Sensors for Diagnostics and Monitoring

Devices, Circuits, and Systems

Series Editor Krzysztof Iniewski

Wireless Technologies
Circuits, Systems, and Devices
Krzysztof Iniewski

Circuits at the Nanoscale
Communications, Imaging, and Sensing
Krzysztof Iniewski

Internet Networks
Wired, Wireless, and Optical Technologies
Krzysztof Iniewski

Semiconductor Radiation Detection Systems
Krzysztof Iniewski

Electronics for Radiation Detection
Krzysztof Iniewski

Radiation Effects in Semiconductors
Krzysztof Iniewski

Electrical Solitons
Theory, Design, and Applications
David Ricketts and Donhee Ham

Semiconductors
Integrated Circuit Design for Manufacturability
Artur Balasinski

Integrated Microsystems
Electronics, Photonics, and Biotechnology
Krzysztof Iniewski

Nano-Semiconductors
Devices and Technology
Krzysztof Iniewski

Atomic Nanoscale Technology in the Nuclear Industry
Taeho Woo

Telecommunication Networks
Eugenio Iannone

For more information about this series, please visit: https://www.crcpress.com/Devices-Circuits-and-Systems/book-series/CRCDEVCIRSYS

Sensors for Diagnostics and Monitoring

Edited by
Kevin Yallup, Laura Basiricò

Managing Editor
Dr. Kris Iniewski

CRC Press is an imprint of the
Taylor & Francis Group, an **informa** business

CRC Press
Taylor & Francis Group
6000 Broken Sound Parkway NW, Suite 300
Boca Raton, FL 33487-2742

First issued in paperback 2020

ISBN 13: 978-0-367-65706-2 (pbk)
ISBN 13: 978-0-8153-7658-3 (hbk)

Visit the Taylor & Francis Web site at
http://www.taylorandfrancis.com

and the CRC Press Web site at
http://www.crcpress.com

Contents

Preface.......... vii
Contributor.......... ix

1. **Nano-carbon Terahertz Sensors and Imagers** 1
   Yukio Kawano

2. **Fiber-Optic Brillouin Distributed Sensors: From Dynamic to Long-Range Measurements** 27
   Alayn Loayssa, Javier Urricelqui, Haritz Iribas, Juan José Mompó, and Jon Mariñelarena

3. **Low-Powered Plasmonic Sensors** 43
   Sasan V. Grayli, Xin Zhang, Siamack V. Grayli, and Gary W. Leach

4. **Ultra-fast Photodiodes under Zero- and Forward-Bias Operations** 65
   Jin-Wei Shi

5. **Semiconductor Sensors for Direct X-Ray Conversion** 97
   Kris Iniewski and Toby Astill

6. **Organic Imagers** 129
   Dario Natali

7. **Tactile Sensors for Electronic Skin** 149
   Fabrizio A. Viola and Pierro Cosseddu

8. **Sensor Systems for Label-Free Detection of Biomolecular Interactions: Quartz Crystal Microbalance (QCM) and Surface Plasmon Resonance (SPR)** 179
   Şükran Şeker, M. Taner Vurat, Arın Doğan, A. Eser Elçin, and Y. Murat Elçin

9. **Low-Power Energy Harvesting Solutions for Smart Self-Powered Sensors** 217
   Albert Álvarez-Carulla, Jordi Colomer-Farrarons, and Pere Ll. Miribel

10. **RFID Supporting IoT in Health and Well-Being Applications** 253
    Sari Merilampi, Johanna Virkki, Nuno Pombo, and Nuno Garcia

11. Low-Power Biosensor Design Techniques Based on Information Theoretic Principles 267
Nicole McFarlane

12. Modern Application Potential of Miniature Chemical Sensors 283
Paul Hichwa and Cristina E. Davis

13. Optical Flow Sensing and Its Precision Vertical Landing Applications 303
Mohammad Al-Sharman, Murad Qasaimeh, Bara Emran, Mohammad A. Jaradat, and Mohammad Amin Al-Jarrah

Index 331

Preface

One of the key developments in recent years has been the emergence of novel technologies enabling smaller, lower-power sensors that can be used to sense a wide range of environmental variables. These new sensor devices are the gatekeepers to the connection between the ever-increasing processing power that is available in modern computers and systems that we want to monitor, diagnose, and ultimately control. In this context, *systems* refer to anything including machinery, automobiles, industrial manufacturing lines, the environment, and even living organisms such as ourselves. With the advent of the Internet of Things (IoT) this trend will accelerate as data from ever-wider ranges of sensors are combined using both traditional processing approaches and artificial intelligence (AI)-based synthesis to enable us to better understand each system and improve the desired outcome, be it better performance, minimized environmental impact, or better human health. Such as approach already has a precedent in the natural world, where advanced organisms such as human beings support a vast array of sensors, all of which are combined and processed by a highly efficient neural network (the biological brain) to enable the organism to interact with and thrive in that environment.

Engineered sensors are a vital part of systems that can help us address many of the challenges facing us as a society. For example, human health will be positively impacted by the shifting approach to preventative medicine, which in turn is enabled by more sophisticated biosensors able to sense key diagnostic variables on a day-to-day basis and report that data back to systems that are capable of interpreting it and proposing preventative "treatments" that can avert problems.

Improved sensors will also have a role to play in limiting the environmental impact of human activity. One aspect is that the sheer number of sensors will represent an increase in power requirements that can be mitigated by the development of more power-efficient sensors. Energy harvesting will have an important role to play in this as well as enabling sensors to operate without the requirement for external power supplies, making it easier to deploy them at the point of need in the environment. Such autonomy supports another role for sensors in the endeavor to address environmental issues that is, monitoring environmental variables to provide feedback on the impact of human activity. Sensors powered by energy harvested from the environment can also be used to monitor the operation of energy-consuming devices in order to optimize their use of energy, thereby reducing their impact on the environment.

One of the areas of that promises to produce significant disruption is the combination of cybernetics and AI, leading to smart machines and robots that can autonomously learn and perform an ever-increasing range of tasks. Such machines heavily rely on sensors to interact with the world around them in a way analogous to humans. There have been tremendous strides in developing certain sensors such as vision systems; however, there are still challenges, such as developing a sensor system that will enable a machine to handle a wide range of objects with the delicacy and precision of a human.

Sensors for Diagnostics and Monitoring brings together a collection of chapters from authors around the world who are working on sensor technologies and the components needed to integrate sensors into intelligent, aware systems. The book showcases some of the latest ideas in sensor development and sensor integration to give the reader an overview of the state of the art. While the book is not exhaustive, the approach is to include topics such as fundamental sensor technology, interface electronics, and the systems needed to make use of sensor information. As many of the ideas can be reused across a wide range of applications, the intent is to provide material that can help to cross-fertilize the development of sensor systems across a whole range of applications.

Contributor

Mohammad Amin Al-Jarrah
College of Engineering
Higher Colleges of Technology (HCT)
Abu Dhabi, UAE

Mohammad Al-Sharman
Aerospace Research Innovation Center (ARIC)
Khalifa University for Science and Technology
Abu Dhabi, UAE

Albert Álvarez-Carullaa
Department of Engineering: Electronics
University of Barcelona
Barcelona, Spain

Toby Astill
Senior Business Development Manager
PerkinElmer
Vancouver, Canada

Laura Basiricò
Department of Physics and Astronomy
University of Bologna
Bologna, Italy

Jordi Colomer-Farraronsa
Department of Engineering: Electronics
University of Barcelona
Barcelona, Spain

Pierro Cosseddu
Department of Electrical and Electronic Engineering
University of Cagliari
Cagliari, Italy

Cristina E. Davis
Department of Mechanical and Aerospace Engineering
University of California
Davis, CA

Arin Doğan
Department of Biochemistry
Ankara University Faculty of Science
Ankara, Turkey

A. Eser Elçin
Department of Biochemistry
Ankara University Faculty of Science
Ankara, Turkey

Y. Murat Elçin
Department of Biochemistry
Ankara University Faculty of Science
Ankara, Turkey

Nuno Garcia
Universidade da Beira Interior
Covilhã, Portugal,
and
Universidade Lusófona de Humanidades e Tecnologias
Lisbon, Portugal

Sasan V. Grayli
School of Engineering Science
Simon Fraser University
Burnaby, Canada

Siamack V. Grayli
School of Engineering Science
Simon Fraser University
Burnaby, Canada

Paul Hichwa
Department of Mechanical and Aerospace Engineering
University of California
Davis, CA

Haritz Iribas
Departamento de Ingenería Eléctrica y Electrónica
Universidad Públia de Navarra
Pampalona, Spain

Kris Iniewski
Head of R&D
Redlen Technologies Inc.
Vancouver, Canada

Mohammad A. Jaradat
Mechanical Engineering Department
American University of Sharjah
Sharjah, UAE
and
Mechanical Engineering Department
Jordan University of Science and Technology
Irbid, Jordan

Yukio Kawano
Department of Electrical and Electronic Engineering
Tokyo Institute of Technology
Tokyo, Japan

Gary W. Leach
School of Engineering Science
Simon Fraser University
Burnaby, Canada

Alayn Loayssa
Departamento de Ingenería Eléctrica y Electrónica
Universidad Públia de Navarra
Pampalona, Spain

Jon Mariñelarena
Departamento de Ingenería Eléctrica y Electrónica
Universidad Públia de Navarra
Pampalona, Spain

Nicole McFarlane
Department of Electrical Engineering and Computer Science
University of Tennessee
Knoxville, TN

Sari Merilampi
Faculty of technology
Satakunta University of Applied Sciences
Pori, Finland

Pere Ll. Miribela
Department of Engineering: Electronics
University of Barcelona

Juan José Mompó
Departamento de Ingenería Eléctrica y Electrónica
Universidad Públia de Navarra
Pampalona, Spain

Dario Natali
Dipartimento di Elettronica, Informazione e Bioingegneria
Istituto Italiano di Tecnologia
Milano, Italy

Nuno Pombo
Universidade da Beira Interior
Covilhã, Portugal,
and
Universidade Lusófona de Humanidades e Tecnologias
Lisbon, Portugal

Murad Qasaimeh
Electrical and Computer Engineering Department
Iowa State University
Iowa, USA

Şükran Şeker
Department of Biochemistry
Ankara University Faculty of Science
Ankara, Turkey

Jin-Wei Shi
Department of Electrical Engineering
National Central University
Taiwan

Javier Urrielqui
Departamento de Ingenería Eléctrica y Electrónica
Universidad Públia de Navarra
Pampalona, Spain

Fabrizio A. Viola
Department of Electrical and Electronic Engineering
University of Cagliari
Cagliari, Italy

Johanna Virkki
Department of Electronics and Communications Engineering
Tampere University of Technology
Tampere, Finland

M. Taner Vurat
Department of Biochemistry
Ankara University Faculty of Science
Ankara, Turkey
Barcelona, Spain

Kevin Yallup
Chief Technology Offier
ACAMP
Alberta, Canada

Xin Zhang
College of Engineering
Boston University
Boston, MA

1

Nano-carbon Terahertz Sensors and Imagers

Yukio Kawano

CONTENTS

1.1 Introduction ... 1
1.2 CNT-Based THz Detector ... 2
1.2.1 Bolometric THz Detection ... 2
1.2.2 THz Photon-Assisted Tunneling ... 3
1.2.3 THz-Induced Electrical-Gate Effect ... 5
1.2.4 THz-Thermoelectric Effect: Room-Temperature THz Detection with a Macroscopic CNT Film and Multi-View Flexible Imaging ... 7
1.3 Graphene-Based THz Detector ... 10
1.3.1 Overview ... 10
1.3.2 Graphene-Based THz and IR Spectrometer ... 11
1.4 Near-Field THz Imager ... 13
1.4.1 Near-Field Imaging ... 14
1.4.2 Integrated Near-Field THz Nano-Imager ... 16
1.5 Conclusion ... 20
Acknowledgments ... 21
References ... 22

1.1 Introduction

The terahertz (THz; 10^{12} Hz) frequency region is located between the microwave region and the visible/near-infrared (near-IR) light region. THz technology nowadays is anticipated to have a strong impact in a wide variety of fields, ranging from basic science such as biochemical spectroscopy, astronomy, and condensed-matter physics to practical areas such as high-speed communication, medicine, agriculture, and security [1, 2]. The advantageous properties of THz wavelengths are that they can be transmitted through objects opaque to visible light and that the energy of THz photons, 1–100 meV, is an important energy spectrum for various materials and biomolecules. Furthermore, THz waves are much safer and create less damage than X-rays. These features have various applications in imaging and spectroscopy, hence making THz measurements a very attractive tool for non-destructive inspection.

However, generating and detecting THz waves is not technically mature, compared with other frequency regions. The reasons are that (1) the THz frequency is too high to be handled with conventional high-frequency semiconductor technology and (2) the photon energy of THz waves (in the order of meV) is much lower than the bandgap energy of semiconductors (typically in the order of eV). Therefore, it is not easy to control and manipulate the THz region with existing electronics and optics/photonics technologies; this is often called the *THz gap*.

The applications of nanoscale materials and devices, however, are opening up new opportunities to overcome such difficulties. Progress in nanoelectronic devices based on semiconductors and superconductors has led to significant improvement in THz devices—for example, quantum cascade lasers with multiple semiconductor quantum wells [3], THz oscillators with resonant tunneling diodes [4], semiconductor quantum dot (QD) detectors [5, 6], superconductor tunneling junction detectors [7], superconductor nanobolometers [8], coherent emitters with high-temperature superconductors [9, 10], and so on.

Among various types of nanomaterials, nano-carbons such as carbon nano tubes (CNTs) and graphene are promising candidates for future high-performance THz devices, owing to their unique electronic and optoelectronic properties [11, 12]. In this chapter, I present several types of THz detectors based on CNT and graphene devices and discuss their future applications to nano-carbon THz imagers.

1.2 CNT-Based THz Detector

1.2.1 Bolometric THz Detection

In the near-IR region, a sensor using a CNT transistor has been developed [13, 14]. The sensing mechanism is based on bolometric detection, in which the near-IR illumination causes a rise in the temperature of the CNT film. A responsivity of ~1000 V/W and a response time of ~50 ms were reported.

Photon energy is much lower in the THz region, thus making highly sensitive THz detection a difficult task. By using CNT bundles with an antenna and a silicon lens, K. Fu et al. demonstrated bolometric THz detection ranging from 0.69 to 2.54 THz and responsivity of ~10 V/W [15].

Although these groups have opened up new possibilities for CNT-based IR/THz detection, much higher performance is required to compete with existing detectors based on other materials such as semiconductors and superconductors. This suggests that new detection mechanisms and/or device structures are required to obtain higher performances. In the following three sections, I introduce a series of CNT-based THz detectors that we have developed [16–20]. Unlike bolometric detection, our detectors have three detection mechanisms: THz photon-assisted tunneling (PAT) [16], the THz-induced electrical-gate effect [17], and the THz-thermoelectric effect [18–20].

1.2.2 THz Photon-Assisted Tunneling

PAT is based on the theory proposed by Tien and Gordon in 1963 [21]. They discussed the interaction of a nanoscale islands with an electromagnetic wave and theoretically showed that new energy bands, so-called photon sidebands, are formed by the AC electric potential of the electromagnetic wave at intervals of nhf (n: integer number, h: Plank constant, f: frequency of the electromagnetic wave). This phenomenon has been previously observed in superconductor tunneling junctions [21–23] and semiconductor QDs [24, 25] with microwave irradiation. Tucker and Feldman called this phenomenon *quantum detection* and discussed its application to an electromagnetic wave detector [26]. However, most of the earlier works on QDs were done in the microwave region. In order to observe PAT in the THz region with QDs, we used QDs fabricated on CNTs. Compared with conventional QDs based on superconductors and semiconductors, the CNT-QD device has the advantage that the charging energy and energy level spacing due to quantum electron confinement typically reach ~10 meV, corresponding to a THz frequency (~2.4 THz). This energy range is larger than that of conventional QDs by a factor of 10. With this view, we studied the THz response of the CNT-QDs [16].

Figure 1.1a shows a sketch of the CNT-QD device. A single-walled metallic CNT with a diameter of ~1 nm was used, and source and drain electrodes with an interval of ~600 nm and side-gate electrodes were patterned with electron beam lithography. In this device, electrons are confined to a very small area of 1×600 nm^2, forming a QD.

When such a QD structure is connected to source, drain, and gate electrodes, this device works as a single-electron transistor (SET). By sweeping either the gate voltage or the source drain voltage and measuring the source-drain current, spectroscopy of energy states in the QD is possible. The photon sidebands in the QD can be observed as a generation of new current signals via inelastic electron tunneling when electrons exchange photons.

The CNT-QD device was cooled down to a temperature of 1.5 K, and the device was irradiated with THz waves through a THz-transparent window made from a Mylar sheet. As a THz illumination source, a THz gas laser pumped by a CO_2 gas laser was used.

Figure 1.1b shows the source-drain current I_{SD} as a function of the gate voltage V_G for the CNT-QD device. The current versus the gate voltage displays a periodic oscillation, indicating that the device works as a SET. The THz response of the CNT-QDs shows that the THz illumination generated

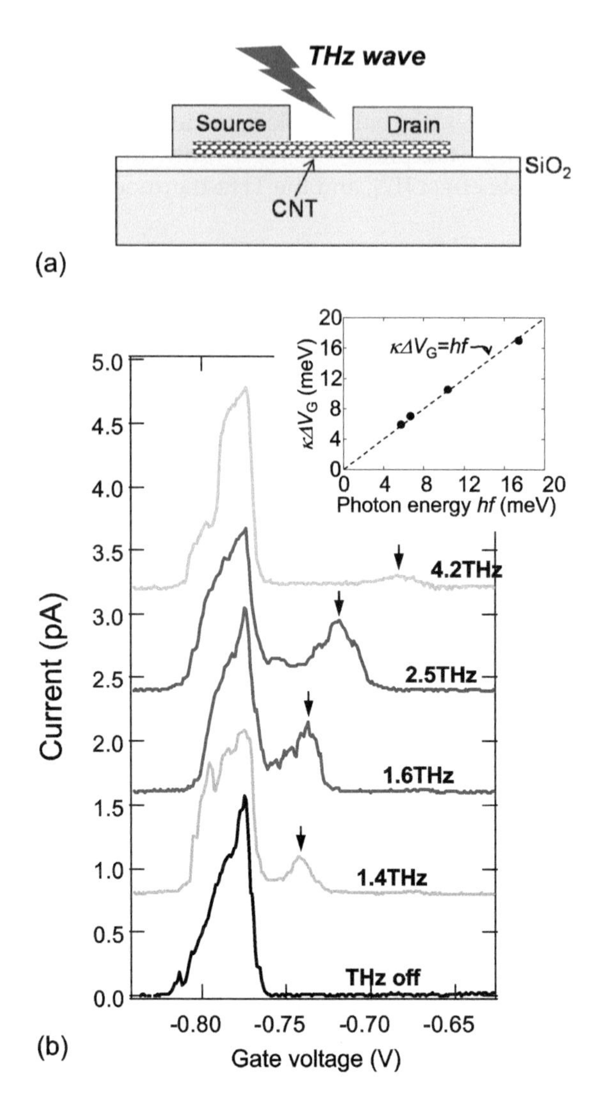

FIGURE 1.1
(a) Sketch of a CNT-QD device. (b) Source-drain current I_{SD} versus gate voltage V_G for the frequency of the incident THz wave, $f = 1.4$, 1.6, 2.4, and 4.2 THz. The experimental curves for the THz irradiation are offset by multiples of 0.8 pA for clarity. The inset shows the energy spacing $\kappa \Delta V_G$ between the original peaks and the satellite peaks as a function of the photon energy hf of the THz wave. The dashed line in the inset is an eye guide corresponding to $\kappa \Delta V_G = hf$. Reprinted with permission from [16].

a new satellite current in the current suppression region (a Coulomb blockade state). It is also seen that its peak position relative to the gate voltage shifts in the positive direction as the frequency, f, of the THz wave increases. As shown in the inset of Figure 1.1b, the energy spacing, $\kappa \Delta V_G$, between the original peak and the satellite peaks linearly depends on the photon energy, *hf*, of the THz wave. From the measurement of the differential conductance dI_{SD}/dV_{SD}, as a function of source drain voltage V_{SD} and gate voltage V_G, we derived the conversion factor $\kappa = 0.18$, which is defined as the conversion ratio of V_G into energy. The good agreement between $\kappa \Delta V_G$ and *hf* provides clear evidence for the THz-PAT in the CNT-QD device. These results demonstrate that the CNT-QD device works as a frequency-tunable THz detector. The noise-equivalent power (NEP) of this detector was estimated to be ~10^{-14} W/Hz$^{1/2}$.

1.2.3 THz-Induced Electrical-Gate Effect

Although the PAT structure has enabled frequency-tunable THz detection, the relatively low detection sensitivity is an issue to be resolved. This is because within the framework of this detection mechanism, the absorption of one photon leads to the generation of just one electron, even if a quantum efficiency of 100% is assumed. In order to obtain a measurable current (e.g., pA–fA), it is necessary to detect 10^4–10^7 photons per second. This limits the detection sensitivity. To resolve this problem, we created a new device design [17]: a CNT-SET is integrated with a GaAs/AlGaAs heterostructure chip containing a two-dimensional electron gas (2DEG) (Figure 1.2a). The hybrid device has two separate components and roles: THz absorption in the 2DEG and signal readout in the CNT-SET. A basic idea of THz detection with this device is that the CNT-SET senses the electrical polarization induced by THz-excited electron–hole pairs in the 2DEG; that is, the current flowing through the CNT-SET is gated with THz-induced polarization in the 2DEG. Since the SET has the detection sensitivity of a single electron, the CNT-2DEG hybrid device is expected to exhibit ultra-high sensitivity.

Figure 1.2b displays the data of the THz response of the CNT-2DEG device. We applied magnetic field *B* in a perpendicular direction to the 2DEG plane. The experimental setup for the THz measurements here is similar to that of the THz-PAT in the previous section. The laser intensity was reduced with THz-attenuating filters, and the intensity of the THz irradiation on the detector was estimated to be 0.75 nW/mm^2.

Figure 1.2b shows that the THz illumination caused a shift in the current peak position in the direction of positive V_G. It was also found that for irradiation at 1.6 THz, the shift was monotonically enhanced by increasing *B* to 3.95 T, then decreased when *B* was further raised beyond 3.95 T. In case of 2.5 THz irradiation, the *B* value for the maximum peak shift changed to 6.13 T.

The physical mechanism behind the preceding features is as follows: When the 2DEG is subject to a magnetic field, the energy state of the 2DEG splits into discrete Landau levels (LLs). When a photon energy, *hf*, of a THz wave is equal to LL separation eB/m^*, the 2DEG resonantly absorbs the THz wave (cyclotron resonance). Here, e is the elementary charge and m^* is the cyclotron-effective mass for the crystal in which the 2DEG is embedded. The preceding experimental data ($B = 3.95$ T for $f = 1.6$ THz and $B = 6.13$ T

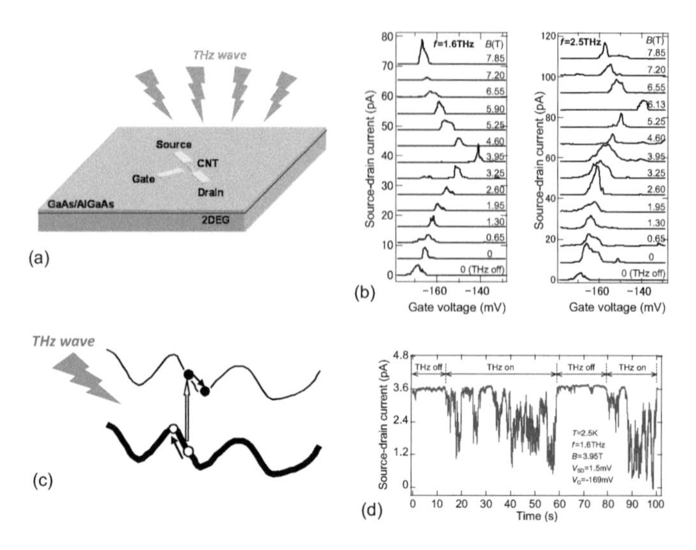

FIGURE 1.2
(a) THz detector with the CNT-2DEG hybrid device. (b) Magnetic field dependence of the source-drain current I_{SD} versus the gate voltage V_G of the CNT with and without the THz irradiation. The magnetic field B was applied to the detector perpendicular to the 2DEG plane; the B values are given on the right-hand side of the figures in units of Tesla (T). The data with the THz irradiation are offset for clarity. (c) Schematic representation of the dynamics of THz-excited carriers in the presence of an impurity potential. (d) Time trace of the THz-detected signal (the CNT-SET current, I_{SD}) as the THz irradiation is cycled on and off. In this measurement, THz cyclotron radiation emitted from another 2DEG was used as an illumination source and the emission intensity was reduced to a very low intensity of ~0.1 fW. (a), (b), (d): Reprinted with permission from [17].

for f = 2.5 THz) indicate that the f value is proportional to the B value for the maximum of the peak shift, which is consistent with the features of cyclotron resonance. In addition, from these data, the associated m^* value is derived to be 0.067 m_0, where m_0 is the free electron mass. This m^* value is in good agreement with the cyclotron-effective mass previously reported for a GaAs-based 2DEG [27]. These facts indicate that the current peak shift of the CNT-SET originates from the THz absorption in the 2DEG.

It is well known that the 2DEG has a random potential with a typical period of 20–100 nm [28], which is mainly due to ionized impurities in dopant layers. It follows that the THz-excited electrons and holes drift in opposite directions, owing to the local electric field gradient arising from the random potential (Figure 1.2c) [29]. As a result, they are spatially separated from each other. The separation of electron–hole pairs generates additional electrical polarization in the 2DEG. This situation is equivalent to the application of an additional gate voltage to the CNT-SET, resulting in the current peak shift.

It should be noted that the CNT-2DEG distance is ~120 nm, the relative dielectric constant of GaAs is 13, and the separation length of the excited electron–hole pairs is 20–100 nm [28, 29]. From these values, the potential change caused by the polarization of a single electron–hole pair is estimated to be 0.4~10 meV. This value is comparable to the width of the current peak of the CNT-SET (~3 meV). This shows that even single-carrier charging via single THz photon absorption in a 2DEG can generate an observable current peak shift.

Based on the preceding discussion, we studied the THz response of the CNT-2DEG device under very weak THz illumination. We expect that in these circumstances, the current of the CNT-SET will switch on and off, according to the excitation and recombination of the electron–hole pairs in the 2DEG. We measured the temporal trace of the THz-detected signal (the I_{SD} change associated with the current peak shift) as the THz irradiation was cycled on and off. Here we used cyclotron radiation [30, 31] from another 2DEG in a different GaAs/AlGaAs chip. We reduced its intensity to an extremely low level, in the order of 0.1 fW. We mounted the detector and the 2DEG-based THz source face to face in the same superconducting magnet so that the 2DEGs of the detector and the source were both in a cyclotron resonance condition. Figure 1.2d displays the temporal behavior of I_{SD} for an on/off sequence of THz irradiation at $V_G = -169$ mV, $V_{SD} = 1.5$ mV, $B = 3.9$ 5 T, and $f = 1.6$ THz. This result shows that the CNT-SET current was stable for THz-off, whereas it repeatedly switched during THz irradiation. This feature means that the CNT-SET detects the temporal processes of the generation and relaxation of the THz-excited carriers in the 2DEG; that is, an ultrasensitive readout of the THz-detected signal has been achieved. The NEP of this detector is 10^{-18}–10^{-19} W/Hz$^{1/2}$, which is far superior to that of the PAT-based THz detector described in Section 1.2.2.

1.2.4 THz-Thermoelectric Effect: Room-Temperature THz Detection with a Macroscopic CNT Film and Multi-View Flexible Imaging

The CNT THz detectors explained in Sections 1.2.2 and 1.2.3 have the advantages of frequency tunability and high sensitivity, but they have to be cooled down. For industrial and/or daily applications, THz detectors that can be operated at room-temperature are strongly desired. Commercially available room-temperature THz detectors include Schottky barrier diodes, pyroelectric detectors, and Golay cell detectors. The CNT is another promising candidate for room-temperature THz detection because even at room temperature, CNTs have the ability to absorb electromagnetic waves in an ultra-wide spectrum range, from nearly DC to ultraviolet [32].

As demonstrated in the previous sections, the use of single-CNT devices is essential for the quantum detection of THz waves. On the other hand, the photon nature of THz waves disappears at room temperature, where the classical nature of electromagnetic waves becomes prominent. In this respect, larger-area CNT devices are desired for room-temperature THz detection so that coupling efficiency with THz waves is enhanced. Nevertheless, a large-area high-quality CNT device suitable for THz detection has been so far unavailable.

In order to achieve room-temperature THz detection with CNT devices, we used macroscopic p-n junction films of highly aligned CNT arrays (Figure 1.3a) [18, 19]. As shown in the THz and IR spectrum of the CNT arrays (Figure 1.3b), this film exhibits strong THz and IR absorption, ranging from 0.1 to 200 THz with polarization anisotropy with respect to the CNT alignment.

Figure 1.3c shows the *I-V* characteristics of this CNT film device with (grey line) and without (black line) illumination at $f = 2.52$ THz. It is found that THz illumination shifts the *I-V* curve and does not change the conductance of the CNT film (i.e. the slope of *I-V* characteristics is unchanged with illumination). This indicates that the THz detection observed here does not originate from the bolometric effect.

In order to elucidate the origin of this THz photoresponse, we studied the THz responses of three samples that have substrates with different thermal conductivities. We found that the THz photosignals remarkably decreased as the thermal conductivity of the substrate became higher. This fact can be reasonably understood in terms of the THz-induced thermoelectric effect.

The CNT film device also has strong polarization dependence with respect to the alignment direction of the CNTs (Figure 1.3d), enabling use as a THz polarizer as well. This CNT array detector and polarizer can also respond to the visible and IR region, demonstrating the use as ultra-broad band photodetectors.

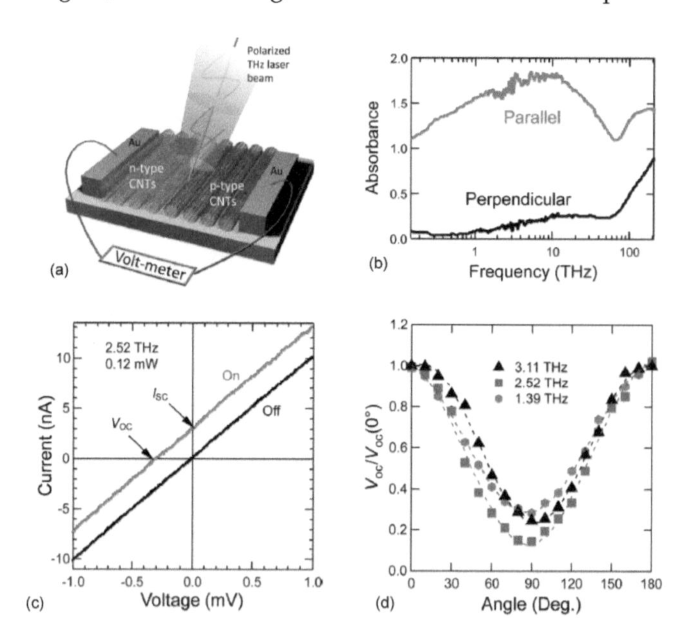

FIGURE 1.3
(a) Sketch of highly aligned p-n junction CNT films. (b) THz and IR absorption spectra of the device in (a). The results for parallel and perpendicular polarization are compared. (c) *I-V* characteristics with and without THz illumination at $f = 2.52$ THz. (d) Polarization dependence of the THz signal at three frequencies ($f = 3.11$, 2.52, and 1.39 THz), normalized by its value of 0 degrees. Reprinted with permission from [18].

We further developed a flexible THz imager by integrating 23-CNT detector elements into an array (Figure 1.4a) [20]. The flexible imager is bendable and thus can match samples with various curved surfaces. The flexible CNT imager can be used to identify a breakage or an impurity in curved samples such as a PET bottle or a syringe, leading to three-dimensional surface scanning with a simple system (Figures 1.4b–d).

Existing flat THz cameras are suitable for two-dimensional scanning [33, 34], whereas most real samples have three-dimensional curvatures. Although THz tomography technology is a powerful method of three-dimensional imaging [35, 36], it requires a complicated and bulky system. Our flexible CNT THz imager is strongly anticipated to serve as a non-invasive inspection tool for applications such as medical care and pharmaceutical inspections and potentially a wearable device for real-time health monitoring.

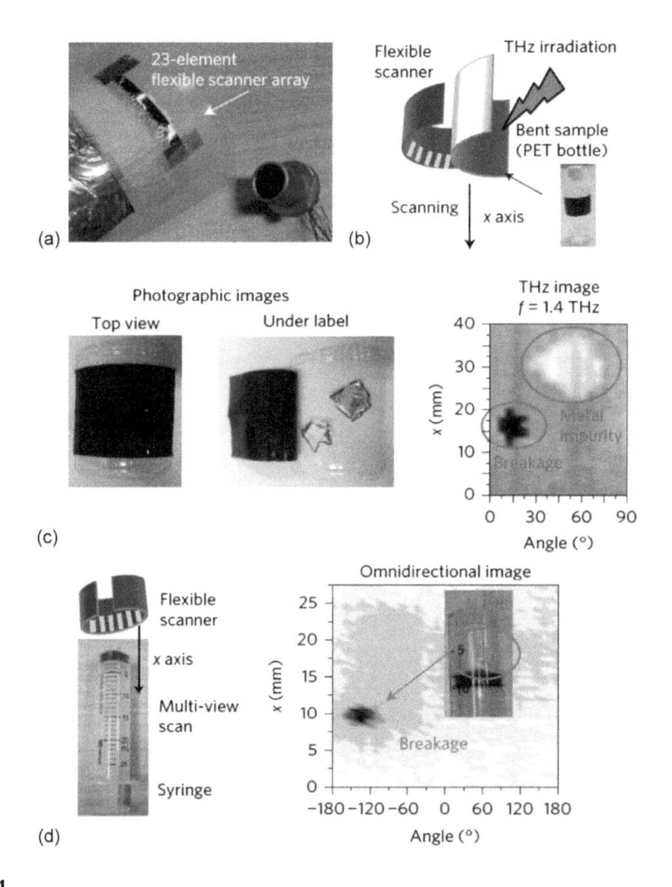

FIGURE 1.4
Flexible THz imager and multi-view THz imaging. (a) Photo of the flexible CNT THz imager. (b) Schematic representation of multi-view THz imaging. (c) Photographic images (left) and THz image (right) of the PET bottle. (d) Multi-view THz image of the syringe. Reprinted with permission from [20].

1.3 Graphene-Based THz Detector

1.3.1 Overview

Since the discovery of graphene, a single atomic layer of graphite, its applications to electronic and optical devices have been intensively studied because of its unique properties, originating from linear dispersion and massless Dirac fermions. Graphene devices have been also attracting much interest as enablers of THz devices. THz detectors with graphene reported so far are categorized into thermal detection [37], nonlinear response [38], and plasmonic detection [39, 40].

Since light illumination onto graphene causes an electron temperature gradient, a large thermoelectric effect induced by THz irradiation is expected. It has been reported that this type of THz detector exhibits excellent NEP performances of $<100\ \mathrm{pW/Hz^{1/2}}$ and response times of ~30 ps [37]. The detector response is much faster than that of conventional silicon-based bolometers (in the order of ms). It has been also demonstrated that graphene field-effect transistors enable THz detection through their nonlinear response to THz electric fields with respect to gate voltages [38]. The NEP was reported to be $\sim 30\ \mathrm{nW/Hz^{1/2}}$ for 0.3 THz irradiation. Another detection method is to use surface plasmons generated on graphene—that is, collective carrier excitation induced via interaction with THz/IR waves [39, 40]. This type of detector is expected to exhibit higher sensitivity due to strong plasmonic resonance.

In spite of the development of a variety of graphene THz detectors, there is an issue to be tackled: these detectors do not have wide-band frequency tunability, which is useful for spectroscopic measurements. THz spectroscopy is utilized as a powerful tool for investigating various materials and structures. This is because the physical/chemical properties of many important materials lie in ~meV energies and a characteristic time constant in the picosecond region, which corresponds to the energy gap of superconductors, the impurity level of semiconductors, phonon energy, LL separation, quantized energy separation due to electron confinement in low-dimensional semiconductors, and so on. This importance is applicable to other research fields such as chemistry, biology, and astronomy, where THz spectroscopy is expected to provide rich information about chemical bonding, higher-order structures of biomolecules, dark matters in space, and so on.

Representative methods for the THz spectroscopy are the time-domain spectroscopy (TDS) and Fourier transform spectroscopy (FTS). However, the signal-to-noise ratio of commercially available TDS systems reduces above ~5 THz, while that of FTS systems degrades below ~10 THz. In the following section, I introduce our wide-band THz spectrometer based on a graphene device that allows frequency-tunable detection ranging from the THz to the IR regions (0.76–33.0 THz) [41].

1.3.2 Graphene-Based THz and IR Spectrometer

Tunable THz and IR detection utilizes the LL formation of Dirac fermions of graphene under a magnetic field and the resulting cyclotron resonance absorption of the electromagnetic waves. The LLs for graphene are expressed as $E_n = \text{sgn}(n)\tilde{c}(2e\hbar B|n|)^{1/2}$ [42] which is quite different from that of conventional semiconductors:

$$E_n = (n + 1/2)\hbar eB / m^*$$

where:

n is the LL index
$\tilde{c}$ is the effective carrier velocity

As explained in Section 1.2.2, when the photon energy, *hf*, of the incident electromagnetic wave matches the LL energy separation, the cyclotron resonance condition is satisfied. Figure 1.5 schematically compares the LL density of states between the graphene and a typical semiconductor, which is based on the preceding LL expressions. It can be seen that the LL energy separation for the graphene is much larger than that of the semiconductor. The graphene also has electron LLs and hole LLs that meet at the Dirac point, whereas the electron LLs of the semiconductor are separated from the hole LLs by the bandgap energy (typically ~eV). The LL structure of graphene has been experimentally confirmed by IR absorption [43, 44] and scanning tunneling spectroscopy [45]. Therefore, the significant features of the graphene

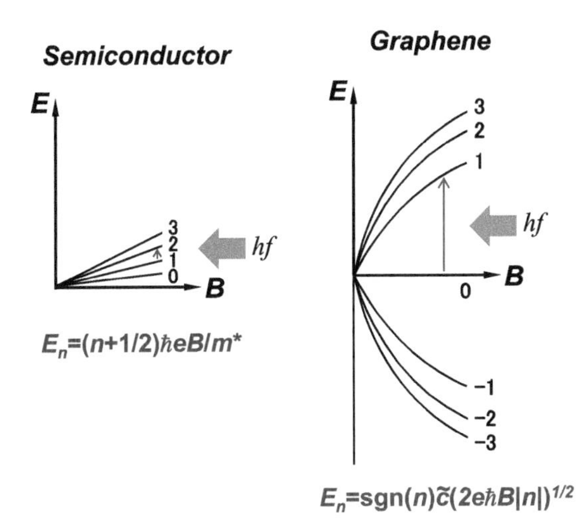

FIGURE 1.5
Schematic view of the LL density of states for a graphene semiconductor (left) and a conventional semiconductor (right).

LL structure make it suitable for use as a magnetically frequency-tunable THz and IR detector over a very broad frequency range.

A monolayer graphene flake was exfoliated from graphite and attached to a SiO_2-Si chip. The graphene device has current electrodes and a back-gate electrode. The graphene device was immersed into a 4.2 K cryostat and the magnetic field B was applied perpendicularly to the graphene surface. A CO_2 gas IR laser (f = 28–33 THz) and a THz gas laser (f = 0.76–4.2 THz) were used as illumination sources. Voltage changes, V_{sig}, with the THz/IR irradiation were measured using a lock-in amplifier, where the intensity of the THz/IR waves was modulated at 24 Hz.

Figure 1.6 displays the experimental results of the THz and IR detection (detected signals V_{sig} versus B) for seven different frequencies of f = 0.76, 1.6, 2.5, 3.1, 4.2, 28, and 33 THz. For comparison, similar measurements were performed for the 2DEG on GaAs/AlGaAs. The data for the graphene show that the peaks of V_{sig} versus B are seen for all the frequencies and that the peak position depends on the frequency f. In contrast, the feature observed for the GaAs-2DEG is that the V_{sig} peaks were seen only at f = 0.76 and 1.6 THz, but their peak positions B = 1.87 and 3.96 T were distinct from B = 0.042 and 0.187 T for the graphene.

These results can be reasonably explained in terms of the large difference in the LL formation between the graphene and the semiconductor, as shown in Figure 1.5. When B = 0–5 T, the corresponding LL energy separation for the GaAs-2DEG is calculated to be 0–8.5 meV using the cyclotron-effective mass of the GaAs-2DEG (0.067 m_0) [27]. This energy range restricts the frequency range of the resonant THz detection to f < 2.1 THz (hf < 8.5 meV). This explains why the GaAs-2DEG device responds only to THz waves with the lower frequencies f = 0.76 and 1.6 THz. In contrast, taking into account the LL expression for the graphene and the Fermi velocity ($\tilde{c} =\sim 10^6$ m/s) [46], we plotted theoretical curves and experimental data of f versus B at the resonant condition (Figure 1.7). From this comparison, the observed THz and IR resonant detection in the graphene device is attributed to carrier transitions between the following LL indexes: n = 12 for 0.76 THz, 12 for 1.6 THz, 12 for 2.5 THz, 12 for 3.1 THz, 12 for 4.2 THz, –11 for 28 THz, and –11 for 33 THz.

The preceding results demonstrate that the graphene device works as a novel type of wide-band THz and IR spectrometer. The tuning range 0.76–33.0 THz is inaccessible via the earlier superconducting hot-electron mixers [47, 48] and 2DEG mixers [49].

The detection bandwidth and sensitivity of the graphene-based spectrometer should depend on the local properties of carrier transport because the signal readout uses THz and IR photoconductivity. In order to investigate local electronic properties, we imaged electric potential distribution in the graphene detector. As depicted in Figure 1.8a, we used a home-made potential mapping system whose detection mechanism is based on capacitive coupling between the tip of a metal-coated cantilever and a sample surface [50]. The detected signal is read out as a voltage change on the electrometer. Local step structures of

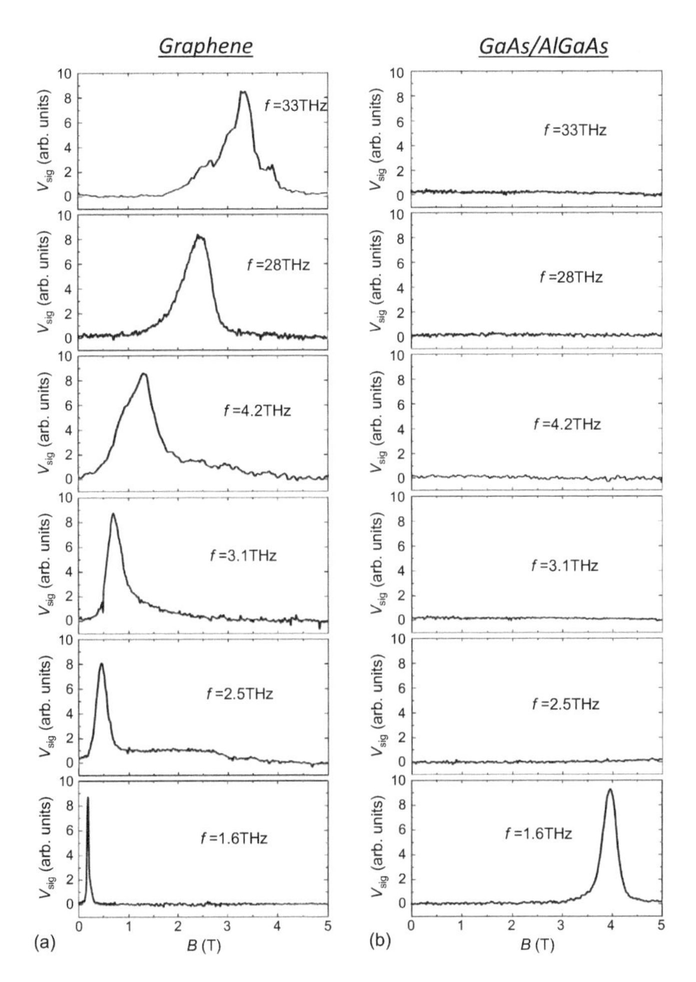

FIGURE 1.6
THz- and IR-detected signal V_{sig} versus the magnetic field B at f = 0.76, 1.6, 2.5, 3.1, 4.2, 28, and 33 THz. Graphene (left) and GaAs-2DEG (right) are compared. Reprinted with permission from [41].

the potential were observed (Figure 1.8b), indicating that the THz and IR photoconductivity of graphene are likely to be sensitive to such potential steps. We are currently taking detailed measurements on the dynamics of THz-excited carriers in terms of the relationship with the potential distribution.

1.4 Near-Field THz Imager

One of important applications of THz detectors is an imaging measurement in which the intensity profile of reflected or transmitted THz waves is mapped. Simultaneous measurements of frequency spectra allow us to

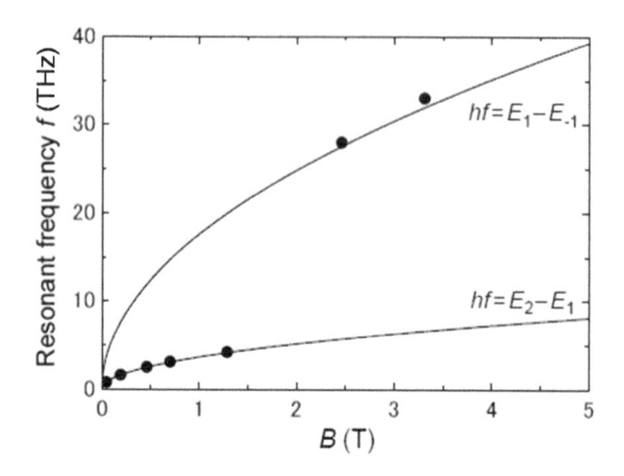

FIGURE 1.7
Resonant frequency f versus magnetic field B for the inter-level transitions in the graphene: $hf = E_2 - E_1, E_1 - E_{-1}$. The theoretical (curves) and experimental (black circles) results were plotted. Reprinted with permission from [41].

characterize the physical/chemical properties. This non-destructive inspection technique can therefore be used in a safer manner than X-ray imaging. However, one concern of the THz imaging is low spatial resolution, which comes from the much longer wavelength compared with that of visible light. This is especially crucial for investigating nanoscale objects whose size is much smaller than the wavelength of the THz waves. A powerful method for overcoming the diffraction limit of the resolution is to use near-field imaging technology [51]. Nevertheless, near-field imaging in the THz region is still developing, partly because of its long wavelength of 30–300 μm.

In this section, I describe the use of a micro-/nano-sized THz detector as a near-field THz imager, which is based on a 2DEG in a GaAs/AlGaAs heterostructure. This is an important step toward carbon-based THz near-field imaging.

1.4.1 Near-Field Imaging

A simplified explanation of near-field imaging is that when an electromagnetic wave is illuminated onto an aperture smaller than a wavelength, an evanescent field (near field) is localized just behind the aperture. The size of the evanescent field is determined by the aperture size and is not influenced by the wavelength. By illuminating and/or detecting the evanescent field, it is possible to obtain a resolution beyond the diffraction limit.

Conventional near-field imaging is categorized into an aperture type and an aperture-less type. In the visible and near-IR regions, either a tapered, metal-coated optical fiber (aperture type) [52] or a metal tip (aperture-less type) [53] is used. In the microwave region, either a sharpened waveguide (aperture type) [54] or a coaxial cable (aperture-less type) [55] is used. Since

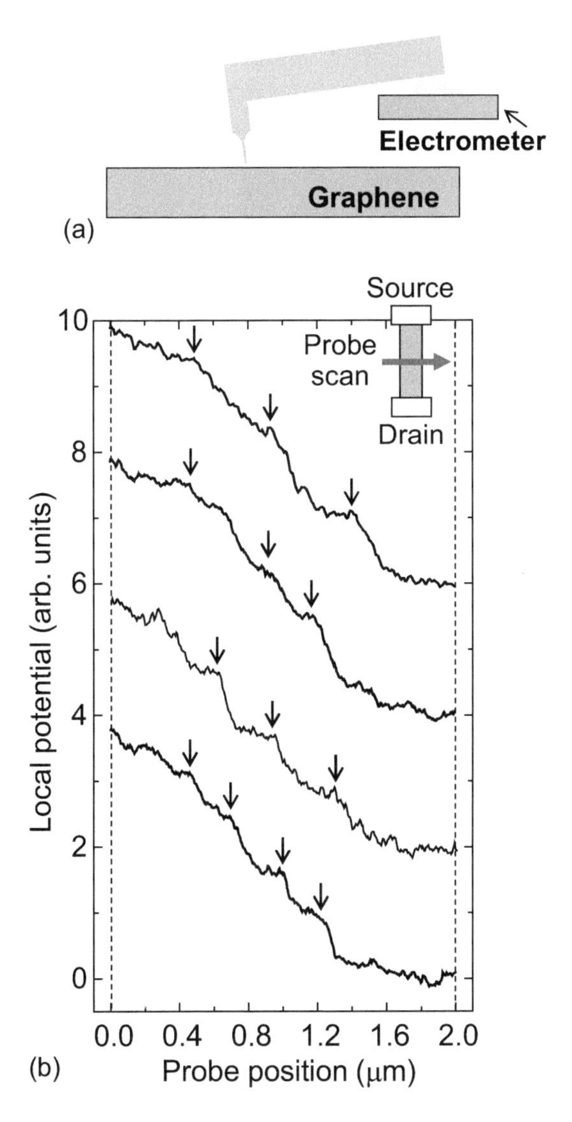

FIGURE 1.8
(a) Potential imaging system with a cantilever-integrated electrometer. (b) Line profiles of the electric potential distribution in the graphene detector at $I = 0.5$ μA and $B = 5$ T. The data are offset for clarity. The two broken lines indicate the channel-width boundaries. Reprinted with permission from [50].

the intensity of the evanescent field is very weak, the near-field imaging system usually needs a highly sensitive detection scheme, such as a high transmission wave line and a highly sensitive detector.

Several methods of near-field imaging in the THz region have been reported. As the aperture type, a metal cone was used to obtain a resolution better than $\lambda/4$ [56]. This method, however, has the drawback of low THz transmission through the small aperture and the spatial resolution is not so high. As the aperture-less type, a sharpened antenna, [57], a metal tip [58], and a cantilever [59] were reported. Although the aperture-less technique allows high spatial resolution, it has a problem separating from the far-field component of an incident wave, which leads to fluctuations of the background signal. For this reason, in most instances the probe position is modulated and a synchronously detected signal is measured. However, this scheme makes the whole system and its operation complicated. Another method is to directly put a sample on a nonlinear crystal surface. By illuminating a femtosecond pulse laser onto the crystals, a THz-evanescent field is generated on the crystal surface due to the nonlinear effect and interacts with the sample. Real-time near-field THz imaging has been reported using this method [60].

1.4.2 Integrated Near-Field THz Nano-Imager

Our idea for near-field THz imaging is that a micro-/nano-sized THz detector is directly coupled with the THz-evanescent field in a near-field region [61–64]. For this purpose, we fabricated a novel integrated device. As shown in Figure 1.9, a sub-wavelength aperture and a planar probe were deposited on a surface of a GaAs/AlGaAs heterostructure chip. The aperture and the

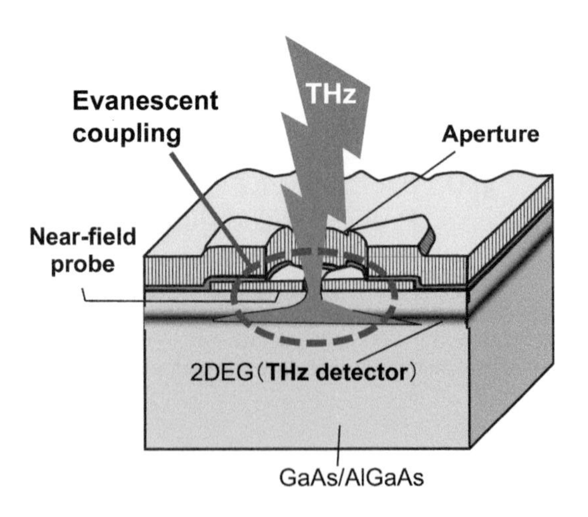

FIGURE 1.9
Near-field THz imager with the 2DEG.

probe are insulated by a 50 nm-thick SiO_2 layer. The 2DEG embedded in the GaAs chip has an electron mobility of 18 m^2/Vs and a sheet electron density of 4.4×10^{15} m^{-2} at 77 K. The 2DEG, located 60 nm below the chip surface, works as a THz detector [65]. The source and drain electrodes were extended to the side surfaces of the chip, each of which was attached to an electrical wire.

In this device, the THz-evanescent field generated just behind the aperture is directly detected by the 2DEG detector owing to the close vicinity to the detector and the aperture. Another advantage of this device structure is that the presence of the planar probe changes the distribution of the evanescent field, enhancing the coupling of the evanescent field to the 2DEG detector. This is expected to increase the detection sensitivity to the THz-evanescent field. Since all the components in this device—an aperture, a probe, and a detector—are integrated on one GaAs/AlGaAs chip, this scheme eliminates any optical and mechanical alignments between each component, leading to an easy-to-use and robust system. This is a collection-mode near-field THz imager, which is in contrast to earlier technologies that are based on sub-wavelength THz sources.

Figure 1.10 shows calculations of THz electric field distributions around the aperture using a finite element method. We compared the device of the aperture alone and that of the aperture plus the probe, where the aperture diameters are the same. In the case of the aperture alone, the electric field is localized near the aperture. This explains why the conventional

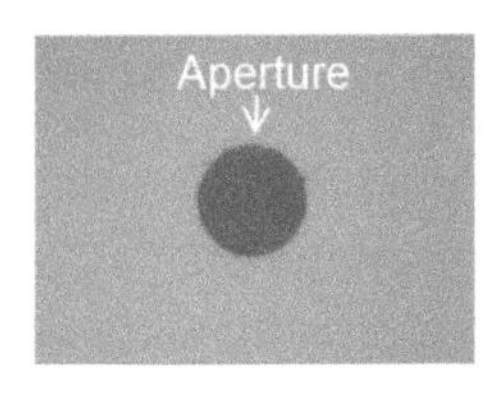

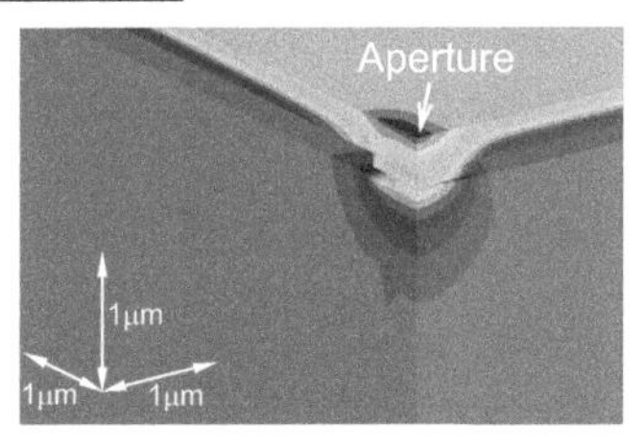

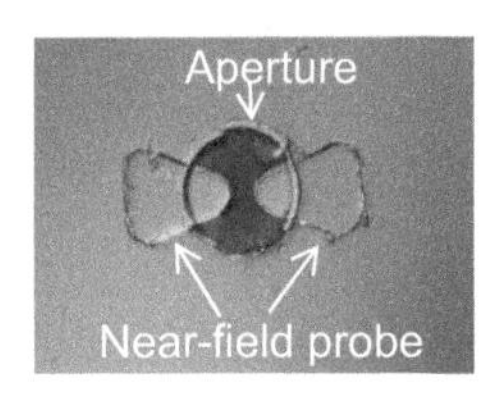

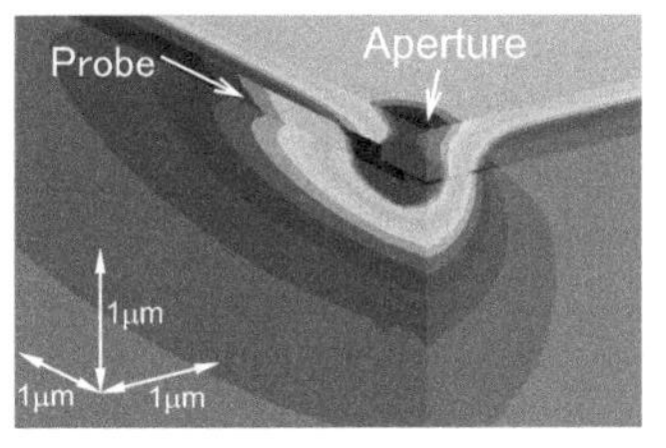

FIGURE 1.10
Calculations of THz electric field distributions in the vicinity of the aperture for the device with the aperture and probe (lower panel) and for the device with the aperture alone (upper panel). Reprinted with permission from [61].

aperture-based technique suffers from low transmission through the aperture. In contrast, when the probe is integrated just behind the aperture, the THz electric field extends into the interior region of the GaAs substrate. This result indicates that the presence of the probe changes the distribution profile of the THz-evanescent field, enhancing the coupling efficiency between the evanescent field and the 2DEG detector.

In order to experimentally confirm the preceding effect and evaluate the spatial resolution, we measured the THz transmission distribution by scanning the device across a sample. In the measurements, the THz gas laser pumped by the CO_2 gas laser was used as a THz source. The THz radiation was cut at 16 Hz and the amplitude modulation of the detected signal (the voltage change of the 2DEG) was measured with a lock-in amplifier. The sample is made up of a THz-transparent substrate, the surface of which is covered at regular intervals by THz-opaque Au films. The widths of the THz-opaque and THz-transparent regions across the scan direction are 80 μm and 50 μm, respectively. Two types of near-field THz devices were used: the aperture integrated with the probe and the aperture alone. Figure 1.11a shows that in the former case, a clear signal is seen, whereas in the latter case, no signal is observed. The enhancement ratio of the signal amplitudes between the two devices was found to be 67. These results clearly demonstrate that the distribution of the THz-evanescent field is largely enhanced thanks to the presence of the probe, resulting in an improvement in the detection sensitivity.

From the decay curves of the data in Figure 1.11a, the spatial resolution of the near-field THz imager is estimated to be 9 μm. The resolution is far beyond the diffraction limit of $\lambda/2$ (107.3 μm) for the wavelength of $\lambda = 214.6$ μm. This feature indicates that the present device works as a near-field THz imager.

Figure 1.11b displays an example of a near-field THz image taken with the device. Here, a THz emission radiating from another 2DEG sample was imaged. The THz emission arises from inter-LL transitions of electrons in the 2DEG. The sample was fabricated on a GaAs/AlGaAs heterostructure wafer, and the length and width of the 2DEG channel was 400 and 200 μm, respectively. The result reveals asymmetric distribution between the source- and drain-contact sides, which does not follow the electric field distribution in quantum Hall devices. This nonlocal feature is due to the macroscopically long scale of the scattering length of the excited electrons associated with the THz emission [66–69].

Figure 1.12a,b displays the improved system of our near-field THz imager [70], in which we integrated a quartz tuning fork. The electrical signal from the tuning fork is fed back to the z-axis translation stage of the THz imager. Because of the very sharp frequency spectrum of the fork oscillation, the distance between the imager and sample surface is accurately kept constant during x–y scanning. With this design and the feedback mechanism, we obtained a much higher resolution of 280 nm (~$\lambda/766$) (Figure 1.12c).

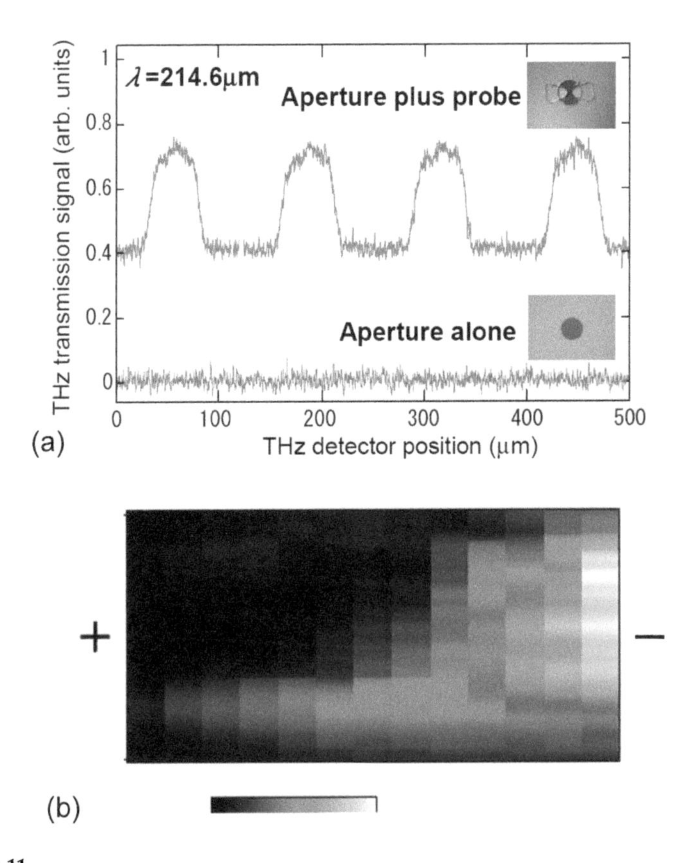

FIGURE 1.11
(a) THz transmission signal as a function of the position of the near-field THz imager. Reprinted with permission from [61]. (b) Near-field image of THz emission distribution in the 2DEG sample without external THz sources (passive near-field THz imaging). Reprinted with permission from [64].

The utilization of this technology enabled us to perform near-field THz emission imaging without external THz sources. We successfully visualized the spatial distribution of a THz emission associated with electron injection from a source electrode in the 2DEG device (Figure 1.13a). The spatial dynamics of non-equilibrium carriers (Figure 1.13b) derived from the THz images is qualitatively consistent with theoretical calculations that showed that the position of the electron injection is separated from one corner as the electronic state deviates from a quantum Hall state to a non-quantum Hall state.

For our near-field THz imager, a low-dimensional structure is essential because this prevents the effects of a far-field component of an incident THz wave. As well as the 2DEG detector, to further improve spatial resolution and perform simultaneous spectroscopy, other low-dimensional devices such as CNT and graphene THz detectors are also promising. When CNT/

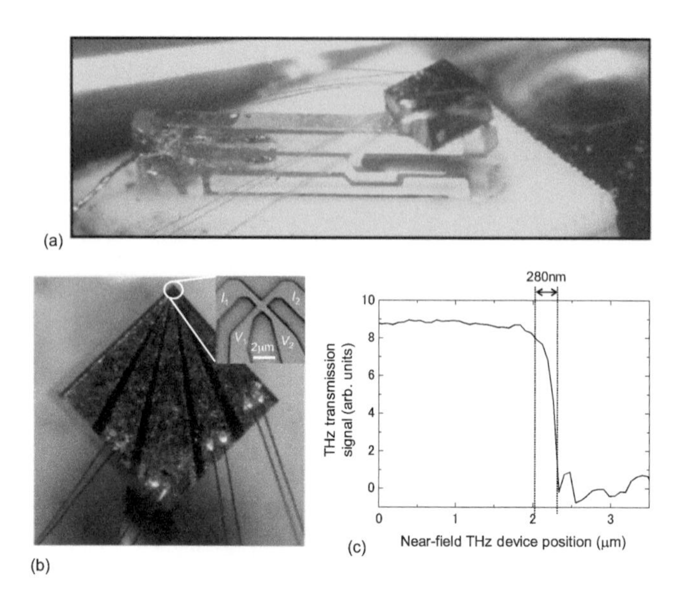

FIGURE 1.12
Photographs of (a) the near-field 2DEG THz imager integrated with a tuning fork and (b) the sensing area of (a). I_1, I_2, V_1, and V_2 denote the current probes and the voltage probes. (c) THz transmission profile at the boundary between the THz-transparent and THz-opaque region, obtained with the THz imager of (a,b). Reprinted with permission from [70].

graphene detectors are fabricated for near-field THz imaging, the resulting devices would exhibit ultra-high sensitivity, nanometer resolution, and a wide tuning frequency range.

1.5 Conclusion

In this chapter, I have described novel THz detectors based on CNTs and graphene. At present, CNT-based THz detectors are categorized into four types: bolometric detection, photon-assisted tunneling, the THz-induced electrical-gate effect, and the THz-thermoelectric effect. These detectors have the advantageous features of frequency selectivity, high sensitivity, and room-temperature detection, respectively. It has been further demonstrated that the use of graphene has made it possible to detect THz and IR waves with frequency tuning. These nano-carbon THz detectors are expected to be used for nanoscale THz imaging and spectroscopy.

As an application of the THz detector, an integrated near-field THz imager was introduced. Using the newly designed device structure, where a micro-sized 2DEG detector is directly coupled with a THz-evanescent field, highly sensitive detection of THz-evanescent fields and near-field THz imaging

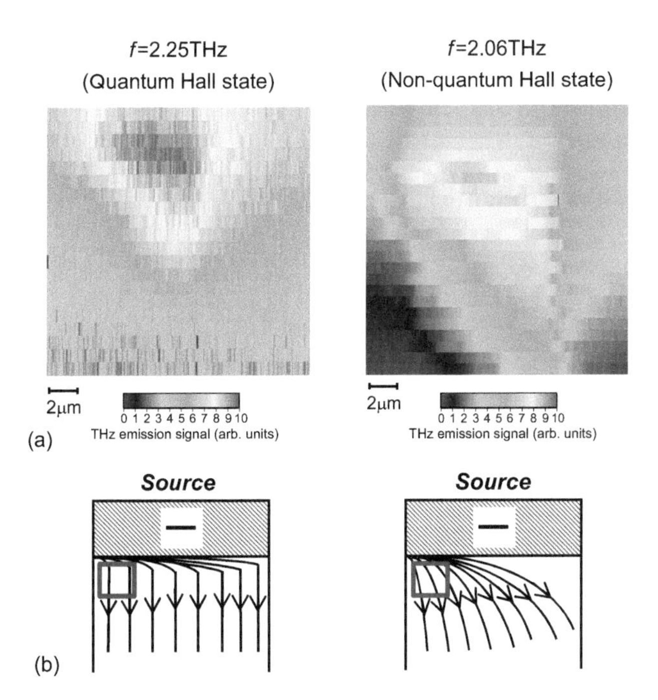

FIGURE 1.13
(a) Near-field THz emission distribution in the vicinity of the source contact of a quantum Hall device. The frequencies of the THz emission observed here are 2.25 and 2.06 THz. The left and right THz images correspond to quantum and non-quantum Hall states, respectively. No external THz sources were used. (b) Sketch of electron injection and transport trajectories near the source electrode of the quantum Hall conductor. The dark gray squares in the center denote the areas of the THz emission images of (a). Reprinted with permission from [70].

with sub-wavelength resolutions were achieved. This technology was successfully applied to the study of non-equilibrium quantum carrier transport in semiconductors. In the future, the use of CNT/graphene THz detectors would enable ultrasensitive and nanometer-resolution THz imaging.

Regarding applications of the THz sensor and imager, materials science and biochemistry are interesting research areas. In these fields, the detection of very weak THz radiation from electrons in materials and biomolecules is expected to reveal much information concealed behind material properties and living activity.

Acknowledgments

We thank the ZEON Corporation for providing CNT film. This work was supported in part by Collaborative Research Based on Industrial Demand, the Matching Planner Program, and the Center of Innovation Program from the Japan Science and Technology Agency, JSPS KAKENHI, Grant Numbers JP17K19026, JP17H02730, JP16H00798, JP16H00906, and JP16J09937, from the Japan Society for the Promotion of Science, the Murata Science Foundation, and Support for Tokyo Tech Advanced Researchers (STAR).

References

1. B. Ferguson and X-C. Zhang, "Materials for terahertz science and technology," *Nature Mater.* **1**, 26 (2002).
2. M. Tonouchi, "Cutting-edge terahertz technology," *Nature Photon.* **1**, 97 (2007).
3. B. S. Williams, "Terahertz quantum-cascade lasers,"*Nature Photon.* **1**, 517 (2007).
4. S. Suzuki, M. Asada, A. Teranishi, H. Sugiyama, and H. Yokoyama, "Fundamental oscillation of resonant tunneling diodes above 1 THz at room temperature," *Appl. Phys. Lett.* **97**, 242102 (2010).
5. S. Ariyoshi, C. Otani, A. Dobroiu, H. Matsuo, H. Sato, T. Taino, K. Kawase, and H. Shimizu, "Terahertz imaging with a direct detector based on superconducting tunnel junctions," *Appl. Phys. Lett.* **88**, 203503 (2006).
6. J. Wei, D. Olaya, B. S. Karasik, S. V. Pereverzev, A. V. Sergeev, and M. E. Gershenson, "Ultrasensitive hot-electron nanobolometers for terahertz astrophysics," *Nature Nanotechnol.* **3**, 496 (2008).
7. S. Komiyama, O. Astafiev, V. Antonov, H. Hirai, and T. Kutsuwa, "A single-photon detector in the far-infrared range," *Nature* **403**, 405 (2000).
8. P. Kleinschmidt, S. Giblin, A. Tzalenchuk, H. Hashiba, V. Antonov, and S. Komiyama, "Sensitive detector for a passive terahertz imager," *J. Appl. Phys.* **99**, 114504 (2006).
9. L. Ozyuzer, A. E. Koshelev, C. Kurter, N. Gopalsami, Q. Li, M. Tachiki, K. Kadowaki, et al., "Emission of Coherent THz Radiation from Superconductors," *Science* **318**, 1291 (2007).
10. M. Tsujimoto, T. Yamamoto, K. Delfanazari, R. Nakayama, T. Kitamura, M. Sawamura, T. Kashiwagi, H. Minami, M. Tachiki, K. Kadowaki, and R. A. Klemm, *Phys. Rev. Lett.* **108**, 107006 (2012).
11. R. Saito, G. Dresselhaus, and M. S. Dresselhaus, *Physical Properties of Carbon Nanotubes*; London: Imperial College Press, 1998.
12. M. I. Katsnelson, *Graphene: Carbon in Two Dimensions*; Cambridge, UK: Cambridge University Press, 2012.
13. P. W. Barone, S. Baik, D. A. Heller, and M. S. Strano, "Near-infrared optical sensors based on single-walled carbon nanotubes," *Nature Mater.* **4**, 86 (2005).
14. M. E. Itkis, F. Borondics, A. Yu, and R. C. Haddon, "Bolometric infrared photoresponse of suspended single-walled carbon nanotube films," *Science* **312**, 413 (2006).
15. K. Fu, R. Zannoni, C. Chan, S. H. Adams, J. Nicholson, E. Polizzi, and K. S. Yngvesson, "Terahertz detection in single wall carbon nanotubes," *Appl. Phys. Lett.* **92**, 033105 (2008).

16. Y. Kawano, T. Fuse, S. Toyokawa, T. Uchida, and K. Ishibashi, "Terahertz photon-assisted tunneling in carbon nanotube quantum dots," *J. Appl. Phys.* **103**, 034307 (2008).
17. Y. Kawano, T. Uchida, and K. Ishibashi, "Terahertz sensing with a carbon nanotube/two-dimensional electron gas hybrid transistor," *Appl. Phys. Lett.* **95**, 083123 (2009).
18. X. He, N. Fujimura, J. M. Lloyd, K. J. Erickson, A. A. Talin, Q. Zhang, W. Gao, et al., "Carbon nanotube terahertz detector," *Nano Letters* **14**, 3953–3958 (2014).
19. K. Erickson, X. He, A. A. Talin, B. Mills, R. H. Hauge, T. Iguchi, N. Fujimura, Y. Kawano, J. Kono, and F. Léonard, "Figure of merit for carbon nanotube photothermoelectric detectors," *ACS Nano* **9**, 11618–11627 (2015).
20. D. Suzuki, S. Oda, and Y. Kawano, "A flexible and wearable terahertz scanner," Nature Photonics **10**, 809–814 (2016).
21. P. K. Tien and J. P. Gordon, "Multiphoton process observed in the interaction of microwave fields with the tunneling between superconductor films," *Phys. Rev.* **129**, 647 (1963).
22. J. M. Hergenrother, M. T. Tuominena, J. G. Lua, D. C. Ralpha, and M. Tinkham, "Charge transport and photon-assisted tunneling in the NSN single-electron transistor," *Physica B* **203**, 327 (1994).
23. B. Leone, J. R. Gao, T. M. Klapwijk, B. D. Jackson, W. M. Laauwen, and G. de Lange, "Electron heating by photon-assisted tunneling in niobium terahertz mixers with integrated niobium titanium nitride striplines," *Appl. Phys. Lett.* **78**, 1616 (2001).
24. T. H. Oosterkamp, L. P. Kouwenhoven, A. E. A. Koolen, N. C. van der Vaart, and C. J. P. M. Harmans, "Photon sidebands of the ground state and first excited state of a quantum dot," *Phys. Rev. Lett.* **78**, 1536 (1997).
25. T. H. Oosterkamp, T. Fujisawa, W. G. van der Wiel, K. Ishibashi, R. V. Hijman, S. Tarucha, and L. P. Kouwenhoven, "Microwave spectroscopy of a quantum-dot molecule," *Nature* **395**, 873 (1998).
26. J. R. Tucker and M. J. Feldman, "Quantum detection at millimeter wavelengths," *Rev. Mod. Phys.* **57**, 1055 (1985).
27. F. Thiele, U. Merkt, J. P. Kotthaus, G. Lommer, F. Malcher, U. Rossler, and G. Weimann, "Cyclotron masses in n-GaAs/$Ga_{1-x}Al_xAs$ heterojunctions," *Solid State Commun.* **62**, 841 (1987).
28. S. H. Tessmer, P. I. Glicofridis, R. C. Ashoori, L. S. Levitov, and M. R. Melloch, "Subsurface charge accumulation imaging of a quantum Hall liquid," *Nature* **392**, 51 (1998).
29. Y. Kawano, Y. Hisanaga, H. Takenouchi, and S. Komiyama: "Highly sensitive and tunable detection of far-infrared radiation by quantum Hall devices," *J. Appl. Phys.* **89**, 4037 (2001).
30. Y. Kawano, Y. Hisanaga, and S. Komiyama, "Cyclotron emission from quantized Hall devices: Injection of nonequilibrium electrons from contacts," *Phys. Rev. B* **59**, 12537 (1999).
31. Y. Kawano and S. Komiyama, "Spatial distribution of non-equilibrium electrons in quantum Hall devices: Imaging via cyclotron emission," *Phys. Rev. B* **68**, 085328 (2003).
32. S. Nanot, E. H. Hároz, J.-H. Kim, R. H. Hauge, and J. Kono, "Optoelectronic properties of single-wall carbon nanotubes," *Advanced Mater.* **24**, 4977 (2012).

33. N. Oda, "Uncooled bolometer-type terahertz focal plane array and camera for real-time imaging," *C. R. Phys.* **11**, 496–509 (2010).
34. R. Han, Y. Zhang, Y. Kim, D.Y Kim, H. Shichijo, E. Afshari, K.O. Kenneth, "Active terahertz imaging using Schottky diodes in CMOS: Array and 860-GHz pixel," *IEEE J. Solid-State Circ.* **48**, 2296–2308 (2013).
35. J. P. Guillet, B. Recur, L. Frederique, B. Bousquet, L. Canioni, I. Manek-Hönninger, P. Desbarats, P. Mounaix, "Review of terahertz tomography techniques," *J. Infrared Millim. Terahertz Waves* **35**, 382–411 (2014).
36. K. Kawase, T. Shibuya, S. Hayashi, and K. Suizu, "THz imaging techniques for nondestructive inspections," *C. R. Phys.* **11**, 510–518 (2010).
37. X. Cai, A. B. Sushkov, R. J. Suess, M. M. Jadidi, G. S. Jenkins, L. O. Nyakiti, R. L. Myers-Ward, et al., "Sensitive room-temperature terahertz detection via the photothermoelectric effect in graphene," *Nature Nanotechnol.* **9**, 814 (2014).
38. L. Vicarelli, M. S. Vitiello, D. Coquillat, A. Lombardo, A. C. Ferrari, W. Knap, M. Polini, V. Pellegrini, and A. Tredicucci, "Graphene field-effect transistors as room-temperature terahertz detectors," *Nature Mater.* 11, 865 (2012).
39. T. Low and P. Avouris, "As a general review of graphene plasmonics," *ACS Nano* **8**, 1086 (2014).
40. A. V. Muraviev, S. L. Rumyantsev, G. Liu, A. A. Balandin, W. Knap and M. S. Shur, "Plasmonic and bolometric terahertz detection by graphene field-effect transistor," *Appl. Phys. Lett.* **103**, 181114 (2013).
41. Y. Kawano, "Wide-band frequency-tunable terahertz and infrared detection with graphene," *Nanotechnology* **24**, 214004-1-6 (2013).
42. J. W. McClure, "Diamagnetism of Graphite," *Phys. Rev.* **104**, 666 (1956).
43. M. L. Sadowski, G. Martinez, M. Potemski, C. Berger, and W. A. de Heer, "Landau Level Spectroscopy of Ultrathin Graphite Layers," *Phys. Rev. Lett.* **97**, 266405 (2006).
44. Z. Jiang, E. A. Henriksen, L.-C. Tung, Y.-J. Wang, M. E. Schwartz, M. Y. Han, P. Kim, and H. L. Stormer, "Infrared Spectroscopy of Landau Levels of Graphene," *Phys. Rev. Lett.* **98**, 197403 (2007).
45. G. Li, A. Luican, and E. Y. Andrei, "Scanning Tunneling Spectroscopy of Graphene on Graphite," *Phys. Rev. Lett.* 102, 176804 (2009).
46. K. S. Novoselov, A. K. Geim, S. V. Morozov, D. Jiang, M. I. Katsnelson, I. V. Grigorieva, S. V. Dubonos, and A. A. Firsov, "Two-dimensional gas of massless Dirac fermions in graphene," *Nature* **438**, 197 (2005).
47. A. D. Semenov, H.-W. Hübers, H. Richter, M. Birk, M. Krocka, U. Mair, Y. B. Vachtomin, et al., "Superconducting hot-electron bolometer mixer for terahertz heterodyne receivers," *IEEE Trans. Appl. Supercond.* **13**, 168–171 (2003).
48. W. Zhang, P. Khosropanah, J. R. Gao, E. L. Kollberg, K. S. Yngvesson, T. Bansal, R. Barends, and T. M. Klapwijk, "Quantum noise in a terahertz hot electron bolometer mixer," *Appl. Phys. Lett.* **96**, 111113 (2010).
49. K. S. Yngvesson, "Ultrafast two-dimensional electron gas detector and mixer for terahertz radiation," *Appl. Phys. Lett.* **76**, 777–779 (2000).
50. Y. Kawano and K. Ishibashi, "Scanning nanoelectrometer based on a two-dimensional electron gas transistor with a probe-integrated gate electrode," *Appl. Phys. Lett.* **96**, 142109 (2010).
51. M. Ohtsu, ed., *Near-Field Nano/Atom Optics and Technology*; New York: Springer, 1998.

52. T. Saiki, S. Mononobe, M. Ohtsu, N. Saito, and J. Kusano, "Tailoring a high-transmission fiber probe for photon scanning tunneling microscope," *Appl. Phys. Lett.* **68**, 2612 (1996).
53. F. Zenhausern, Y. Martin, and H. K. Wickramasinghe, "Scanning interferometric apertureless microscopy: Optical imaging at 10 Angstrom resolution," *Science* **269**, 1083 (1995).
54. W. C. Symons III, K. W. Whites, and R. A. Lodder, "Theoretical and experimental characterization of a near-field scanning microwave microscope (NSMM)," *IEEE Trans. Microwave Theory Tech.* **51**, 91 (2003).
55. M. Tabib-Azar and Y. Wang, "Design and fabrication of scanning near-field microwave probes compatible with atomic force microscopy to image embedded nanostructures," *IEEE Trans. Microwave Theory Tech.* **52**, 971 (2004).
56. S. Hunsche, M. Koch, I. Brener, and M. C. Nuss, "THz near-field imaging," *Opt. Commun.* **150**, 22 (1998).
57. N. C. J. van der Valk and P. C. M. Planken, "Electro-optic detection of subwavelength terahertz spot sizes in the near field of a metal tip," *Appl. Phys. Lett.* **81**, 1558 (2002).
58. H. T. Chen, R. Kersting, and G. C. Cho, "Terahertz imaging with nanometer resolution," *Appl. Phys. Lett.* **83**, 3009 (2003).
59. A. J. Huber, F. Keilmann, J. Wittborn, J. Aizpurua, and R. Hillenbrand, "Terahertz near-field nanoscopy of mobile carriers in single semiconductor nanodevices," *Nano Lett.* **8**, 3766 (2008).
60. A. Doi, F. Blanchard, H. Hirori, and K. Tanaka, "Near-field THz imaging of free induction decay from a tyrosine crystal," *Opt. Exp.* **18**, 18419–18424 (2010).
61. Y. Kawano and K. Ishibashi, "An on-chip near-field terahertz probe and detector," *Nature Photon.* **2**, 618–621 (2008).
62. Y. Kawano, "Highly Sensitive Detector for On-Chip Near-Field THz Imaging," *IEEE J. Selec. Top. Quantum Electron.* **17**, 67 (2011).
63. Y. Kawano, "Terahertz sensing and imaging based on nanostructured semiconductors and carbon materials," *Laser Photon. Rev.* **6**, 246 (2012).
64. Y. Kawano and K. Ishibashi, "On-chip near-field terahertz detection based on a two-dimensional electron gas," *Physica E*, **42**, 1188–1191 (2010).
65. D. Suzuki, S. Oda, and Y. Kawano, "GaAs/AlGaAs field-effect transistor for tunable terahertz detection and spectroscopy with built-in signal modulation," *Appl. Phys. Lett.* **102**, 122102–1–4 (2013).
66. S. Komiyama, Y. Kawaguchi, T. Osada, and Y. Shiraki, "Evidence of nonlocal breakdown of the integer quantum Hall effect," *Phys. Rev. Lett.* **77**, 558 (1996).
67. I. I. Kaya, G. Nachtwei, K. von Klitzing, and K. Eberl, "Spatially resolved monitoring of the evolution of the breakdown of the quantum Hall effect: Direct observation of inter-Landau-level tunneling," *Europhys. Lett.* **46**, 62 (1999).
68. Y. Kawano and T. Okamoto, "Imaging of intra- and inter-Landau-level scattering in quantum Hall systems," *Phys. Rev. B* **70**, 081308(R) (2004).
69. Y. Kawano and T. Okamoto, "Macroscopic channel-size effect of nonequilibrium electron distributions in quantum Hall conductors," *Phys. Rev. Lett.* **95**, 166801 (2005).
70. Y. Kawano, "Chip-based near-field Terahertz microscopy," *IEEE Trans. Terahertz Sci. Technol.* **6**, 356–364 (2016).

2

Fiber-Optic Brillouin Distributed Sensors: From Dynamic to Long-Range Measurements

Alayn Loayssa, Javier Urricelqui, Haritz Iribas, Juan José Mompó, and Jon Mariñelarena

CONTENTS

2.1 Introduction....27
2.2 Fundamentals of Brillouin Distributed Sensors....28
2.3 Research Trends in BOTDA Sensors....31
  2.3.1 Dynamic Measurements....31
  2.3.2 Long-Range Measurements....35
2.4 Conclusions....39
References....39

2.1 Introduction

Brillouin distributed sensors (BDSs) have been extensively researched and developed over the last three decades due to their ability to measure the distribution of temperature and strain along an optical fiber, providing thousands of measurement locations along a structure in a cost-effective manner. The applications of the technology are multiple and in diverse fields—for instance, in the oil and gas industry, where BDSs have been applied to measure temperature and strain along the umbilical cables used for subsea wells. They have also been extensively deployed for more than a decade to assess the integrity of oil and gas pipelines regarding the detection of leaks as well as geohazard threats such as erosion, landslide, and subsidence [1]. In the electric grid, BDSs have been deployed in the ampacity and thermal rating of high voltage cables, particularly in underground or subsea installations [2]. In aerospace and other industries such as wind power, they can be used to measure the dynamic strain of composite structures during operation or in fatigue tests. In geotechnical engineering, BDSs can be installed directly in the soil to monitor slopes and landslides or in levees and embankments.

They can also be fitted to structures such as tunnels or piles to assess their behavior in response to soil characteristics or to nearby construction works [3]. BDSs can also be deployed to detect seepage in hydraulic structures such as dams, to measure the water content of soil for agricultural applications, or to regulate the irrigation systems of parks and gardens [3].

In the following, we review the fundamentals of BDSs, focusing on their most successful variety: Brillouin optical time-domain analysis (BOTDA) sensors. In addition, we discuss recent progress in the technology, focusing first on the fast measurements that are needed in applications that require dynamic characterization and then on techniques for long-range measurement.

2.2 Fundamentals of Brillouin Distributed Sensors

The nonlinear Brillouin scattering effect results from the interaction in a material between optical photons and acoustic phonons. In optical fibers, spontaneous Brillouin scattering takes place when a narrowband pump light interacts with thermally excited acoustic waves. Due to the acousto-optic effect, the acoustic wave generates a periodic perturbation of the refractive index that reflects part of the energy of the pump wave by Bragg diffraction. The reflected light (Stokes wave) experiences a shift in frequency due to the Doppler effect. This is the so-called Brillouin frequency shift (BFS), which is related to the velocity of the acoustic waves in the fiber. Spontaneous Brillouin scattering can become stimulated Brillouin scattering (SBS) if a Stokes wave whose optical frequency is downshifted by the BFS from that of the pump is injected into the fiber. In SBS, the counterpropagation of the pump and the Stokes waves generates a moving interference pattern which induces an acoustic wave due to the electrostriction effect, reflecting part of the energy of the pump wave by Bragg diffraction. The frequency of the light reflected as a consequence of this process is downshifted by the BFS, which is given by

$$BFS = \frac{2n\nu_A}{\lambda_P} \tag{2.1}$$

where:

ν_A is the velocity of the acoustic wave
n is the refractive index
λ_P is the wavelength of the pump wave

This shift is around 10.8 GHz at a wavelength of 1550 nm for silica fibers. As a result of this interaction, there is a transfer of energy between the pump and Stokes waves that simultaneously enhances the amplitude of the acoustic waves, leading to a stimulation of the process. Therefore, as the signals propagate along the fiber, the Stokes wave is amplified while the pump wave loses energy and is attenuated. In terms of spectral response, this interaction can be described by the Brillouin gain coefficient, given by

$$g_B(\Delta\nu) = g_B \frac{(\Delta\nu_B/2)^2}{(\Delta\nu)^2 + (\Delta\nu_B/2)^2} \tag{2.2}$$

where:

g_B is the peak gain coefficient

$\Delta\nu$ is the detuning from the center of the Brillouin resonance

$\Delta\nu_B$ is the Brillouin line width, given by the inverse of the phonon lifetime and usually in the order of a few tens of MHz for standard single-mode fibers

When using the Brillouin scattering effect for optical fiber sensing purposes, the dependence of Brillouin parameters on temperature and strain in the fiber is exploited. This linear dependence of the BFS with the applied strain $\delta\varepsilon$ or temperature change δT is given by

$$BFS - BFS_0 = A \cdot \delta\varepsilon + B \cdot \delta T \tag{2.3}$$

where BFS_0 is the reference BFS value, measured at room temperature and in the loose state of the fiber—that is, laid freely in order to avoid any artificial disturbances. A and B are the strain and temperature coefficients given in units of MHz/με and MHz/°C, respectively. Typical values for these strain and temperature coefficients are around 0.05 MHz/με and 1 MHz/°C for most standard fibers at 1550 nm.

Therefore, the BFS in the fiber can be found by measuring the Brillouin gain spectrum (BGS) and then translating the results to temperature or strain. In order to perform this spectral measurement, a pump wave can be injected at one end of an optical fiber and a probe wave at the other, and the gain experienced by the latter can be recorded as the wavelength separation between the two waves is scanned. Finally, the strain and temperature of the fiber can be obtained from the measured Brillouin spectrum peak gain using the coefficients obtained from a previous calibration of the deployed sensing fiber. Nevertheless, this procedure measures the global gain spectrum of the total length of the fiber, which is given by the integration of the local gain (or loss) at each position along the fiber. In order to implement a distributed sensor featuring spatially resolved measurements, the gain spectra at individual locations in the fiber must be isolated. This can be done in the time, coherence, or frequency domains, giving rise to the three main analysis BDS types: Brillouin optical time-domain analysis (BOTDA), Brillouin optical correlation-domain analysis (BOCDA), and Brillouin optical frequency-domain analysis (BOFDA). This chapter is focused on BOTDA sensors because they are the most successful BDS type in terms of performance and practical applications.

The fundamentals of BOTDA sensors are highlighted in Figure 2.1. A continuous wave (CW) probe is counterpropagated with a pump pulse. The pulse propagates along the fiber and at each location imparts Brillouin gain to the probe. Finally, the received probe signal is detected in the time domain, so that the position-dependent gain can be calculated by the classical time-of-flight method. If we consider t_0 the time when the trailing end of the pump pulse enters the fiber, then the probe signal received at $t = t_0 + 2\,z/c$ has interacted with the pulse between positions z and $z + u$, where c is the speed of light in the fiber and $u = \tau\, c/2$ is the interaction length, with τ being the temporal duration of the pulse. Therefore, the spatial resolution of the measurement is given by u. For instance, in order to achieve a resolution of 1 m, a pulse duration of around 10 ns must be deployed. Notice that the division by two in the expression for u is due to the fact that the pump pulse and the CW probe are counterpropagating. From the discussion above, it is clear that the time-dependent BOTDA signal can be translated to position-dependent gain by the simple relation $z = (t - t_0)\, c/2$. Therefore, as shown in the figure, it is possible to reconstruct the position-dependent Brillouin spectra distributed along the fiber by sweeping the wavelength separation between pump and probe and measuring multiple BOTDA time-domain traces. Finally, strain or temperature can be quantified from the measured BFS at each position.

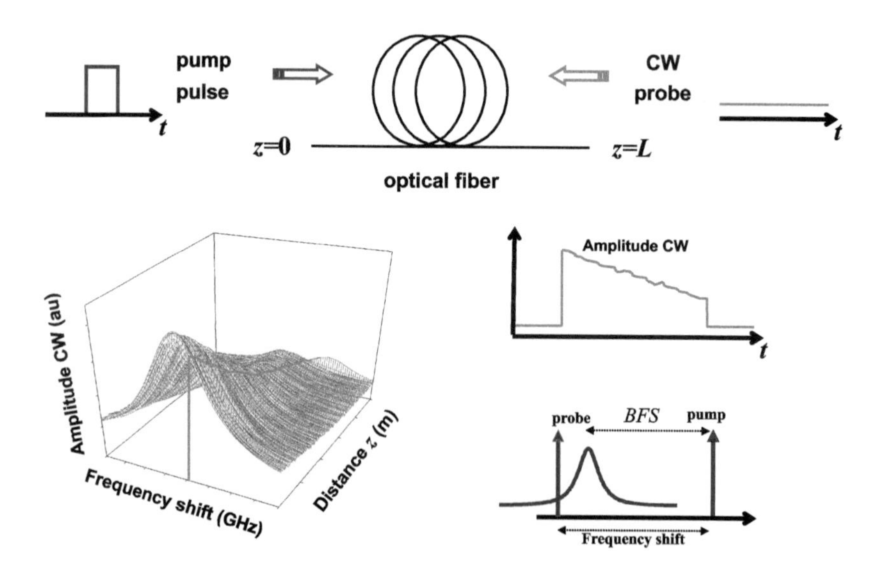

FIGURE 2.1
Fundamentals of BOTDA.

2.3 Research Trends in BOTDA Sensors

BOTDA sensor technology has reached a high degree of maturity in the last decade, coming from a time when the technology fundamentals were developed and reaching a stage in which research has been increasingly focused on enhancing the performance of the sensor system so that it can became suitable for deployment in practical applications and contribute to solving a number of industrial and societal challenges. One of the main research topics in recent years has been reducing the measurement time to provide dynamic measurements, which are needed for a number of applications in which there is a rapid variation of the measurand. Typically, this is for assessing the structural health of structures that are experiencing dynamic loads and the quantity of interest is strain. Paradigmatic examples of these applications can be found in the aerospace industry, where the performance of the wings, fuselage, and other elements of an aircraft need to be assessed. Another example would be wind turbines, particularly those in harsh environments such as offshore wind farms. Continuous structural monitoring of the blades and other components of those turbines can lead to significant savings in maintenance and extended operational service life.

Another area of research where a number of contributions can be found is related to developing methods to extend the distance range of the sensor. This is driven by applications in which very large structures or systems need to be measured. For instance, monitoring pipelines, high voltage power lines, or railways. These require sensing lengths that can be in the hundreds of kilometers.

2.3.1 Dynamic Measurements

As explained in Section 2.2, the measurement procedure followed in BOTDA sensors involves sweeping the frequency difference between the pump and probe waves so as to obtain the BGS along every location of the fiber. Then, the BFS is obtained after performing a regression to the experimentally measured data in order to find the frequency of the peak of these spectra. However, this is a time-consuming process that hinders the ability of BOTDA sensors to provide dynamic measurements in the hundreds or even thousands of measurements per second that are required for applications such as the fatigue testing of industrial structures or the detection of vibrations.

Several techniques have been proposed to circumvent this limitation in the measurement time of BOTDA sensors—for instance, the use of arbitrary waveform generation in conventional BOTDA setups to speed up the measurement process [4] or the use of pump and probe waves containing multiple spectral components so that several frequencies within the Brillouin spectrum can be measured simultaneously [5, 6]. However, the simplest solution to performing real-time measurements in a cost-effective manner without increasing the complexity of the BOTDA setup is probably the so-called slope-assisted method. This is based on the frequency discriminator principle, in which the optical frequency of the probe wave is fixed on the slope of the BGS and the amplitude of the detected signal is interrogated, so that variations of the BFS due to temperature or strain changes are translated to variations in the amplitude of the detected probe wave [7]. This measurement principle is schematically depicted in Figure 2.2a, where it can be seen that changes in the detuning of the probe wave within the BGS slope lead to changes in the detected probe power. Therefore, the time-consuming frequency sweep process required by conventional BOTDA sensors is removed altogether.

However, a major issue with the slope-assisted method lies in its reliance on amplitude measurements. In order to obtain correct measurements, the BFS to probe wave power transfer characteristics provided by the slope of the BGS must remain stable. However, the amplitude of the Brillouin spectrum depends on the pump pulse power. Hence, any variation in the attenuation of the fiber being tested, which is highly probable when it is affixed to a structure that is experiencing dynamic deformation, would be misinterpreted as a strain change and lead to measurement error. This is highlighted in Figure 2.2a, which shows the effect of a reduction in Brillouin gain on the measurement (dark gray line). This effect can be mitigated with a slight modification of the slope-assisted method in which sequential readings on both the positive and negative slopes of the BGS are performed [8]. This has been shown to increase the tolerance of the method to power variations of up to 6 dB, at the cost of doubling the measurement time.

Nevertheless, an alternative slope-assisted BOTDA method for dynamic measurements has been proposed that greatly increases the tolerance to variations of the pump pulse power [9]. Figure 2.3 depicts the fundamentals

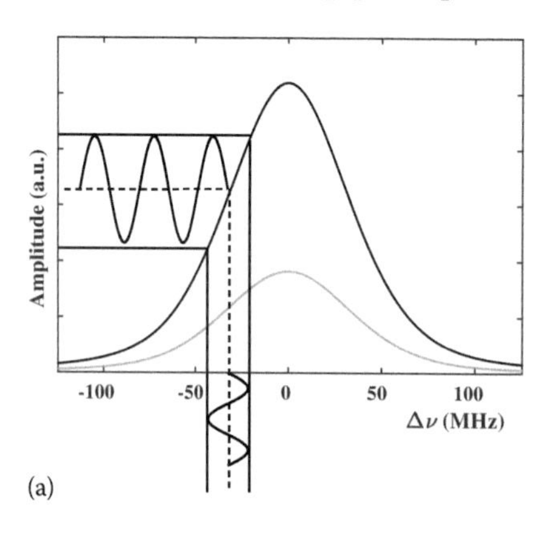

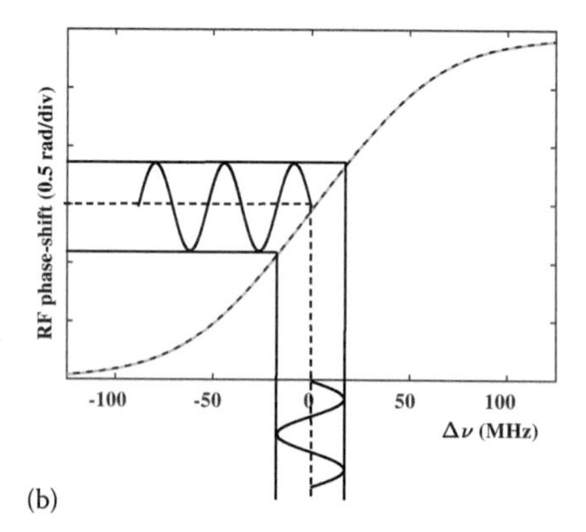

FIGURE 2.2
Fundamentals of slope-assisted methods for dynamic measurements in BOTDA sensors: (a) conventional slope-assisted BOTDA based on the amplitude of the BGS, (b) BOTDA using PM and self-heterodyne detection to measure RF phase shift.

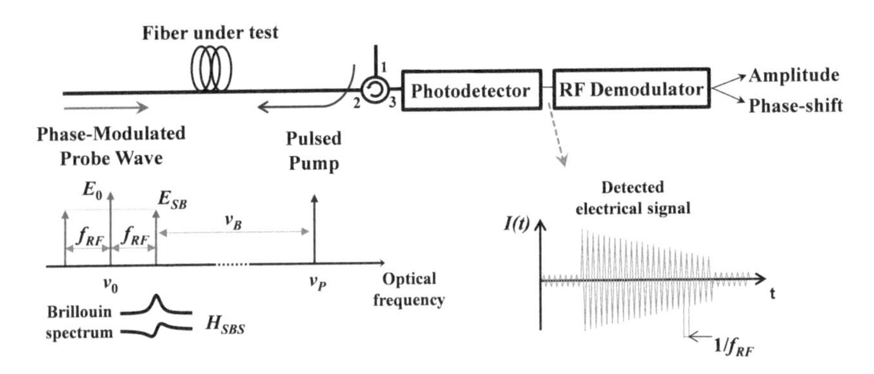

FIGURE 2.3
Fundamentals of BOTDA RF. (From J. Urricelqui et al., *Opt. Express*, 20, 26942–26949, 2012. Copyright 2012 Optical Society of America.)

of the method, which is based on using a phase-modulated (PM) probe wave and self-heterodyne detection. As it is schematically depicted in the figure, the pump wave induces a Brillouin spectrum (amplitude and phase) that is experienced by just one of the sidebands of the PM probe wave. Then, the probe wave is detected by a photo-receiver generating an radio frequency (RF) signal at the frequency of the PM signal. Finally, RF demodulation is deployed to obtain the amplitude and phase shift of the RF signal. For dynamic measurements, we focus on the detected RF phase shift. Figure 2.2b depicts the RF phase shift measured at a particular location in the fiber as a function of the detuning of the sideband of the PM wave from the center of the BGS. Dynamic measurements are performed by setting the pump and probe frequency difference to the center of the spectra ($\Delta\nu = 0$) so that any change in BFS due to strain in the fiber is then translated to a change in the measured RF phase shift. A remarkable feature of this RF phase shift spectrum is that its shape does not depend on the peak Brillouin gain or the pump pulse power [9]. This is due to the special characteristics of the detection process, which involves interference between two RF signals originating in the beating between the optical carrier of the PM wave and each sideband. Moreover, the phase shift in addition to the amplitude of the Brillouin interaction spectrum intervenes, in contrast to conventional BOTDA sensors that just take advantage of the amplitude response.

Figure 2.4 depicts a sample dynamic measurement performed with self-heterodyne BOTDA. A 1-m section at the end of a 930 m-length fiber was fixed to a 1-m cantilever beam that was made to vibrate. It can be seen in the measurement that only the fiber section in the cantilever beam experiences dynamic strain, while the rest of the fiber remains static. This measurement was performed with a setup that implemented a slight modification of the principle in Figure 2.3, in which two pump pulses with orthogonal polarizations are deployed to simultaneously induce a gain spectrum and a loss spectrum on the upper and lower frequency sidebands, respectively, of the

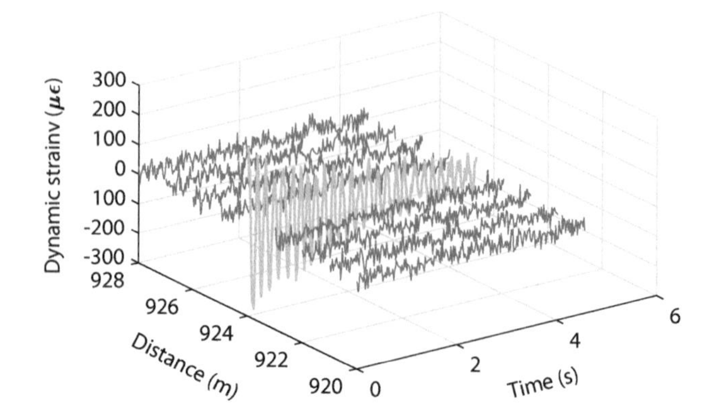

FIGURE 2.4
Dynamic measurement example. (From J. Urricelqui et al., *J. Lightwave Technol.*, 33(12), 2633–2638, 2015. Copyright 2015 IEEE.)

PM probe wave [10]. This setup provided a polarization diversity measurement of the Brillouin interaction independent of polarization-induced fading. This leads to a halving of the measurement time and negates the use of the polarization switch or scrambler that is usually deployed in BOTDA sensors.

Another limitation of slope-assisted BOTDA setups is the dynamic measurement range—that is, the range of BFS values that can be measured with a specified precision. In conventional slope-assisted BOTDA, this is roughly given by the Brillouin spectrum line width. Considering a typical value for a standard single-mode fiber of $\Delta\nu_B \approx 30$ MHz and a strain coefficient of 20 με/MHz, this translates to a strain measurement range of 600 με. This is insufficient for many applications in which the dynamic strain of industrial parts or structures is measured. For instance, the maximum strain experienced by the blade of a wind turbine in operation is in the order of 10,000 με. Moreover, the actual measurement range required in strain measurements can be even larger because the use of different fiber types or the installation process of the fiber in the structure can lead to variations in the BFS along the fiber even before dynamic loads are experienced.

Therefore, it is necessary to extend the measurement range for the practical applications of slope-assisted BOTDA in dynamic scenarios related to structural health monitoring. One option is switching the probe wave frequency in several steps within a given range, so that multiple BGS slopes can be interrogated [11] [12]. However, the measurement time also increases proportionally to the number of frequency steps. Another option is to deploy short pump pulses with the aim of broadening the BGS shape, which has been applied in conventional [13] as well as RF-based slope-assisted BOTDA [14]. Figure 2.5 depicts the measurement of the RF amplitude and RF phase shift spectra in self-heterodyne BOTDA as the pump pulse duration is reduced. Notice that the measurement range for dynamic measurements using the RF

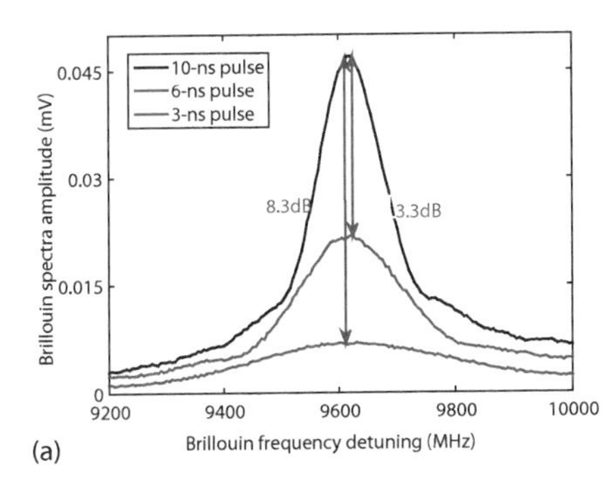

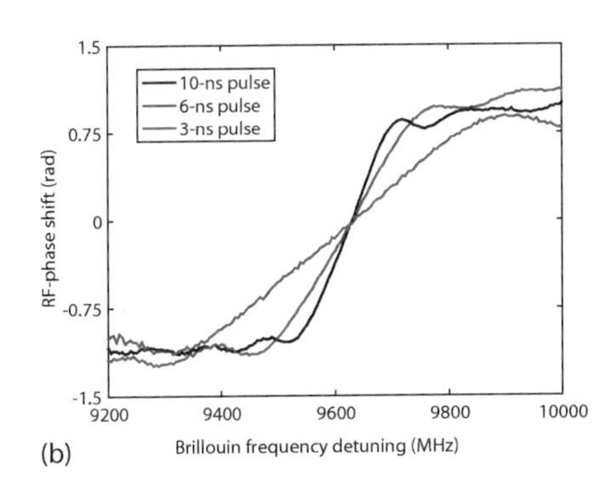

FIGURE 2.5
Experimental (a) amplitude and (b) RF phase shift spectra when short pulses are employed. (From J. Mariñelarena et al., *IEEE Photon. J.*, 9(3), 1–10, 2017. Copyright 2017 IEEE.)

phase shift is enhanced for shorter pulses. However, as shown in Figure 2.5a, the price to pay for the increased dynamic range comes in the form of a reduction of the magnitude of the gain spectra, which leads to an equal reduction of the signal-to-noise ratio (SNR) of the measurement. In addition, it is also possible to extend the dynamic range of the RF-based slope-assisted BOTDA by launching pump pulses containing multiple frequency components [14]. This causes the Brillouin spectra generated by each component to overlap, providing a wider linear region of the detected RF phase shift spectrum and allowing the larger Brillouin frequency shift variations to be measured. Notice that this method is exclusive to self-heterodyne BOTDA and is not applicable to conventional BOTDA (Figure 2.5).

Finally, another issue that needs to be tackled with slope-assisted BOTDA sensors for dynamic operation is the recently found dependence of the Brillouin spectrum line width on the pump power [15]. This unexpected finding means that BFS measurement errors can appear in techniques such as slope-assisted BOTDA that rely on the shape of the BGS [9]. However, it has been found that in self-heterodyne BOTDA, this effect is reduced and only becomes significant at extremely high pump powers [16].

2.3.2 Long-Range Measurements

The distance range of measurements performed using a BOTDA sensor is given by the length of sensing fiber that the system is able to measure with a specified performance in terms of measurement precision and time. This performance ultimately depends on the SNR of the probe signal received from far-away locations, which in turn depends on the power of the probe and pulse pump waves that it is possible to inject into the sensing fiber.

The maximum power of the pump pulse that can be injected into the sensing fiber is constrained by the onset of nonlinear effects: mainly modulation instability and Raman scattering [17]. As for the probe wave power, its maximum power in the fiber is limited firstly by the onset of so-called nonlocal effects and ultimately by the Brillouin threshold of the fiber link.

Nonlocal effects are intrinsically linked to the nature of the Brillouin interaction in a BOTDA sensor. The pump pulse gives energy to the probe wave as they counterpropagate in the sensing fiber, leading to an amplification of the former that is used for sensing. However, the pump pulse is also depleted in its journey due to the energy that it loses in the interaction at each location. Moreover, this depletion depends on the BFS at every position at which the pump pulse has interacted with the probe wavefronts prior to its arrival at a given location. This distorts the measured spectrum and introduces a bias in the obtained BFS that entails a systematic error in the measurement [18]. Notice that this measurement impairment is a nonlocal effect in the sense that the BFS measurement at a particular location in the fiber is affected by the Brillouin interaction between pump and probe at other locations.

Several methods have been proposed to mitigate nonlocal effects. An early proposal was the use of a post-processing technique for BFS reconstruction using a multidimensional minimization algorithm to determine the coefficients of a parameterized unknown profile [19]. However, this is a time-consuming process and its applicability to real-world measurement scenarios seems rather difficult. Another proposal for nonlocal effects mitigation was a time-multiplexing method based on pulsing the probe wave so as to limit the interaction length with the pump pulse [20][21]. However, this entails a measurement time penalty.

Altogether, the simplest and most successful means to mitigate the nonlocal effects of pump depletion is the use of a dual-probe setup in which two probe waves with identical separation in optical frequency from the pump are deployed [22]. These two waves induce two simultaneous Brillouin interactions (gain and loss) on the pulse that compensate each other, hence suppressing pulse depletion. However, this technique only mitigates nonlocal effects of the first order, but there is a second-order nonlocal effect related to the appearance of linear distortion in the pump pulse spectrum due to its interaction with the two probe waves [23]. This is due to the fact that the gain and loss spectra induced by the two probe waves in the pump only overlap when their optical frequency separation equals the BFS of the fiber. However, during the spectral scan required to characterized the full BGS, the frequency separation between both waves is swept, so that for certain frequency differences the gain and loss spectra fall at different frequencies and distort the pulse spectrum, leading to measurement errors.

One method to overcome second-order nonlocal effects is the use of an alternative scanning method in the dual-probe BOTDA sensor configuration in which the frequency difference between pump and probe waves is maintained so as to ensure that the gain and loss spectra induced in the pulse continuously overlap [24] [25]. We have also introduced a method to compensate second-order nonlocal effects that also enables dual-probe BOTDA setups to overcome the Brillouin threshold of the fiber [26]. The basis of this technique is to introduce modulation or *dithering* to the optical frequency of the probe waves, as schematically depicted in Figure 2.6. This causes the pump pulse to experience interaction with probe wavefronts that have a different frequency detuning, the net effect being that the integrated gain and loss spectrum generated by both probe waves broadens so that no distortion is introduced in the pump pulse spectrum and second-order nonlocal effects are suppressed [26]. A side benefit of this probe dithering method is that the optical frequency modulation of the probe waves overcomes the Brillouin threshold limit of the fiber. Altogether, the deployment of the probe dithering method has demonstrated the largest probe wave power ever injected in a long-range BOTDA sensing link, hence enhancing the SNR of the measurements.

The probe dithering method can be used with little modification to amplify the pump pulses as they travel along the fiber, hence increasing the sensor range. Figure 2.6 hints at the way to achieve this amplification. Notice that the pump pulse can be made to experience gain simply by removing the lower optical frequency probe component. Then the higher-frequency probe wave induces a broad and flat gain spectrum that equally amplifies all frequency components of the pump pulse. This technique has been demonstrated in a 100 km BOTDA sensor incorporating distributed Brillouin gain provided by the probe [27]. Figure 2.7a depicts the BGS distribution measured along the fiber when a dual-probe BOTDA configuration with dithering is deployed. Notice that for far-away locations, the BGS becomes very small. Figure 2.7b shows the measurement of the same link once the lower optical frequency is removed. In this case, the amplitude starts to increase and recover mid-link when the gain provided by the probe wave starts to become significant. This amplification

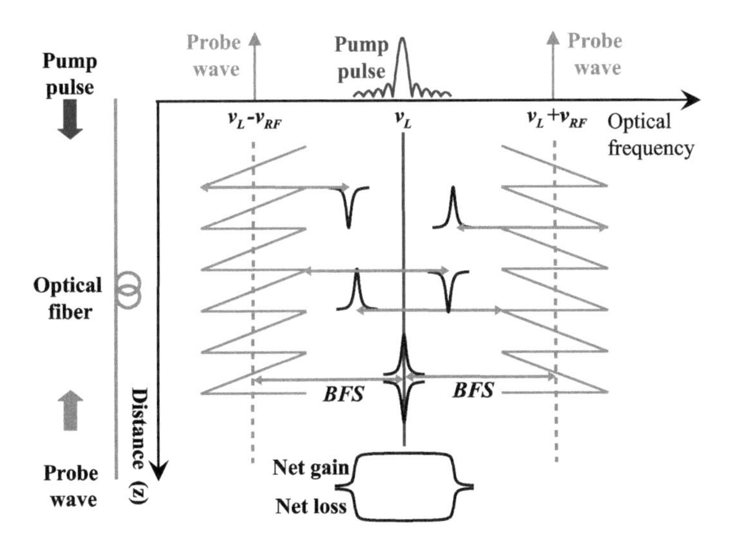

FIGURE 2.6
Fundamentals of the probe dithering technique for the mitigation of second-order non-local effects (From R. Ruiz-Lombera et al., *IEEE Photon. J.*, 7(6), 1–9, 2015. Copyright 2015 IEEE.)

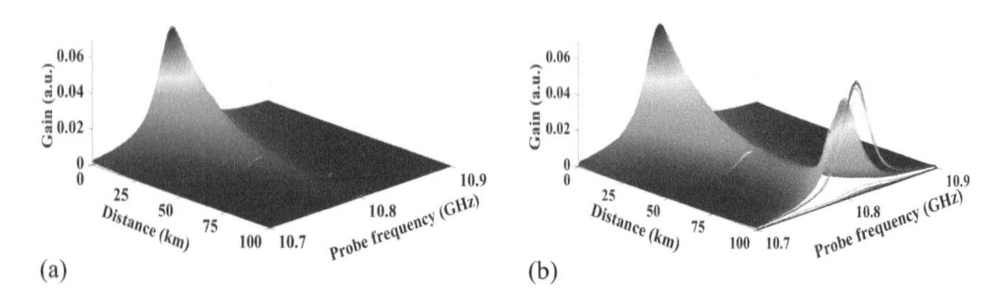

FIGURE 2.7
Brillouin gain distribution measured with (a) a dual-probe-sideband BOTDA sensor using the frequency modulation of the probe wave and (b) a novel BOTDA with pulse amplification. A pulse duration of 45 ns was deployed in both measurements. (From J. J. Mompó et al., *Opt. Express*, 24, 12672–12681, 2016. Copyright 2016 Optical Society of America.)

of the pump pulse can also be obtained by deploying a separated Brillouin pump source with dithering to generate distributed Brillouin gain [28].

Another method of extending the measurement range in BOTDA sensors is the use of pulse compression coding. The idea behind pulse coding is to launch coded sequences of pulses into the sensing fiber so that an improvement in the SNR of the retrieved single-pulse response is obtained after detection and processing. The first proposal for BOTDA sensors deployed the unipolar simplex coding imported from optical time-domain reflectometers [29]. This concept was later improved, deploying cyclic unipolar simplex codes, which significantly reduced the measurement time because successive code words are obtained by time delaying the cyclic sequence [30]. The SNR enhancement brought by unipolar simplex codes is equal to $(L + 1)/(2\sqrt{L})$, where L is the code length. Further refinements of pulse coding have been proposed, such as using bipolar codes that make simultaneous use of the gain and loss spectra [31] or deploying so-called color coding [32], in which the optical frequency of each pulse in the sequence is changed.

An important issue to take into account when deploying coding in BOTDA is that the successive pulses of a sequence experience an increased depletion factor, since the nth pulse in the sequence meets a probe wavefront that has previously been amplified by $n - 1$ pulses [33]. Hence, significant levels of pump depletion will appear at lower powers in pulse-coded BOTDA setups than in conventional single-pulse BOTDA sensors. Furthermore, pulse-coded BOTDA sensors are less tolerant to a given level of pump depletion because proper decoding requires that all pulses in a sequence have identical energy. Altogether, this translates to a constraint on the length of the pulse sequences that can be deployed and hence on the maximum SNR enhancement provided by these methods. A way to mitigate this problem is to deploy probe dithering. This has been demonstrated in a 164 km fiber loop link using a dual-probe BOTDA setup with probe dithering, deploying a 79-bit cyclic unipolar coding sequence and providing 3-MHz BFS measurement precision

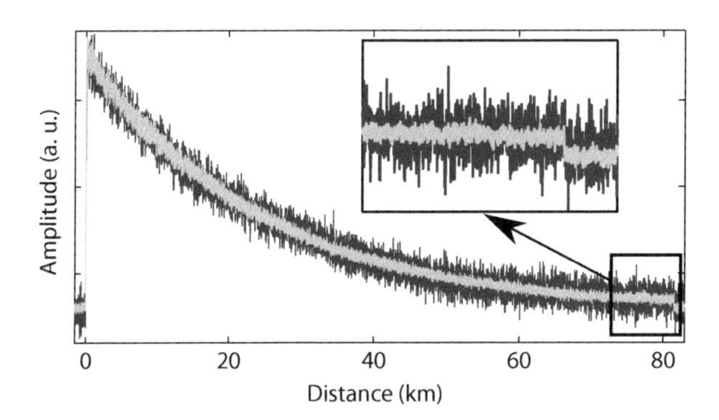

FIGURE 2.8
BOTDA traces at BFS obtained for the analyzer with a 79-bit cyclic coding (light gray) and without applying coding (dark gray) (From H. Iribas et al., *Opt. Express*, vol. 25, pp. 8787–8800, 2017. Copyright 2017 Optical Society of America.)

at 1 m spatial resolution. Figure 2.8 compares single-pulse to coded-pulse distributed probe gain measurements, where the 6.5 dB SNR enhancement associated with the code length is clearly noticeable.

2.4 Conclusions

In this chapter, we have reviewed the fundamentals and the latest research directions in BDSs. This technology is bound to experience increased adoption in practical applications due to its unique features, particularly after the latest research efforts have greatly enhanced its performance. Future developments need to focus on developing novel application fields and standardizing the methods for the deployment of the sensors and the interpretation of the measurement results.

References

1. S. S. C. Madabhushi, M. Z. E. B. Elshafie, and S. K. Haigh, "Accuracy of distributed optical fiber temperature sensing for use in leak detection of subsea pipelines," *J. Pipeline Sys. Engr. Practice*, vol. 6, no. 2, p. 04014014, May 2015.
2. A. Ukil, H. Braendle, and P. Krippner, "Distributed temperature sensing: Review of technology and applications," *IEEE Sensor. J.*, vol. 12, pp. 885–892, May 2012.
3. L. Schenato, "A review of distributed fibre optic sensors for geo-hydrological applications," *Appl. Sci.*, vol. 7, no. 9, Sep 2017.

4. I. Sovran, A. Motil, O. Danon, and M. Tur, "An ultimately fast frequency-scanning Brillouin optical time domain analyzer," in *Optical Fiber Communication Conference*, p. W2A.44, Optical Society of America, Mar 2015.
5. A. Voskoboinik, D. Rogawski, H. Huang, Y. Peled, A. E. Willner, and M. Tur, "Frequency-domain analysis of dynamically applied strain using sweep-free Brillouin time-domain analyzer and sloped-assisted FBG sensing," *Opt. Express*, vol. 20, pp. B581-F586, Dec 2012.
6. C. Jin, N. Guo, Y. Feng, L. Wang, H. Liang, J. Li, Z. Li, C. Yu, and C. Lu, "Scanning-free BOTDA based on ultra-fine digital optical frequency comb," *Opt. Express*, vol. 23, pp. 5277–5284, Feb 2015.
7. R. Bernini, A. Minardo, and L. Zeni, "Dynamic strain measurement in optical fibers by stimulated Brillouin scattering," *Opt. Lett*, vol. 34, pp. 2613–2615, Sep 2009.
8. A. Motil, O. Danon, Y. Peled, and M. Tur, "Pump-power-independent double slope-assisted distributed and fast Brillouin fiber-optic sensor," *IEEE Photon. Technol. Lett*, vol. 26, pp. 797–800, April 2014.
9. J. Urricelqui, A. Zornoza, M. Sagues, and A. Loayssa, "Dynamic BOTDA measurements based on Brillouin phase-shift and RF demodulation," *Opt. Express*, vol. 20, pp. 26942–26949, Nov 2012.
10. J. Urricelqui, F. López-Fernandino, M. Sagues, and A. Loayssa, "Polarization diversity scheme for BOTDA sensors based on a double orthogonal pump interaction," *J. Lightwave Technol*, vol. 33, pp. 2633–2638, June 2015.
11. D. Ba, B. Wang, D. Zhou, M. Yin, Y. Dong, H. Li, Z. Lu, and Z. Fan, "Distributed measurement of dynamic strain based on multi-slope assisted fast BOTDA," *Opt. Express*, vol. 24, pp. 9781–9793, May 2016.
12. D. Zhou, Y. Dong, B. Wang, T. Jiang, D. Ba, P. Xu, H. Zhang, Z. Lu, and H. Li, "Slope-assisted BOTDA based on vector SBS and frequency-agile technique for wide-strain-range dynamic measurements," *Opt. Express*, vol. 25, pp. 1889–1902, Feb 2017.
13. Q. Cui, S. Pamukcu, W. Xiao, and M. Pervizpour, "Truly distributed fiber vibration sensor using pulse base BOTDA with wide dynamic range," *IEEE Photon. Technol. Lett*, vol. 23, pp. 1887–1889, Dec 2011.
14. J. Mariñelarena, J. Urricelqui, and A. Loayssa, "Enhancement of the dynamic range in slope-assisted coherent Brillouin optical time-domain analysis sensors," *IEEE Photon. J*, vol. 9, pp. 1–10, June 2017.
15. A. Motil, R. Hadar, I. Sovran, and M. Tur, "Gain dependence of the linewidth of Brillouin amplification in optical fibers," *Opt. Express*, vol. 22, pp. 27535–27541, Nov 2014.
16. J. Mariñelarena, J. Urricelqui, and A. Loayssa, "Gain dependence of the phase-shift spectra measured in coherent Brillouin optical time-domain analysis sensors," *J. Lightwave Technol*, vol. 34, pp. 3972–3980, Sept 2016.
17. L. T. Stella and M. Foaleng, "Impact of Raman scattering and modulation instability on the performances of Brillouin sensors," *Proc. SPIE*, vol. 7753, p. 77539V, May 2011.
18. L. Thévenaz, S. F. Mafang, and J. Lin, "Effect of pulse depletion in a Brillouin optical time-domain analysis system," *Opt. Express*, vol. 21, pp. 14017–14035, Jun 2013.
19. A. Minardo, R. Bernini, L. Zeni, L. Thévenaz, and F. Briffod, "A reconstruction technique for long-range stimulated Brillouin scattering distributed fibre-optic sensors: Experimental results," *Measure. Sci. Technol.*, vol. 16, no. 4, p. 900, Feb 2005.

20. Y. Dong, L. Chen, and X. Bao, "Time-division multiplexing-based BOTDA over 100 km sensing length," *Opt. Lett.*, vol. 36, pp. 277–279, Jan 2011.
21. A. Zornoza, A. Minardo, R. Bernini, A. Loayssa, and L. Zeni, "Pulsing the probe wave to reduce nonlocal effects in Brillouin optical time-domain analysis (BOTDA) sensors," *IEEE Sensor. J.*, vol. 11, pp. 1067–1068, April 2011.
22. A. Minardo, R. Bernini, and L. Zeni, "A simple technique for reducing pump depletion in long-range distributed Brillouin fiber sensors," *IEEE Sensor. J.*, vol. 9, pp. 633–634, June 2009.
23. A. Dominguez-Lopez, X. Angulo-Vinuesa, A. Lopez-Gil, S. Martin-Lopez, and M. Gonzalez-Herraez, "Non-local effects in dual-probe-sideband Brillouin optical time domain analysis," *Opt. Express*, vol. 23, pp. 10341–10352, Apr 2015.
24. A. Dominguez-Lopez, Z. Yang, M. A. Soto, X. Angulo-Vinuesa, S. Martin-Lopez, L. Thevenaz, and M. Gonzalez-Herraez, "Novel scanning method for distortion-free BOTDA measurements," *Opt. Express*, vol. 24, pp. 10188–10204, May 2016.
25. X. Hong, W. Lin, Z. Yang, S. Wang, and J. Wu, "Brillouin optical time-domain analyzer based on orthogonally-polarized four-tone probe wave," *Opt. Express*, vol. 24, pp. 21046–21058, Sep 2016.
26. R. Ruiz-Lombera, J. Urricelqui, M. Sagues, J. Mirapeix, J. M. López-Higuera, and A. Loayssa, "Overcoming nonlocal effects and Brillouin threshold limitations in Brillouin optical time-domain sensors," *IEEE Photon. J.*, vol. 7, pp. 1–9, Dec 2015.
27. J. J. Mompó, J. Urricelqui, and A. Loayssa, "Brillouin optical time-domain analysis sensor with pump pulse amplification," *Opt. Express*, vol. 24, pp. 12672–12681, Jun 2016.
28. J. Urricelqui, M. Sagues, and A. Loayssa, "Brillouin optical time-domain analysis sensor assisted by Brillouin distributed amplification of pump pulses," *Opt. Express*, vol. 23, pp. 30448–30458, Nov 2015.
29. M. A. Soto, G. Bolognini, F. D. Pasquale, and L. Thévenaz, "Simplex-coded BOTDA fiber sensor with 1 m spatial resolution over a 50 km range," *Opt. Lett.*, vol. 35, pp. 259–261, Jan 2010.
30. M. Taki, Y. Muanenda, C. J. Oton, T. Nannipieri, A. Signorini, F. D. Pasquale, "Cyclic pulse coding for fast BOTDA fiber sensors," *Opt. Lett.*, vol. 38, pp. 2877–2880, Aug 2013.
31. M. A. Soto, S. L. Floch, and L. Thévenaz, "Bipolar optical pulse coding for performance enhancement in BOTDA sensors," *Opt. Express*, vol. 21, pp. 16390–16397, Jul 2013.
32. S. L. Floch, F. Sauser, M. Llera, M. A. Soto, and L. Thévenaz, "Colour simplex coding for Brillouin distributed sensors," *Proc. SPIE*, vol. 8794, May 2013.
33. H. Iribas, A. Loayssa, F. Sauser, M. Llera, and S. L. Floch, "Cyclic coding for Brillouin optical time-domain analyzers using probe dithering," *Opt. Express*, vol. 25, pp. 8787–8800, Apr 2017.

3

Low-Powered Plasmonic Sensors

Sasan V. Grayli, Xin Zhang, Siamack V. Grayli, and Gary W. Leach

CONTENTS

3.1 Introduction to Plasmonics 43
3.1.1 History and Development of Plasmonics 44
3.1.2 Fabrication of Plasmonic Devices 46
3.1.3 Mathematical Simulation in Studying Plasmonics 48
3.2 Low-Powered Plasmonic Sensors 49
3.2.1 State-of-the-Art Plasmonic Sensors in Biomedical Technology 49
3.2.2 State-of-the-Art Plasmonic Sensors in Communication Industries 50
3.2.3 Research and Development in the Area of Low-Power Plasmonic Sensing 51
3.3 Mechanisms of Hot Electron–Based Plasmonic Sensors 52
3.4 Controlling Nanostructure and Local Field Enhancements 54
3.5 Wavelength Selectivity with Nanostructures 55
3.6 Electroless Approach to Deposit Crystalline Films 56
3.7 Current and Future Market Drivers for Plasmonic Sensors 58
3.8 Alternative Materials to Cope with Challenges 58
3.9 Outlook 59
References 60

3.1 Introduction to Plasmonics

A plasmon is a quantum of the collective oscillation of a conductive electron gas in a metal, or generally speaking in a conductor (Berini et al. 2016) or a semiconductor where a conductive electron gas can form under appropriate conditions, as plasmonic phenomena have been observed and studied with diverse materials far beyond metals and alloys (Zhong et al. 2015). To excite a plasmon on a metal surface, a photon needs to be coupled to the surface either by a carefully selected combination of incident medium (a coupling prism) and angle or with properly designed fine surface structures. It is worth mentioning that an accelerated electron can also be used to excite a plasmon on a metal surface, but the following context will focus on optically excited plasmons.

Plasmonics is the science and technology that makes use of excited plasmons and allows the manipulation, including confining, focusing, guiding, converting, and propagating, of the electromagnetic field with sub-wavelength capability, using structured conductor–dielectric interfaces. It is now a major subject in the field of nanophotonics, as well as a very popular topic in many other fields, such as sensors, clinical cancer treatment, energy conversion, signal processing, lighting and displays, and so on. Some commercialized examples of plasmonics in sensors and sensing materials include those by Biacore (Lofas 1995), now under GE Healthcare, BioNavis, IBIS Technologies (Lofas and Johnsson 1990), Renishaw Diagnostics, Cabot Corporation, and Causeway Sensors ("Commercializing Plasmonics" 2015) and demonstrate the usefulness and potential of plasmonics in real-life applications and its potential for much more commercialization in other diverse applications in the near future. The capacity of the biosensor market alone is estimated to reach $22.68 billion (Lopez et al. 2017) by 2020, where the current contribution from plasmonics can't be overlooked and will most certainly increase in coming years. The market share of plasmonics will further increase, not only in the field of biosensors but also in other applications such as plasmonic-enhanced hard drives and integrated photonics, where Seagate and Intel have ongoing research projects, respectively ("Commercializing Plasmonics" 2015).

3.1.1 History and Development of Plasmonics

Despite its popularity in current science and technology research, modern plasmonics did not begin until 1998, when Ebbesen and coworkers ("Commercializing Plasmonics" 2015) demonstrated an enhanced light transmission effect using a 200 nm silver film perforated with a square array of sub-wavelength circular holes. The explosion of recent interest is in part due to the fact that nanofabrication methods have become much more accessible across various fields, beyond the needs of the semiconductor industry. Since then, the study of surface plasmons and surface structuring by nanofabrication have been closely combined, and plasmonic structures have overcome difficulties laid down by the *diffraction limit* of light, enabling progress in spectroscopy, sensing, energy conversion, display, nanoelectronics, and many other fields. The work by Ebbesen and coworkers is recognized as a milestone in the history of learning about photons (Pile 2010). It was immediately followed by the work from Pendry and coworkers (Heber 2010) in 1999 and 2000, demonstrating the concept of a negative refractive index using fabricated devices with engineered sub-wavelength structures and pitches—man-made materials now named *metamaterials* that can't be found in nature. This work has inspired eye-opening research interest in optical magnetism, perfect lenses for imaging and lithography beyond the diffraction limit, invisible cloaks, and ultimately, controlling electromagnetic waves at will.

Historically, the plasmonic phenomena from metals have been used to generate colors in stained glass, including the famous dichroic Lycurgus Cup in ancient times, and the documented scientific work on such phenomena began with a paper by Sommerfeld in 1899. At the beginning of the twentieth century, scientists started to give mathematical descriptions and predictions of how electromagnetic waves propagate along a conductor surface, reporting the observation of "abnormal" intensity loss in the reflection spectra of visible light by metallic gratings, as well as providing theoretical explanations of the coloring effects of metallic particles. The observed phenomena due to the excitation of surface plasmons, however, was not associated with the concurrent theoretical work until the middle of the twentieth century. In the 1960s, the loss phenomena by diffracting electron beams with metallic foils and by diffracting visible light with metallic gratings was linked to the excitation of plasmons—that is, Sommerfeld's surface waves—and the excitation of such waves by visible light had been demonstrated by both the Kretschmann and Otto prism-coupling methods. In the 1970s, observations of the enhanced Raman scattering effects of light by different molecules adsorbed on roughened or structured silver surfaces led to rapid experimental expansion and technological development of surface-enhanced Raman scattering (SERS) (Kneipp et al. 2006; Moskovits and Suh 1984; Fleischmann et al. 1974; Wang and Kong 2015), allowing devices for detection at the single-molecule level (Taylor and Zijlstra 2017) as well as the earliest commercialized plasmonic device. Following the pioneering work by Ebbesen and coworkers in 1998 and Pendry and coworkers in 1999 and 2000, the sub-wavelength manipulation of light opened up many application possibilities, including bridging photonics and electronics at the nanometer scale (for fast and large data transfer interconnects between chips of integrated circuits), nanolasing or laser enhancing and tuning, creating magnetism and thus a negative refractive index with fine structures of nonmagnetic materials, invisible cloaking, sub-100 nm photolithography using a broad beam illumination, and even various functional components for a complete plasmonic circuit (Ozbay 2006; Fang and Sun 2015). Free-space and wide spectral range coupling using regular surface nanostructures had become a popular alternative method for the excitation of plasmons, without the restrictions of coupling prisms, expanding the choice of possible plasmonic materials and structures, as well as extending plasmonic applications beyond the visible range. As attractive as the applications in ancient times, color generation by plasmonic nanostructures has recently gained huge research interest within the last few years for applications in paint-free printing, security coding, imaging, and display technology.

We provide a timeline of some of the important scientific and technological developments in the history of plasmonic phenomena and modern plasmonics in Table 3.1.

TABLE 3.1

Historical Development and Progress in Plasmonics

Time	Major Development and Progress in Plasmonics	References
Ancient times	Color formation in stained glass using plasmonic particles	(Barchiesi 2015)
1899	First scientific investigation on plasmonic effects: theoretical study of propagating radio waves on conductor surface	(Pile 2010; Maier 2007)
1900s	First experimental observation of plasmonic loss phenomena with visible light reflected off metal gratings Theoretical explanation on the color formation in ancient stained glass	(Pile 2010; Maier 2007)
1957	First experimental observation of plasmonic loss phenomena with accelerated electrons diffracted through metal foils	(Maier 2007)
1960s	Established connection between observed plasmonic loss phenomena with the earlier theoretical studies Discovery of plasmon excitation by Kretschmann and Otto configurations Theoretical study on a material with a negative refractive index	(Pile 2010; Maier 2007)
1970s	Discovery of enhanced Raman scattering for molecules attached to silver electrodes	(Pile 2010)
1990	The first commercial surface plasmon resonance sensor by Biacore	(Lofas 1995)
1998	Start of modern plasmonics; sub-wavelength manipulation of light	(Pile 2010)
1999	Demonstration of a structure for magnetism by nonmagnetic materials	(Heber 2010)
2000	Metamaterials: negative refractive index for making perfect lenses	(Heber 2010)
2001	Birth of the first metamaterial with a negative refractive index	(Heber 2010)

3.1.2 Fabrication of Plasmonic Devices

As free-space coupling using plasmonic nanostructures has made its way into many applications, the realization of nanostructures in device fabrication has become diverse. For proof of concept and fundamental research interests, many fabrication routines can be borrowed or created to achieve one specific design, regardless of complexity, cost, or scale-up feasibility. However, for making practical devices, the realization of the device structure is critical to optimal functionality, long-term device stability, compatibility with microelectronics, cost-effective manufacturability, and potential commercialization. In general, fabrication routines for nanostructures can be grouped into three categories: bottom-up, top-down, and a combination of the two.

The predominant bottom-up fabrication technique is the solution phase chemical synthesis of solution-suspended nano-sized structures and the assembly of these structures onto a surface to build plasmonic devices. Many SERS-based sensors have been built this way using synthetic Ag, Au, and other diverse types of colloids. It should be mentioned that surface roughening by processes such as anodization to achieve plasmonic structures also belongs to this category. However, the synthesis of various nanoparticles is more widely used as a simple and cost-effective approach, as it allows a wide range of size, composition, and shape control, and more importantly, scale-up production capability. In addition to simple metallic nanoparticles, core shell nanoparticles such as SiO_2- or Al_2O_3-coated Ag or Au have been used in SERS applications to provide enhanced nanoparticle stability and functionality control. Furthermore, nanowires, nanorods, nanoprisms, nanocubes, nanostars, nanosheets, and more are available to provide directionality, sharp tips and edges for added field enhancement, and control based on structure shapes. However, it is widely recognized that the manipulation or simple assembly of these tiny structures onto substrate surfaces to make high-fidelity devices remains a significant technical challenge for scale-up production. Although not frequently mentioned, the use of organic surfactants, which are necessary and important for size and shape control and solution stability in the chemical synthesis of these nanoparticles, can sometimes create contamination and compatibility constraints, as well as limit electrical contact to these nanostructures, presenting major hurdles in device fabrication.

Top-down fabrication, by comparison, refers primarily to the micro- and nanomachining of structures from bulk materials. Because top-down fabrication processes such as photolithography, electron beam lithography (EBL), focused ion beam (FIB) milling, shadow masking, and imprint lithography have been developed and tested to satisfy the needs of the semiconductor industry, the nanostructures and devices successfully demonstrated by top-down processes can easily be scaled up for production due to their compatibility with semiconductor production line processes. Top-down processes provide great repeatability, shape control, and prototype flexibility, but at a very high cost per device. In addition to the relatively high processing costs, top-down fabrication facilities are not easily accessible to many researchers. Also, the choice between bottom-up and top-down processes can also depend on the application. For example, it is very difficult to chemically synthesize aluminum nanoparticles, so the fabrication of aluminum nanopillars or nanodiscs would certainly be undertaken through top-down fabrication. When core shell structures are needed, bottom-up fabrication would be considered more appropriate. For applications with high-precision requirements or specialized structure assemblies, such as plasmonic focusing arrays and various metamaterials, top-down methods provide for more straightforward and faster fabrication.

When integrating plasmonic devices with nanoelectronics (e.g., for point-of-care diagnostic tools), top-down fabrication steps can't be avoided for the device integration, regardless of functional nanostructures fabricated by bottom-up approaches. In many plasmonic applications, such as solar cells, imaging and displays, and circuit interconnects, plasmonic structures are employed for functional enhancement. Introducing nanostructures and special material combinations for such purposes is not always practical with exclusively bottom-up or top-down techniques. Furthermore, the achievable fabrication results by the bottom-up and top-down fabrication techniques are not always highly overlapped or consistent. Appropriate combinations of both techniques should allow more structural options and bring improved results. For example, one such combination could be using anodized alumina as the shadow mask to deposit fine structures on substrate surfaces, eliminating the use of photoresist and the lift-off or etch step by adding a bottom-up step. Another example could be using lithography patterns for guided nanostructure growth by chemical processes, for materials that are not readily available by physical vapor deposition (PVD).

3.1.3 Mathematical Simulation in Studying Plasmonics

Plasmonics has showed an inspiring potential in any application where the confinement and manipulation of electromagnetic fields play a role, making use of the advantages of photonics at the miniaturized size range of electronics. Such confinement and manipulation provide opportunities for device miniaturization, high density and high efficiency in energy conversion and data transfer, and low-power driven devices. In the rapid development and progress of plasmonics, numerical analyses have greatly assisted research in the design and discovery of plasmonic materials, structures, and devices. With appropriate boundary conditions and using bulk material parameters, simulation results by finite-difference time domain (FDTD), finite element method (FEM), finite-difference method (FDM), discrete dipole approximation (DDA), and other approaches have validated numerous experimental data from plasmonic structures and devices, as well as predicted the possibility of new plasmonic designs and their required optimal material combinations and structural shapes and sizes, reducing the exploratory labor and costs for new discoveries and providing proper theoretical understanding (Murray and Barnes 2007). Many commercial software packages are now readily available and comprise an integral component of almost all plasmonic device development efforts.

In the following sections, our discussions are narrowed to provide a brief view of plasmonic sensors in various applications. In what follows, we describe their market potential, their underlying theory of operation, the nature of the plasmonic field enhancements responsible for their sensory response, the selection of materials and structures, and a model fabrication example for achieving desired plasmonic structures.

3.2 Low-Powered Plasmonic Sensors

The rapid development of modern plasmonics has benefited significantly from top-down microelectronics fabrication methods, and microelectronics fabrication materials and strategies remain indispensable for the development of state-of-the-art plasmonic sensors. Although silicon has many suitable characteristics that allow for a majority of silicon-based micro-devices and components to reach optimum performance levels, it falls somewhat short in its photoconductive and photovoltaic applications. The nature of silicon's indirect bandgap impairs carrier generation through limited photon absorption, and thus, detectors that use silicon junctions require thicker layers of photoactive material in order to compensate for low carrier density generation. The incorporation of plasmonic structures into silicon-based devices can improve photon absorption and consequent carrier generation by trapping the incident photons into the silicon layer and inducing stronger near-field photon–plasmon coupling (Maier 2007; Shahbazyan and Stockman 2013; Stewart et al. 2008). Beyond the use of plasmonic structures to enhance silicon's modest photoresponse, existing microelectronics approaches to signal coupling, readout, and the amplification of plasmonic response remain important components of plasmonics-based sensor design and application.

3.2.1 State-of-the-Art Plasmonic Sensors in Biomedical Technology

Following the achievements in SERS detection at the single-molecule level in recent years (Taylor and Zijlstra 2017), there has been an extensive body of research focusing on the use of surface plasmons to detect or enhance the detection of biological tissues. Many biomedical applications today are reaching their natural limitations in terms of optical penetration depth and resolution for imaging techniques (Shahbazyan and Stockman 2013; Maier 2007; Ouchi et al. 2014; Brongersma et al. 2015). The driving force is the need for the development of detectors that allow for higher resolution as well as more power-efficient signal generation. This is specifically important since concise differentiation between healthy and tumorous tissues through non-invasive methods such as imaging can result in the early detection of cancerous tissues and help to save many cancer-stricken patients. The use of terahertz radiation for medical imaging has garnered great interest in recent years (Siegel 2004; Ouchi et al. 2014). In other related fields, use of terahertz radiation in cellular biology, biochemistry, and protein dynamics has become an enabling way to study structural changes without damaging or denaturing the target biological compounds. However, due to low quantum efficiencies, conventional photoconductors have low responsivity when exposed to terahertz radiation. The incorporation of plasmonic contact electrodes into the photoconductive material both reduces the average transport length of the photocarriers and increases electron injection by plasmonic coupling (Berry et al. 2013). As a result, the output and response time of the plasmonically enhanced devices increases significantly in comparison with conventional detectors without the incorporation of plasmonic structures into the photoconductor's detection.

3.2.2 State-of-the-Art Plasmonic Sensors in Communication Industries

With recent rapid increases in computation capacity and constant market demand for more, further improvement of very-large-scale integration (VLSI) design is required. Industry and academia alike are in search of new techniques to both miniaturize and speed up semiconductor devices (Maier 2007; Shahbazyan and Stockman 2013; Stewart et al. 2008). Current VLSI circuits are approaching fundamental limits mainly due to limits associated with scaling, signal delay, and parasitic heat dissipation. Simultaneously, in order to enhance faster data transfer rates, there is a need for a paradigm shift in the current computational state variable from *electrical charge* to a *post-complementary metal oxide semiconductor (CMOS) era parameter.* Many believe that this future will involve technologies that incorporate aspects of photonics, excitonics, and/or plasmonics (Stewart et al. 2008; Maier 2007; Shahbazyan and Stockman 2013; Berry et al. 2013; Watanabe et al. 2012). Of the proposed regimes, photonics offers advantages such as broader bandwidth, shorter signal delay time, less signal cross-talk and, most importantly, lower power consumption. However, while photonics is currently being used in long-range and multiband communication, the size mismatch between electronic and diffraction-limited photonic components remains an enormous hurdle (Maier 2007; Shahbazyan and Stockman 2013). The use of surface plasmons (SPs) in the form of traveling wave surface plasmon polaritons (SPPs) and/or localized surface plasmon resonant (LSPR) structures offers the possibility of circumventing this issue. SPs allow one to couple and confine the extended fields of electromagnetic waves to nanoscale dimensions, making plasmonic and electronic components size compatible and facilitating the integration of photonics with current CMOS processes. The general categorization of plasmonic components in microelectronics differentiates between passive and active devices. The most common example of a passive device is the on-chip waveguide that is used to create signal links between different components of a chip. Common active plasmonic devices of interest include modulators and detectors, both of which need to meet the requirements of high extinction ratio and low power consumption (Maier 2007; Shahbazyan and Stockman 2013; Berry et al. 2013). Amid the frequent use of silicon-based photodetectors in many optoelectronic applications, silicon-based detectors are transparent to infrared data communication signals. The integration of other lower bandgap semiconductors such as germanium and gallium as active layers in silicon-based detectors is an expensive solution that also renders larger chip sizes less practical due to the need for additional external components. Furthermore, it does not resolve chip industry requirements for monolithic process flows (Maier 2007; Shur 2015).

Recently, researchers have investigated the use of internal photoemission to allow the sub-bandgap detection of communication signals by silicon detectors. The latter is done via the introduction of a Schottky barrier to the detection arrangement. The Schottky contact in semiconductor photodetectors is achieved via the deposition of a thin metal layer on the surface of the semiconductor material (Maier 2007). Since the Schottky barrier potential is lower than the bandgap energy, when sub-bandgap photons reach the detecting structure, they overcome the Schottky potential and convert into charge carriers that are then injected into the semiconductor material's channel (Figure 3.1; Maier 2007). However, even in the presence of a Schottky interface, the small photon–electron interaction volume at the Schottky interface results in small output currents, which are not ideal for device applications. The incorporation of plasmonic corrugated structures into the Schottky barrier confines the incident optical power at the boundary between metal and the semiconductor, hence increasing the probability of photon–electron interaction.

3.2.3 Research and Development in the Area of Low-Power Plasmonic Sensing

As described, simple Schottky contacts and plasmonically enhanced Schottky contacts have been reported to enhance the response of photodetectors for sub-bandgap incident radiation. However, recent work in the field has focused on more complex thin film structures in an effort to enhance the generation of hot electrons and increase the internal photoemission probability.

Scales et al. have developed phenomenological models to predict the internal quantum efficiency of Schottky barrier photodetectors for sub-bandgap wavelength radiation in various geometries. The models demonstrated

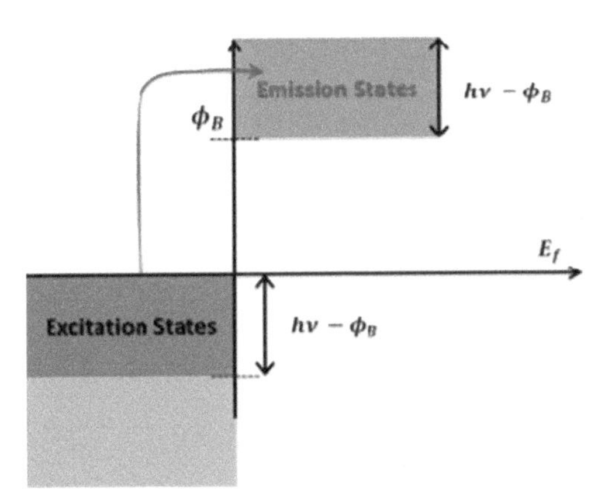

FIGURE 3.1
Hot-electron injection in a Schottky contact where the excited electron enters the semiconductor's channel after overcoming the Schottky barrier's potential ϕ_B. (S. A. Maier. *Plasmonics: Fundamentals and Applications.* Berlin, Germany: Springer.)

enhanced efficiencies for a dual Schottky configuration consisting of a thin metal film sandwiched between two semiconducting layers and attributed the enhanced response to multiple internal reflections (Scales and Berini 2010; Brongersma et al. 2015). Although the system does not have a direct link to lower power consumption, the enhancement of the output photocurrent can provide grounds for low-power plasmonically enhanced Schottky contacts for sub-bandgap photodetection. Other novel designs focus on the use of periodic plasmonic structures to both enhance the hot electron–generated photocurrent and define a narrow bandgap for selective photodetection applications. Since the frequency of the resonant plasmon-coupled photon absorption is defined by the plasmonic structure's periodicity, the photodetectors of this configuration offer high wavelength selectivity via the definition of a narrow bandgap for resonant absorption and lower power consumption through plasmon-induced hot-electron generation (Sobhani et al. 2013). Other novel applications of plasmonics in microelectronics are the plasmonic modulators that use a plasmonic waveguide to invoke different plasmon waveguide modes in a metal oxide semiconductor (MOS) geometry. In such configurations, the MOS transistor's gate voltage controls a near-infrared (NIR) optical transmission through the transistor's channel, whereby the NIR optical signal is selectively switched between a photonic mode with a cut-off frequency and a plasmonic mode with wavelengths on the order of half a cubic wavelength of the propagating NIR signal. The use of the plasmonic mode therefore allows for smaller device dimensions and lower power consumption per transmission (Maier 2007; Berry et al. 2013).

As a rule of thumb, designers of optoelectronic components such as photodetectors aim to lower the device power consumption as much as the technology cutoffs permit. Consequently, and in the course of doing so, the use of CMOS-based Schottky contacts, as previously discussed, offers power consumption as low as 70 mW across the entire circuit. The use of plasmonics can improve the detector's signal-to-noise ratio as well as reduce the power consumption to an even lower rate for applications such as interchip communication (Tavernier and Steyaert 2008; Akbari and Berini 2009; Shahbazyan and Stockman 2013). An estimated energy consumption of 20 fJ/bit was reported by simulating a Franz–Keldysh effect plasmonic modulator, much lower than the energy consumption of photonic Franz–Keldysh effect modulators (Abadía et al. 2014).

3.3 Mechanisms of Hot Electron–Based Plasmonic Sensors

Much of the recent work on plasmon-enhanced internal photoemission has focused on the development of plasmon-induced hot-electron generation in photovoltaic and photocatalytic systems. Given these systems are ideally conceived to function in the absence of an applied bias, the extension of the principles responsible for their operation to low-power sensing applications is both natural and extremely promising. Optimizing the requirements for efficient hot-electron generation at plasmonic metal–dielectric interfaces is thus a key element in the development of efficient plasmonic photovoltaics as well as low-power plasmonic sensors. Many of the important factors affecting plasmonic photovoltaics have been reviewed previously in great detail, and readers are referred there for more detail (Jang et al. 2016). Here, we provide a brief summary of some of the important concepts and device requirements.

The basis for device operation relies on the capture of electromagnetic waves through the excitation of SPs. The conditions required to achieve efficient SP absorption are highly dependent on the local metal–dielectric interface structure and range from prism-coupling methods for planar traveling wave SPP geometries, to narrowband (wavelength-specific) free space–coupled excitation of LSPRs in size-selected nanostructures, to broadband-absorbing LSPR structures comprised of nanostructures with a polydispersity in size and shape. Each of these coupling situations has its benefits and challenges regarding device practicality, wavelength sensitivity, and ultimate device efficiency.

Planar Schottky devices are designed for implementation with a narrow range of wavelengths and restricted to the Kretschmann and Otto prism-coupling geometries or other evanescent coupling waveguide strategies. The resulting SPPs are localized to the metal–dielectric plane, with fields that decay exponentially into both the adjacent metal and dielectric materials. The decay of the SPPs is dominated by non-radiative mechanisms and results in the formation of excited (hot) electrons and holes at the metal–dielectric interface. A well-formed Schottky interface drives the hot electrons into the adjacent dielectric material, effecting charge separation and the generation of a photocurrent via internal photoemission (Moskovits 2015; Brongersma et al. 2015; Knight et al. 2013). The quality of the interface and the nature of the metal–dielectric material pair define the characteristics of the Schottky barrier, where the barrier is nominally defined by the work function of the plasmonic metal and the electron affinity of the dielectric semiconductor.

The extension of this idealized planar SPP-based geometry to a nanostructured interface capable of the free-space coupling of radiation relies on the enhancement of the local field through locally resonant surface plasmons (LRSPs). These SP excitations are characterized by plasmonic fields that are much more tightly localized to the interface and enhanced in magnitude than SPPs, with correspondingly higher probabilities of generating hot electrons. These free carriers can lead to enhanced rates of internal photoemission or transfer their energy by plasmonic resonant energy transfer (PRET; Jang et al. 2016) to adjacent materials. However, in addition to these near-field enhancements, the nanostructured nature of the interface also leads to the radiative decay of the pathways of the LRSPs not experienced by SPPs and increases in their far-field scattering cross-sections. The resonant properties of the LRSPs are dictated by the shape, aspect ratio, composition, and crystallinity of the nanostructures.

While the discussion thus far has focused on the plasmonic enhancement of local fields, it is important to recognize that the fate of hot electrons is intimately connected to their local environment. For example, the fate of SPPs and LRSPs is expected to be highly dependent on the degree of crystallinity of the surface or nanostructure that is supporting the excitation. Non-radiative decay of the excited carriers can occur through electron–electron scattering processes on timescales of 100 fs to 1 ps, and electron–phonon scattering processes that occur on 1 ps to 10 ns timescales, providing only a very short time window in which to capture the hot electrons (Brongersma et al. 2015). SP scattering at grain boundaries and defects can accelerate plasmon decay, resulting in hot electrons with insufficient energy to overcome the barrier to internal photoemission. Further, polycrystalline nanostructures are expected to have compromised thermal and mechanical stabilities, making them less desirable for long-term sensing applications. Finally, the ultimate efficiency of hot electron–based devices is expected to be highly dependent on the electronic characteristics of the interface through the internal fields of the Schottky barrier. The quality of this interface is therefore of utmost importance in determining device efficiency and power consumption.

3.4 Controlling Nanostructure and Local Field Enhancements

As described in the previous section, the excitation of LSPRs is characterized by field enhancements that can be categorized as either far-field or near-field in nature. The far-field enhancements are typically associated with scattering effects and are well described by Mie scattering theory for isolated non-interacting spherical particles. As the particle density, level of interaction, and complexity of the nanostructure increases, modeling by more sophisticated approaches is required.

The nature of the interaction of incident light with plasmonic nanoparticles is the resonant excitation of electron density oscillations. For light interacting with very small particles (diameters below 10 nm), only a single absorbing dipolar mode is supported. As the particle diameter grows, it can support other higher-order modes (e.g., quadrupolar, octopolar, etc.). For small particles supporting predominantly dipolar modes, the total extinction is primarily due to absorption with a smaller scattering cross-section. For larger particles with diameters greater than approximately 100 nm, the opposite is true: the scattering cross-sections are much larger than the absorption cross-sections. Depending on the size, shape, composition, and distribution of nanoparticles, far-field scattering effects can play an important role in hot electron–based sensor devices. The size and shape of the nanostructures (or nanoparticles) of these materials are important factors that control the characteristics of the scattering (Jang et al. 2016). Nanostructure sizes below 50 nm are ideal for the forward scattering of incident light and by modifying the shape as well as the geometry of nanoparticles, the scattered wavelengths can be selectively controlled (Jang et al. 2016). As the structure size increases, the scattering effects will mainly be dominated by backward (reflection) scattering (Jang et al. 2016; Eustis and El-Sayed 2006). A combination of these scattering effects can be used in the fabrication of devices to tailor device performance by creating regions in which photons of certain wavelengths can be trapped, excluded, or preferentially directed to regions where they have an enhanced probability of absorption (Jang et al. 2016; Sorger et al. 2012; Veenkamp et al. 2014; Eustis and El-Sayed 2006).

Near-field enhancements at planar metal–dielectric interfaces can be driven by SPPs, the propagating mode of surface plasmons (Jang et al. 2016; Shokri Kojori et al. 2015; Atwater 2007). The oscillating electric field at the metal–dielectric interface can extend into the semiconductor layer and lead to the injection of the generated hot electrons into the conduction band of the adjacent material. The one-dimensional propagating SPP wave increases the probability of this process; however, the exponential decay of the SPP across the surface (Jang et al. 2016) should be considered in the design of devices that take advantage of the field enhancement of propagating surface plasmons.

Likewise, plasmonic near-field enhancement relies on the LSPR excitation of the nanostructure electron density, which provides a locally excited electric field. The resonant excitation mode structures for the near-field enhancement of LSPRs occur in response to the incident exciting field wavevector and are determined by the particle shape and dimension. Spherical particles generally support simple mode structure; however, complex high-order modes can be supported as the shape deviates from that of a simple sphere and the particle grows in size. The proximity of nanostructures to each other can also lead to the formation of collective modes of interaction that can modulate single-particle modes. Careful nanostructure design and fabrication can in some cases be tailored to provide significantly enhanced near fields or *hot spots*. The control of local, near-field enhancements can provide regions of enhanced hot-electron density and therefore higher internal photoemission yield.

Often, far-field enhancements are used to improve the efficiencies of conventional Si-based solar cells. However, the combination of both far-field and near-field enhancements can be used in the types of devices described here.

3.5 Wavelength Selectivity with Nanostructures

The excitation of LSPR structures will result in the generation of a localized, confined and highly energetic electric field (Jang et al. 2016; Moskovits 2015; Clavero 2014). The resonating field of the LSPR depends greatly on the size, shape, and geometry of the nanostructures at different wavelengths (Jang et al. 2016; Knight et al. 2011; Clavero 2014), and therefore, controlling these variables is a path to designing more efficient and selective sensors. It has been demonstrated that the absorption of plasmonic nanoparticles (or nanostructures) can be tailored by changing the shape and size of the structures, and this property can be used to create regions that absorb specific wavelengths of light (Eustis and El-Sayed 2006; Knight et al. 2013, 2011). The near-field enhancement of LSPRs, excited at the selected frequencies, can give rise to hot carrier generation and transfer for those specific wavelengths, acting as both an active region and color filter in a photosensor. Selective photon absorption/coupling of plasmonic nanostructures in such configurations promises a new generation of opto-plasmonic sensors with tunable wavelength capability. Figure 3.2 illustrates some of the many possible plasmonic nanostructures that can be made on a surface in different shapes and geometries.

3.6 Electroless Approach to Deposit Crystalline Films

It is well understood that the deposition quality of plasmonic materials plays an important role in the efficiency of plasmonic devices (Park et al. 2012;

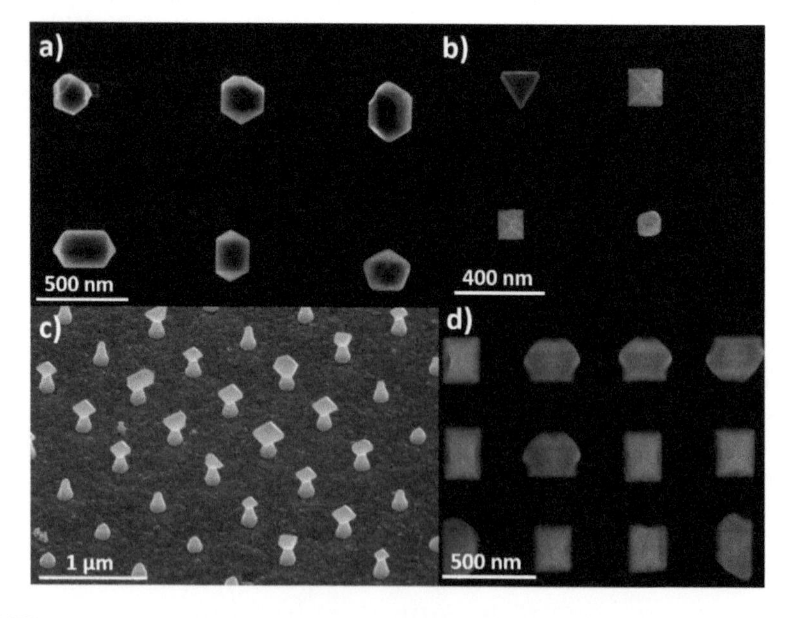

FIGURE 3.2
Deposition of shape-specific plasmonic nanostructures: (a) growth of highly faceted Ag nanostructures in different sizes and shapes, (b)–(d) growth of Au nanostructures in different shapes, sizes, and orientations in a periodic and ordered pattern.

High et al. 2015). The existence of defects and grain boundaries can lead to lossy plasmonic excitation, where the intensity of the plasmon will rapidly decay through defect scattering (Park et al. 2012). In most cases, the deposition of plasmonic materials leads to the formation of polycrystalline films and structures, resulting in inefficient Schottky junctions. To avoid this problem, plasmonic materials with high levels of crystallinity are needed. Unfortunately, the majority of the commonly used tools and techniques for the deposition of plasmonic metals yield polycrystalline materials, limiting their efficiency in these types of sensors and potentially compromising their commercial viability.

Leach and coworkers at Simon Fraser University have recently demonstrated an alternative approach for the deposition of ultra-smooth plasmonic metal surfaces. This approach takes advantage of solution phase chemistry enabling the cost-effective fabrication of high-quality crystalline thin-film plasmonic devices. The technique employs the electroless reduction of metal ions from solution and enables metal deposition with highly controllable thickness.

It was demonstrated that commonly known plasmonic metals such as gold, silver, and copper can be grown epitaxially with ultra-smooth surfaces (Figure 3.3). It has also been demonstrated that nanostructures made using this chemistry are grown with a high degree of crystallinity, making it ideal for the formation of low–defect density metal surfaces for metal semiconductor Schottky junctions. The nanostructures demonstrated in Figure 3.2 are grown using this deposition technique. This chemistry appears to be a promising path forward to achieving high-efficiency, low-cost, plasmonic, hot electron–based sensors.

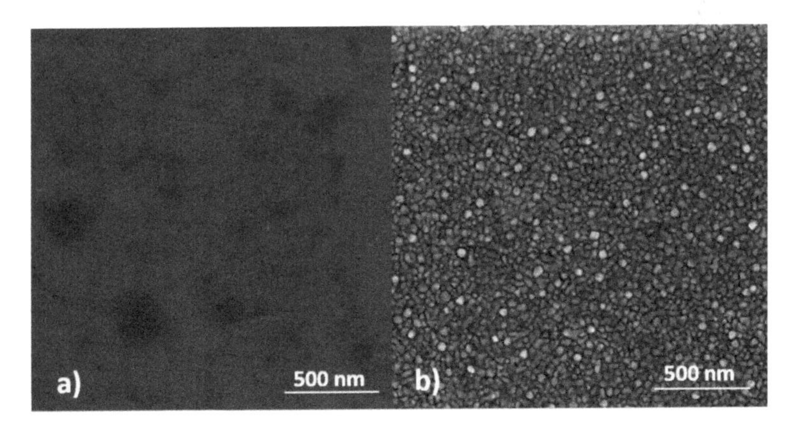

FIGURE 3.3
Scanning electron micrograph of (a) epitaxially grown single-crystal and ultra-smooth Au film using a novel electroless deposition technique, (b) thermally evaporated multicrystalline Au film.

3.7 Current and Future Market Drivers for Plasmonic Sensors

The semiconductor industry represents an enormous market with a capitalization size of nearly $300 billion (SIA, 2014). Predicted growth trends estimate continued levels of interest in the development of new semiconductor-based devices that can maintain the market size at and beyond $300 billion for the decade to come. With such promise, semiconductor-based photodetectors will continue to find their way into the market by virtue of industry-desired features such as fast conversion rate, energy efficiency, linearity, and smaller dimensions (Park et al. 2012; High et al. 2015). With recent advances in nanophotonics and plasmonics, nanoscale confinement and micron-scale propagation lengths make it possible to scale optoelectronic components down to market-valued dimensions. Together with lower power consumption, further device developments will continue to secure a place for low-power plasmonic devices in this fast-paced growing market. Another very promising market for low-power plasmonic photodetectors is the high-speed processing chip industry. With insatiable need for computing power enhancement and design specifications that keep pushing the limits of large-scale integration, companies such as Intel® and IBM® are dedicating significant resources in order to lower the power consumption of nanophotonic detectors integrated in silicon chip interconnects and intrachip data transfer couplers. The commercial projection of photonically enhanced VLSI is in the supercomputing market, with applications in machine learning, artificial intelligence, and large data computation. These markets have each been predicted to grow exponentially as industry and the public sector target such areas for electronic commerce, predictive analysis, and ease of access (Soref 2008; Feldman 2016).

3.8 Alternative Materials to Cope with Challenges

With the revolution in plasmonic device design, discovery, and function, and the promise of future device development, plasmonics represents a burgeoning research and development opportunity with many exciting sensor applications. Nevertheless, there remain some general and well-recognized obstacles in the commercialization of plasmonic devices. For instance, commonly used plasmonic materials such as silver and gold, which are ideal for visible and short wavelength infrared applications, are not considered compatible with industry fabrication routines for the integration of plasmonics with electronic circuits (Naik et al. 2013). Interband absorption losses in the metals is another limitation imposed by such materials that can lead to a loss of information or energy in plasmonic-driven devices (Murray and Barnes 2007). The direct coupling of electromagnetic waves to the plasmonic components of an apparatus can be very complex for applications that require high coupling efficiencies (Atwater and Polman 2010; Jang et al. 2016). Localized plasmonic heating at the nanoscale and photothermal effects (Naik et al. 2013), which can be useful for some applications such as heat-assisted magnetic recording and cancer therapy, are, however, quite wasteful and destructive in SPP-assisted optical circuits, energy conversion devices, and sensors (Galvez 2017). The integration of different plasmonic functional units into a complex plasmonic circuit is far more difficult without sophisticated design considerations. Despite the enhancement achieved with plasmonic components in some applications such as light-emitting diodes (LEDs) and photovoltaics, the improved efficiency is still not sufficient to justify the cost incurred for the added fabrication steps and materials (Jang et al. 2016). To overcome the current limitations, it is crucial to know more about plasmonics and find alternative plasmonic materials that are compatible with the industry fabrication protocols at low cost. In addition to plasmonic pure metals such as Ag, Au, Cu, Al, Ga, In, Cr, Pd, Pt, Ni, Ti, and W, where many have limited applications due to either their instability in air or their very high absorption loss, alloys, metal germanides, silicides, oxides, nitrides, bromides, graphene, and other two-dimensional materials, as well as engineered metamaterials (Boltasseva and Atwater 2011; Zhong et al. 2015), are now beginning to be widely explored as alternative plasmonic materials that are compatible with semiconductor fabrication methods and also fit applications in the broader range of the electromagnetic spectrum.

As we expand our knowledge about plasmonics and its application in various fields, some low-powered plasmonic devices (Feldman 2016) such as sensors (the first commercial surface plasmon resonance instrument by Biacore in 1990), full-color displays, and anti-counterfeiting coding have matured and moved from research labs into real-life applications. Other plasmonic devices (e.g., those in heat-assisted data recording and plasmonic interconnects), will likely soon follow. However, many more in other plasmonic applications remain in need of further breakthroughs in materials, structures, designs, and fabrication.

3.9 Outlook

Since its inception only a relatively short time ago, modern plasmonics has undergone revolutionary growth. The number of research articles and publications related to plasmonics is increasing exponentially and reflects the enormous excitement and creativity of plasmonics researchers. As well as expanding our understanding of its fundamental underpinnings and defining its limitations, these contributions describe the novel designs and functions of plasmonic elements and how they relate to meaningful applications of plasmonic-based technology, including low-powered sensors. Together with advances in and the increasing availability of modern patterning methods and new bottom-up fabrication strategies, the rate of growth of plasmonic-based technologies will continue to accelerate. Some of the current limitations to this growth remain factors related to device cost and manufacturability. The requirements to produce high-fidelity nanostructures are often accompanied by high pattern transfer costs. As these costs continue to fall and new lower-cost strategies emerge, low-power plasmonic sensors will increasingly find their place in the technological marketplace. In view of the inherent limits to patterning density in the electronics industry and the necessity for a new paradigm to drive chip density further, the use of plasmonics should not be ruled out. The ability to capture and confine electromagnetic waves at the nanoscale offers a means of marrying larger-scale, high-speed photonics-based approaches with nanometer-scale electronic elements. Thus, plasmonics should not be viewed as a competitor to the photonics or electronics industries but rather a bridge between the two that offers many new opportunities.

References

Abadía, N., T. Bernadin, P. Chaisakul, S. Olivier, D. Marris-Morini, R. Espiau de Lamaëstre, J. C. Weeber, and L. Vivien. 2014. "Low-Power Consumption Franz–Keldysh Effect Plasmonic Modulator." *Optics Express* 22 (9): 11236–11243.

Akbari, A., and P. Berini. 2009. "Schottky Contact Surface-Plasmon Detector Integrated with an Asymmetric Metal Stripe Waveguide." *Applied Physics Letters* 95 (2): 021104.

Atwater, H. A. 2007. "The Promise of Plasmonics." *Scientific American* 296 (4): 56–62.

Atwater, H. A., and A. Polman. 2010. "Plasmonics for Improved Photovoltaic Devices." *Nature Materials* 9 (3): 205.

Barchiesi, D. 2015. "Lycurgus Cup: Inverse Problem Using Photographs for Characterization of Matter." *JOSA A* 32 (8): 1544–1555.

Berini, P., S. Bozhevolnyi, and D.-S. Kim. 2016. "Plasmonics (Scanning the Issue)." *Proceedings of the IEEE* 104 (12): 2267–2269

Berry, C. W., N. Wang, M. R. Hashemi, M. Unlu, and M. Jarrahi. 2013. "Significant Performance Enhancement in Photoconductive Terahertz Optoelectronics by Incorporating Plasmonic Contact Electrodes." *Nature Communications* 4: 1622.

Boltasseva, A., and H. A. Atwater. 2011. "Low-Loss Plasmonic Metamaterials." *Science* 331 (6015): 290–291.

Brongersma, M. L., N. J. Halas, and P. Nordlander. 2015. "Plasmon-Induced Hot Carrier Science and Technology." *Nature Nanotechnology* 10 (1): 25–34

Clavero, C. 2014. "Plasmon-Induced Hot-Electron Generation at Nanoparticle/Metal-Oxide Interfaces for Photovoltaic and Photocatalytic Devices." *Nature Photonics* 8 (2): 95–103

"Commercializing Plasmonics." 2015. *Nature Photonics* 9 (8): 477–477

Eustis, S., and M. A. El-Sayed. 2006. "Why Gold Nanoparticles Are More Precious than Pretty Gold: Noble Metal Surface Plasmon Resonance and Its Enhancement of the Radiative and Nonradiative Properties of Nanocrystals of Different Shapes." *Chemical Society Reviews* 35 (3): 209–217.

Fang, Y., and M. Sun. 2015. "Nanoplasmonic Waveguides: Towards Applications in Integrated Nanophotonic Circuits." *Light: Science & Applications* 4 (6): e294.

Fleischmann, M., P. J. Hendra, and A. J. McQuillan. 1974. "Raman Spectra of Pyridine Adsorbed at a Silver Electrode." *Chemical Physics Letters* 26 (2): 163–166.

Galvez, F., D. Pérez de Lara, J. Spottorno, M. A. García, and J. L. Vicent. 2017. "Heating Effects of Low Power Surface Plasmon Resonance Sensors." *Sensors and Actuators B: Chemical* 243: 806–811.

Heber, J. 2010. "The Masters of Light." *Nature Materials* 9: S18–S19.

High, A. A., R. C. Devlin, A. Dibos, M. Polking, D. S. Wild, J. Perczel, N. P. de Leon, M. D. Lukin, and H. Park. 2015. "Visible-Frequency Hyperbolic Metasurface." *Nature* 522 (7555): 192.

Jang, Y. H., Y. J. Jang, S. Kim, L. N. Quan, K. Chung, and D. H. Kim. 2016. "Plasmonic Solar Cells: From Rational Design to Mechanism Overview." *Chemical Reviews* 116 (24): 14982–15034.

Kneipp, K., M. Moskovits, and H. Kneipp. 2006. *Surface-Enhanced Raman Scattering: Physics and Applications*. Berlin, Germany: Springer.

Knight, M. W., H. Sobhani, P. Nordlander, and N. J. Halas. 2011. "Photodetection with Active Optical Antennas." *Science* 332 (6030): 702–704.

Knight, M. W., Y. Wang, A. S. Urban, A. Sobhani, B. Y. Zheng, P. Nordlander, and N. J. Halas. 2013. "Embedding Plasmonic Nanostructure Diodes Enhances Hot Electron Emission." *Nano Letters* 13 (4): 1687–1692.

Lofas, S. 1995. "Dextran Modified Self-Assembled Monolayer Surfaces for Use in Biointeraction Analysis with Surface Plasmon Resonance." *Pure and Applied Chemistry* 67 (5): 829–834.

Lofas, S., and B. Johnsson. 1990. "A Novel Hydrogel Matrix on Gold Surfaces in Surface Plasmon Resonance Sensors for Fast and Efficient Covalent Immobilization of Ligands." *Journal of the Chemical Society: Chemical Communications*, no. 21: 1526–1528.

Lopez, G. A., M.-C. Estevez, M. Soler, and L. M. Lechuga. 2017. "Recent Advances in Nanoplasmonic Biosensors: Applications and Lab-on-a-Chip Integration." *Nanophotonics* 6 (1): 123–136.

Maier, S. A. 2007. *Plasmonics: Fundamentals and Applications*. Berlin, Germany: Springer.

Feldman, M. 2016. "Market for Artificial Intelligence Projected to Hit $36 Billion by 2025." August 30. Top500. Online: https://www.top500.org/news/market-for-artificial-intelligence-projected-to-hit-36-billion-by-2025 (accessed September 12, 2017).

Moskovits, M. 2015. "The Case for Plasmon-Derived Hot Carrier Devices." *Nature Nanotechnology* 10 (1): 6–8.

Moskovits, M., and J. S. Suh. 1984. "Surface Selection Rules for Surface-Enhanced Raman Spectroscopy: Calculations and Application to the Surface-Enhanced Raman Spectrum of Phthalazine on Silver." *The Journal of Physical Chemistry* 88 (23): 5526–5530.

Murray, W. A., and W. L. Barnes. 2007. "Plasmonic Materials." *Advanced Materials* 19 (22): 3771–3782.

Naik, G. V., V. M. Shalaev, and A. Boltasseva. 2013. "Alternative Plasmonic Materials: Beyond Gold and Silver." *Advanced Materials* 25 (24): 3264–3294.

Ouchi, T., K. Kajiki, T. Koizumi, T. Itsuji, Y. Koyama, R. Sekiguchi, O. Kubota, and K. Kawase. 2014. "Terahertz Imaging System for Medical Applications and Related High Efficiency Terahertz Devices." *Journal of Infrared, Millimeter, and Terahertz Waves* 35 (1): 118–130.

Ozbay, E. 2006. "Plasmonics: Merging Photonics and Electronics at Nanoscale Dimensions." *Science* 311 (5758): 189–193

Park, J. H., P. Ambwani, M. Manno, N. C. Lindquist, P. Nagpal, S.-H. Oh, C. Leighton, and D. J. Norris. 2012. "Single-Crystalline Silver Films for Plasmonics." *Advanced Materials* 24 (29): 3988–3992.

Pile, D. 2010. "Small and Beautiful." *Nature Materials* 9: S18–S19.

Scales, C., and P. Berini. 2010. "Thin-Film Schottky Barrier Photodetector Models." *IEEE Journal of Quantum Electronics* 46 (5): 633–643.

Shahbazyan, T. V., and M. I. Stockman. 2013. *Plasmonics: Theory and Applications*. Berlin, Germany: Springer.

Shokri Kojori, H., J.-H. Yun, Y. Paik, J. Kim, W. A. Anderson, and S. J. Kim. 2015. "Plasmon Field Effect Transistor for Plasmon to Electric Conversion and Amplification." *Nano Letters* 16 (1): 250–254.

Shur, M. 2015. "Plasmonic Terahertz Detectors." *ECS Transactions* 66 (7): 139–144.

SIA, Semiconductor Industry Association. 2014. "The US Semiconductor FactBook 2014." Online Report. The US Semiconductor FactBook 2014.

Siegel, P. H. 2004. "Terahertz Technology in Biology and Medicine." *IEEE Transactions on Microwave Theory and Techniques* 52 (10): 2438–2447.

Sobhani, A., M. W. Knight, Y. Wang, B. Zheng, N. S. King, L. V. Brown, Z. Fang, P. Nordlander, and N. J. Halas. 2013. "Narrowband Photodetection in the Near-Infrared with a Plasmon-Induced Hot Electron Device." *Nature Communications* 4: 1643.

Soref, R. 2008. "The Achievements and Challenges of Silicon Photonics." *Advances in Optical Technologies* 2008. doi:10.1155/2008/472305.

Sorger, V. J., R. F. Oulton, R.-M. Ma, and X. Zhang. 2012. "Toward Integrated Plasmonic Circuits." *MRS Bulletin* 37 (8): 728–738.

Stewart, M. E., C. R. Anderton, L. B. Thompson, J. Maria, S. K. Gray, J. A. Rogers, and R. G. Nuzzo. 2008. "Nanostructured Plasmonic Sensors." *Chemical Reviews* 108 (2): 494–521.

Tavernier, F., and M. Steyaert. 2008. "Power Efficient 4.5Gbit/s Optical Receiver in 130nm CMOS with Integrated Photodiode." IEEE 162–165

Taylor, A. B., and P. Zijlstra. 2017. "Single-Molecule Plasmon Sensing: Current Status and Future Prospects." *ACS Sensors* 2 (8): 1103–22.

Veenkamp, R., S. Ding, I. Smith, and W. N. Ye. 2014. "Silicon Solar Cell Enhancement by Plasmonic Silver Nanocubes." Proceedings of SPIE OPTO: 89811B–89811B.

Wang, A., and X. Kong. 2015. "Review of Recent Progress of Plasmonic Materials and Nano-Structures for Surface-Enhanced Raman Scattering." *Materials* 8 (6): 3024–3052.

Watanabe, T., S. B. Tombet, Y. Tanimoto, Y. Wang, H. Minamide, H. Ito, D. Fateev, et al. 2012. "Ultrahigh Sensitive Plasmonic Terahertz Detector Based on an Asymmetric Dual-Grating Gate HEMT Structure." *Solid-State Electronics* 78: 109–114.

Zhong, Y., S. D. Malagari, T. Hamilton, and D. M. Wasserman. 2015. "Review of Mid-Infrared Plasmonic Materials." *Journal of Nanophotonics* 9 (1): 093791.

4

Ultra-fast Photodiodes under Zero- and Forward-Bias Operations

Jin-Wei Shi

CONTENTS

4.1 Introduction 65
4.2 Limitations in Speed of PDs under Zero-Bias Operations 67
4.3 InP-Based Uni-Traveling Carrier PDs 68
4.4 GaAs-Based UTC-PDs and LPCs 73
4.5 Type II InP-Based UTC-PDs 79
4.6 High-Speed Ge-on-Si-Based PDs 87
4.7 Conclusion 90
References 92

4.1 Introduction

The amount of global network data traffic continues to grow, driven primarily by the use of wireless mobile data and internet video. The information and communication technology (ICT) sector is thus gradually coming to take up a larger portion of global electricity consumption (~10%) [1, 2]. In order to minimize the demands of the growth of the ICT sector, a large number of techniques have been adapted for the energy-efficient processing of high-speed data streams in wireless [3] and wired networks [4–10]. Optical interconnect (OI) techniques [4–10], which replace bulky and power-hungry active/passive microwave components with more energy-saving and high-speed optoelectronic devices, provide a revolutionary way to further reduce the carbon footprint of data centers and the wired networks inside them.

Figure 4.1 shows the conceptual picture of a modern OI channel [11] based on multi-mode fiber (MMF). In the transmitter side, it is mainly composed of vertical-cavity surface-emitting lasers (VCSELs) and laser diode drivers (LDDs). Regarding the receiving end, there are photodiodes (PDs), transimpedance amplifiers (TIAs), and limiting amplifiers (LAs) inside. To see the energy consumption breakdown of the whole channel [11], the major consumed power is in the receiver side. It mainly consists of two parts: one

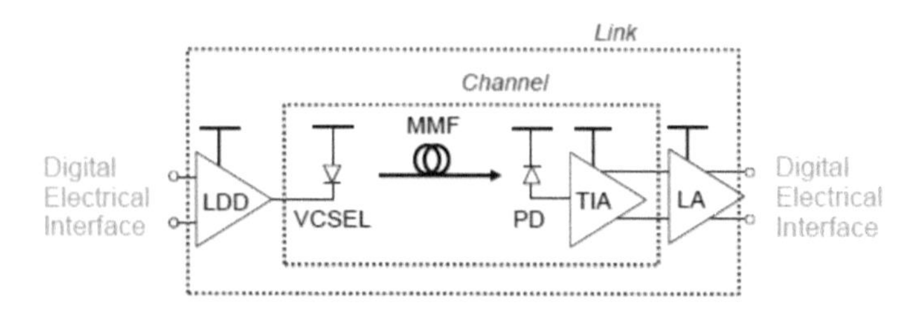

FIGURE 4.1
Conceptual diagram of a modern OI channel. (C. L. Schow, "Power-Efficient Transceivers for High-Bandwidth, Short-Reach Interconnects," *Proceedings of OFC 2012*, Los Angeles, CA, March 2012, pp. OTh1E4. © 2012 IEEE).

is from LAs, which need to have high peak-to-peak (Vpp) driving voltage output to drive the next-stage digital circuit; the other is from the DC component of photocurrent generated from the high-speed optical data stream. This is because in traditional p-i-n PDs, reverse-bias operation is necessary for high-speed optical data detection. It will result in extra power consumption with excess heat generation, which is proportional to the product of DC reverse bias of the PDs and the output photocurrent [12]. Such heating effects could be a serious issue for the next generation of OI systems, which have densely packaged integrated circuits (ICs) with millions of optoelectronic components and optical channels for high-speed linking (>50 Gbit/s) [13]. To have PDs that can sustain high-speed performance under zero-bias operations would be an effective solution to minimize this thermal issue. If the high-speed performance can be sustained even when the operating voltage of PD is further pushed to the forward bias, it can function as a laser power converter (LPC) [14]. We can thus generate (instead of consuming) DC electrical power by using this device during high-speed optical data detection in an OI system.

Figure 4.2 shows a conceptual diagram of a high-speed LPC used in an OI system [15]. As can be seen, by using the bias tee circuits to separate the DC and AC components of detected signal from the laser power converter, we can simultaneously generate the DC electrical power in an OI system and obtain a clear eye pattern [16, 17].

The other possible application for the zero-bias and high-speed PD is in an advanced coherent receiver array, which is used for the detection of optical signals with a complex modulation format for long-haul transmission, where integration with a bulky broadband DC bias network with low electrical cross-talk (less than –30 dB) performance is necessary to handle the increased data rate capacity (>100 Gbit/s) [18]. However, the large geometric size of this bias network and its cross-talk performance will become an issue in the dense package of a high-speed (>100 Gbit/s) coherent photo-receiver array of compact size [18]. Clearly, a high-speed PD that can be operated under the zero-bias condition is highly desired to minimize the footprint of its external bias circuit and electrical cross-talk.

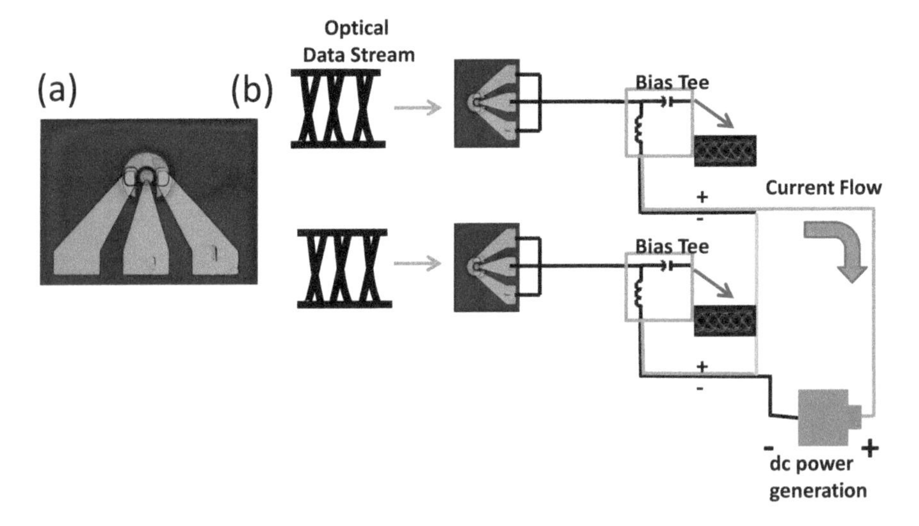

FIGURE 4.2
(a) Top view of a single device and (b) conceptual view of an LPC for OI applications (J.-W. Shi et al., *IEEE Transactions on Electron Devices*, vol. 58, pp. 2049–2056, July 2011. © 2011 IEEE).

In this chapter, we will review several kinds of high-speed PDs, which have the capability in zero-bias or forward-bias operations for the application of OI, coherent detection, and millimeter wave (MMW) photonics [19]. We will introduce their working principles and explain why they can have high-speed performance within an extremely small electric (E-) field inside their active layers.

4.2 Limitations in Speed of PDs under Zero-Bias Operations

There are two major bandwidth-limiting factors for high-speed PDs under zero-bias operations. One is the slow internal transit time of a photo-generated hole due to the extremely small E-field inside the active layer under zero bias. Figure 4.3 shows a conceptual band diagram of a typical p-i-n PD before (dash line) and after (solid line) light illuminations. As can be seen, the photo-generated hole induces the space-charge screening (SCS) field, reduces the net E-field inside the absorption region, and slows down the process of carrier drift [20]. SCS is a fundamental mechanism that limits the output power and electrical bandwidth of PDs under high optical power illumination. Compared with the photo-generated electron, the photo-generated hole more easily resides in the photo-absorption volume due to its lower mobility and the existence of heterostructure barriers in p-i-n PDs, as illustrated in Figure 4.1. This will become a serious issue and limits the

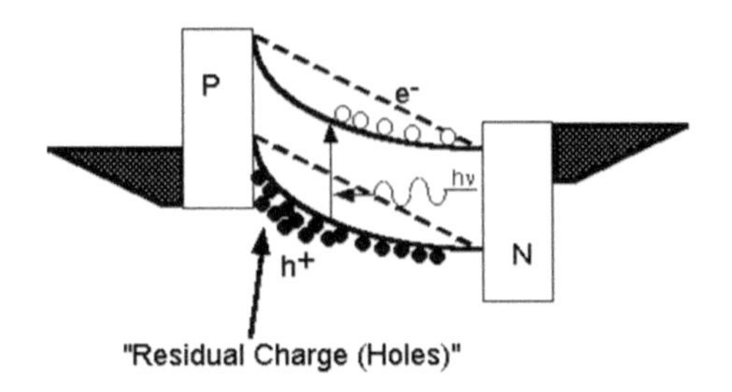

FIGURE 4.3
The serious space-charge screening originates from the hole storage in the photo-absorption volume.

speed performance of PDs when there is no externally applied E-field inside the absorber (e.g., zero-bias operation).

The other possible major bandwidth-limiting factor of PD under zero-bias operation is its large junction capacitance (C_J) and the induced small RC-limited bandwidth. Under the reverse-bias operation of a typical p-i-n PD, the value of C_J can become smaller with the increase of reverse-bias voltage due to the widening of the depletion region. On the other hand, since there is no more externally applied E-field to deplete the intrinsic (i)-layer under zero-bias operation, the background doping in the i-layer becomes a determinant issue in the value of C_J. In order to pursue the ultra-high-speed performance of the PD under zero-bias operation, ultra-low background doping ($\sim 1 \times 10^{14}$ cm^{-3}) in the i-layer is usually necessary to sustain a high RC-limited bandwidth [18, 19].

In the following sections, we will review several techniques that can overcome the aforementioned two problems and sustain the high-speed performance of PD under zero-bias operation.

4.3 InP-Based Uni-Traveling Carrier PDs

One of the most effective ways to fundamentally overcome the limitations of the SCS effect in p-i-n PDs is to remove the hole transport in the active layer of the PD. As discussed in Figure 4.3, the residual hole, which usually has much smaller mobility than that of the electrons in the absorption region, is the root cause for the SCS phenomenon. In order to eliminate the hole-induced SCS effect, the uni-traveling carrier PD (UTC-PD) structure has been demonstrated with excellent performance in terms of speed and output power [21]. Figure 4.4 shows the conceptual band diagram of a typical UTC-PD. As can be seen, the photo-absorption process happens at the p-type $In_{0.53}Ga_{0.47}As$ absorption layer; and the wide-bandgap InP, which has no photo-absorption process, serves as the intrinsic (i) carrier collection and *n*+ contact layers. A diffusion block layer is necessary to block the diffusion of photo-generated

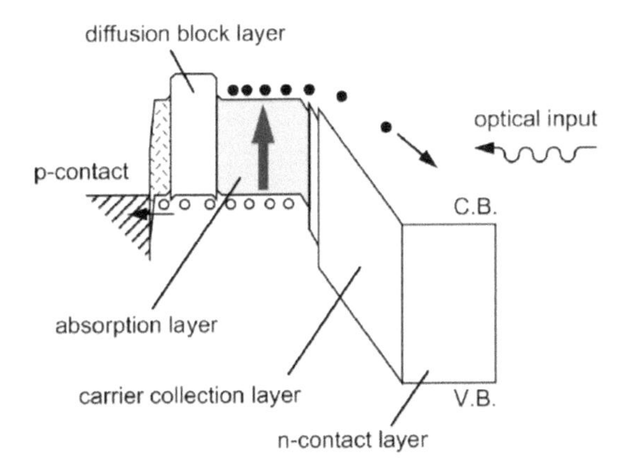

FIGURE 4.4
Conceptual band diagram of an InP-based UTC-PD (H. Ito et al., "High-Speed and High-Output InP-InGaAs Uni-traveling-Carrier Photodiodes," *IEEE Journal of Selected Topics in Quantum Electronics*, vol. 10, pp. 709–727, July/August 2004. © 2004 IEEE).

electron into the p-contact layer. Besides, in order to accelerate the electron diffusion process in the p-type absorber, the step or graded p-type doping profile is usually preferred [22]. Here, the hole transport in the UTC-PD can be eliminated due to the fact that the photo-generated hole will become the majority in the p-type absorber and directly relax to the anode contacts. The excellent high-speed performance of UTC-PD under zero-bias operation can thus be expected due to the superior transport characteristics of electrons at the relatively low E-field in the InP carrier collection layer.

Furthermore, for zero-bias operation, ultra-low background doping in the intrinsic carrier collection layer (or *collector*) is usually preferred to reduce the C_J and increase the built-in E-field inside the collector. Figures 4.5 and 4.6

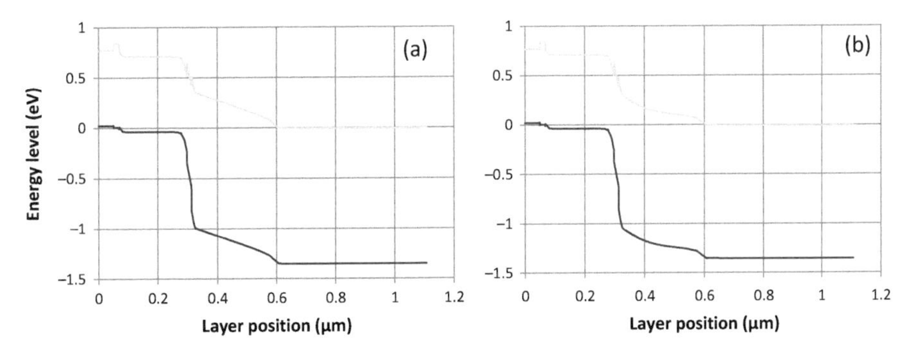

FIGURE 4.5
Energy band diagram comparison between doping levels of (a) 1×10^{14} cm^{-3} and (b) 1×10^{16} cm^{-3} in the carrier collector of a UTC-PD at 0 V. (T. Umezawa et al., "Bias-Free Operational UTC-PD above 110 GHz and Its Application to High Baud Rate Fixed-Fiber Communication and W-Band Photonic Wireless Communication," *IEEE/OSA Journal of Lightwave Technology*, vol. 34, no. 13, pp. 3138–3147, July 2016. © 2016 IEEE).

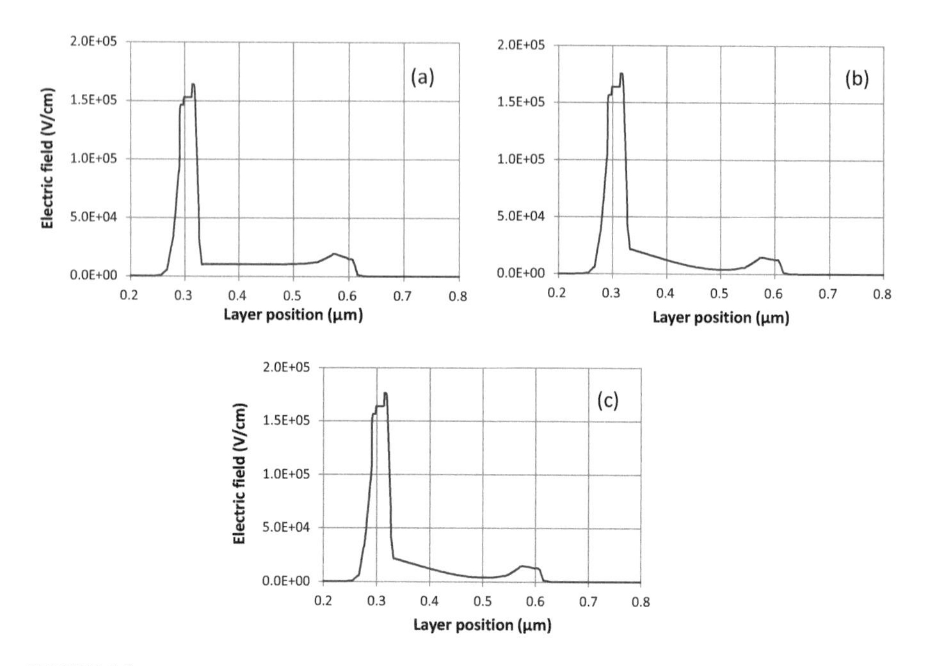

FIGURE 4.6
Internal electric distributions in the carrier collector in a UTC-PD at various doping levels: (a) 1×10^{14} cm^{-3}, (b) 1×10^{15} cm^{-3}, (c) 1×10^{16} cm^{-3}. (T. Umezawa et al., "Bias-Free Operational UTC-PD above 110 GHz and Its Application to High Baud Rate Fixed-Fiber Communication and W-Band Photonic Wireless Communication," *IEEE/OSA Journal of Lightwave Technology*, vol. 34, no. 13, pp. 3138–3147, July 2016. © 2016 IEEE).

show the simulated energy band diagram and E-fields for UTC-PDs with different background dopings in the collector [19]. We can clearly see that compared with the cases of collector background doping at 1×10^{15} and 1×10^{16} cm^{-3}, by lowering the background doping to ~1×10^{14} cm^{-3}, a stronger E-field and a more abrupt electron potential drop in the collector can clearly be observed.

By using the epi-layer structure of the UTC-PD with ultra-low background doping (~1×10^{14} cm^{-3}) in the collector, a 3 dB optical-to-electrical (O-E) bandwidth of over 110 GHz with a saturation current of over 7 mA can be achieved simultaneously under a 25 Ω effective load and zero-bias operation [19]. Figures 4.7 and 4.8 show the measured O-E frequency responses and photo-generated MMW power at 100 GHz versus the output photocurrent, respectively.

Other than using ultra-low background doping in the InP collector to enhance the performance of the UTC-PD under zero-bias operation, another possible way is to series (cascade) two (or several) UTC-PDs, which can increase the total built-in potential in the PD array and benefit its high-speed performance with a larger output voltage swing compared with that

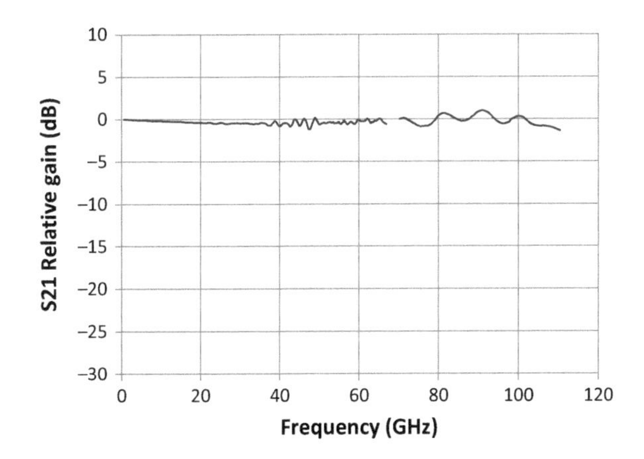

FIGURE 4.7
Measured O-E frequency response of a UTC-PD under zero-bias operation and 25 Ω effective load. (T. Umezawa et al., "Bias-Free Operational UTC-PD above 110 GHz and Its Application to High Baud Rate Fixed-Fiber Communication and W-Band Photonic Wireless Communication," *IEEE/OSA Journal of Lightwave Technology*, vol. 34, no. 13, pp. 3138–3147, July 2016. © 2016 IEEE).

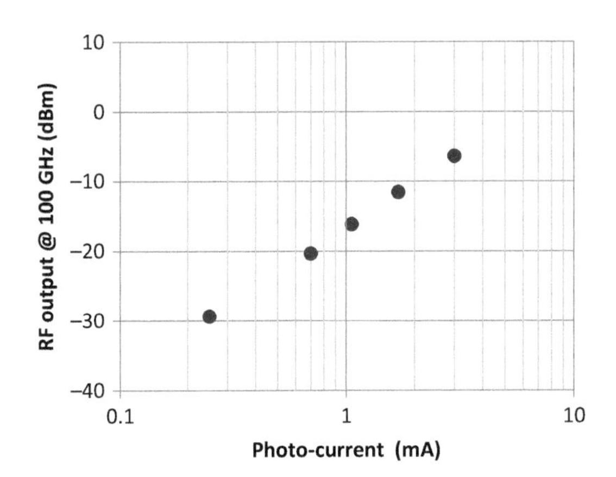

FIGURE 4.8
Measured photo-generated MMW power versus photocurrent of a UTC-PD under zero-bias operation and 25 Ω effective load. (T. Umezawa et al., "Bias-Free Operational UTC-PD above 110 GHz and Its Application to High Baud Rate Fixed-Fiber Communication and W-Band Photonic Wireless Communication," *IEEE/OSA Journal of Lightwave Technology*, vol. 34, no. 13, pp. 3138–3147, July 2016. © 2016 IEEE).

of a single PD [23]. Figure 4.9 shows a conceptual layout of the cascaded UTC-PDs. As can be seen, an optical 3 dB beam splitter is necessary to equally distribute the optical power into two such PDs in series. The unbalanced injected optical power between these two PDs will induce serious speed degradation of the cascade array [24]. Figure 4.10 shows a schematic

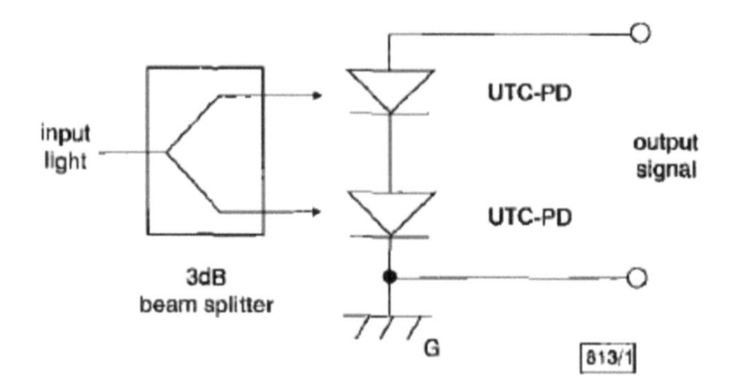

FIGURE 4.9
Circuit diagram of a CT-UTC-PD (H. Ito et al., "Zero-Bias High-Speed and High Output Voltage Operation of Cascade-Twin Unitravelling-Carrier Photodiode," *Electronics Letters*, vol. 36, no. 24, November 2000. © 2000 IEEE).

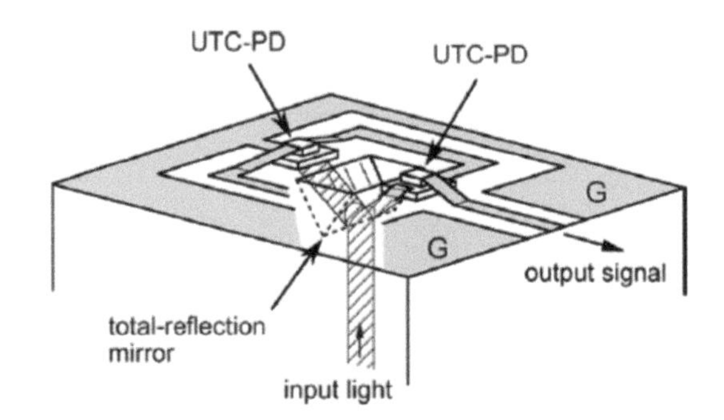

FIGURE 4.10
Schematic drawing of a CT-UTC-PD (H. Ito et al., "Zero-Bias High-Speed and High Output Voltage Operation of Cascade-Twin Unitravelling-Carrier Photodiode," *Electronics Letters*, vol. 36, no. 24, November 2000. © 2000 IEEE).

drawing of the cascaded UTC-PDs. By wet-etching a V-groove between two UTC-PDs, the injected optical power from the substrate side can not only equally split into two arms but also form a refracting facet in each UTC-PD. Such facets will lead to an increase in effective optical absorption paths and enhance responsivity performance [25]. Figure 4.11 shows the measured 3 dB O-E bandwidth versus the peak output voltage of cascade-twin UTC-PDs (CT-UTC-PDs) and the single total-reflection UTC-PD (TR-UTC-PD) under zero-bias operations. We can clearly see that the cascade structure can endure a higher peak optical power illumination and exhibit a faster speed performance under the same peak output voltage swing than those of the reference device due to the doubling of the built-in potential (internal E-field) in the CT-UTC-PDs.

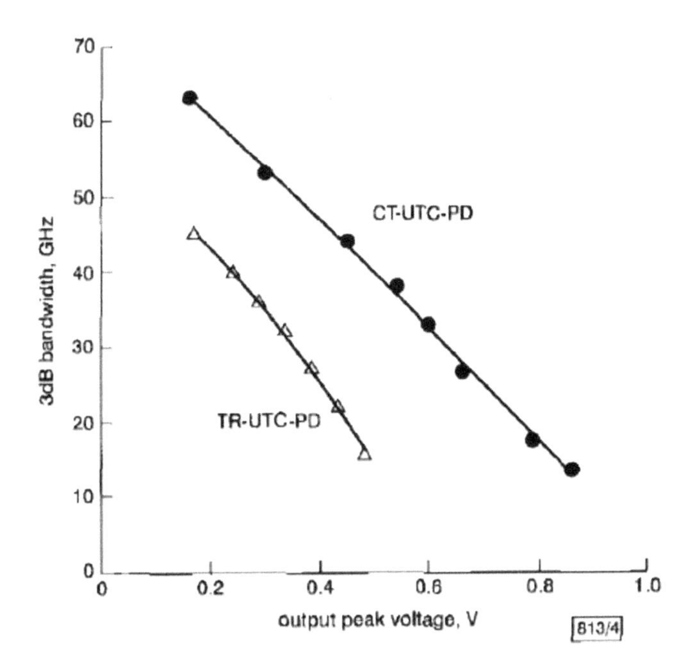

FIGURE 4.11
Relationship between the peak output voltage and 3 dB O-E bandwidth of CT-UTC-PDs and the single TR-UTC-PD under zero-bias and optical pulse excitations (H. Ito et al., "Zero-Bias High-Speed and High Output Voltage Operation of Cascade-Twin Unitravelling-Carrier Photodiode," *Electronics Letters*, vol. 36, no. 24, November 2000. © 2000 IEEE).

4.4 GaAs-Based UTC-PDs and LPCs

Although the InP-based UTC-PD has demonstrated excellent speed performance at a wavelength of 1.55 μm, this device structure may not work under 0.8 μm wavelength excitation, which is the most popular optical window for OI applications. The incident photon at this wavelength will have enough photon energy to induce the absorption process in the InP collector of the UTC-PD, which is supposed to be transparent at operation wavelength to eliminate the hole transport [26]. Thus, in our previous work, we have demonstrated a high-speed GaAs/AlGaAs-based UTC-PD composed of a GaAs-based p-type photo-absorption layer and an $Al_{0.15}Ga_{0.85}As$-based collector to avoid the undesired photo-absorption that occurs in the collector under 800 nm wavelength excitation [26].

Furthermore, by pushing the DC bias of the PD into a forward-bias regime, we can generate the DC electrical power instead of wasting the DC component of the optical power injected into the device. This device will function as an LPC and its generated DC electrical power can be reused in the OI system to improve its overall efficiency. However, forward bias will kill the speed performance of the traditional p-i-n PD because the extremely small E-field under the forward bias will slow down the drift velocity of the carrier (the photo-generated hole) and increase the C_J. As in our previous discussion, we can expect the UTC-PD-based LPC to have superior speed performance to the p-i-n LPC.

In this section, we will review our recent work on a GaAs/AlGaAs-based high-speed LPC for applications in green OI systems [27, 28]. Figure 4.12 shows a cross-section of the demonstrated LPC. We adopted the typical vertically illuminated PD structure with an active circular mesa and a p-type ring contact on the top. The diameter of the whole mesa and the inner circle for light illumination were 28 μm and 20 μm, respectively. As shown in Figure 4.12, the epi-layer structure of our device is similar to that reported for our GaAs/AlGaAs-based UTC-PD [26] at a 850 nm wavelength. It mainly consists of a p-type GaAs-based photo-absorption layer with a 450 nm thickness and an undoped $Al_{0.15}Ga_{0.85}As$-based collector with a thickness of 750 nm. A graded p-doped profile (1×10^{19} cm^{-3} [top] to 1×10^{19} cm^{-3} [bottom]) is used in the absorption layer to accelerate the diffusion velocity of the photo-generated electrons. In addition, the whole epi-layer structure was grown on an n-type distributed Bragg reflector (DBR) to enhance its responsivity performance. The main drawback of the photovoltaic LPC is its low output voltage, which is usually too low to directly power the other active components in the OI system. In order to boost the DC operating voltage of our LPC, several

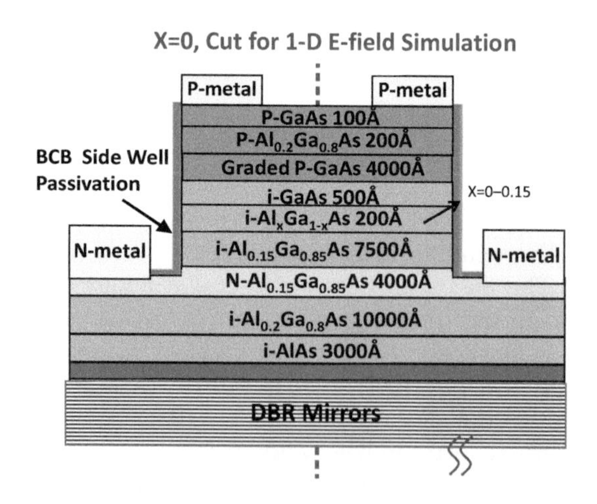

FIGURE 4.12
Cross-sectional view of the demonstrated high-speed GaAs/AlGaAs-based LPC (J.-W. Shi et al., "Dynamic Analysis of Cascade Laser Power Converters for Simultaneous High-Speed Data Detection and Optical-to-Electrical DC Power Generation," *IEEE Transactions on Electron Devices*, vol. 58, pp. 2049–2056, July 2011. © 2011 IEEE).

high-speed LPCs are series wound (linear cascade). The DC operation voltage will thus be proportional to the number of cascaded units. In addition, in order to maximize the net output photocurrent, the output photocurrent from each cascaded unit must be as close as possible [24]. A similar working principle has also been implemented in an InP-based cascade-twin UTC-PD [23], which can effectively improve the output voltage due to its double increase in built-in potential, as shown in Figures 4.9 through 4.11.

The measured DC responsivity of our single device is around 0.41 A/W, which corresponds to an external quantum efficiency of around 60% under zero-bias operation. Under low-power injection (<0.5 mW), there is a slight degradation in this value to around 0.36 A/W when the operating voltage reaches +0.9 V. Figure 4.13a,b shows the measured current-voltage (I-V) curves of the single-LPC and linear-cascade two-LPC devices under different output photocurrents. The value of the injected optical power to each single device is specified in traces (a) and (b). As can be seen, the operating voltage of the cascade device is about twice as high as that of a single cell with around one-half of the responsivity. This is because the optical power needs to be about two times higher to feed each single cell in the cascaded structure to generate the same amount of output photocurrent as that of a single device. Figure 4.14a,b shows the measured O-E power conversion efficiency versus the bias voltage of the single- and cascaded two-LPC devices under different optical pumping powers. As can be seen, the maximum O-E power conversion efficiency of both devices under low pumping power and optimum bias voltage is around 34%. On the other hand, when the injected optical power exceeds 0.5 mW, as shown in Figures 4.13 and 4.14, significant reduction in both the photocurrent and external efficiency can be observed.

Figure 4.15a,b shows the measured O-E and extracted RC/transient time–limited 3 dB bandwidths of a single device and a two-element cascaded device under different forward biases, respectively. As can be seen, when

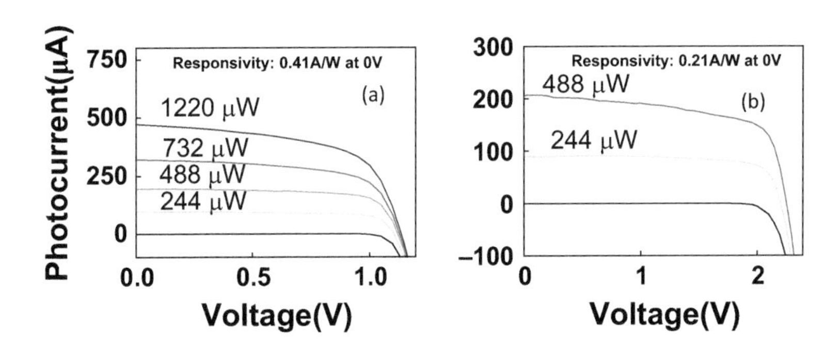

FIGURE 4.13
Measured I-V curves of (a) the single device and (b) the linear-cascade device under different optical pumping powers. (J.-W. Shi et al., "Dynamic Analysis of Cascade Laser Power Converters for Simultaneous High-Speed Data Detection and Optical-to-Electrical DC Power Generation," *IEEE Transactions on Electron Devices*, vol. 58, pp. 2049–2056, July 2011. © 2011 IEEE).

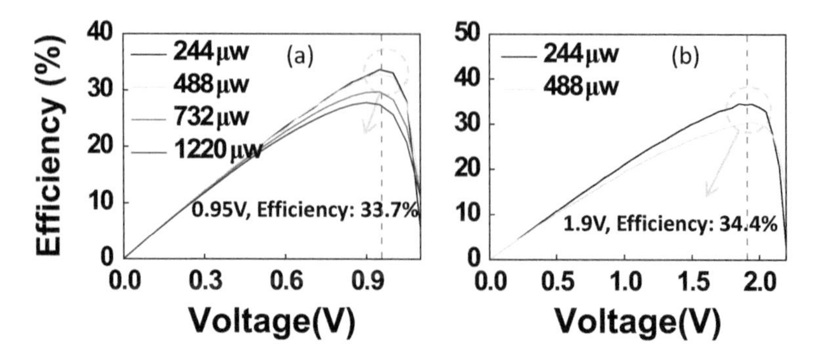

FIGURE 4.14
Measured O-E power conversion efficiency versus bias voltage of (a) the single device and (b) the linear-cascade device under different optical pumping powers. (J.-W. Shi et al., "Dynamic Analysis of Cascade Laser Power Converters for Simultaneous High-Speed Data Detection and Optical-to-Electrical DC Power Generation," *IEEE Transactions on Electron Devices*, vol. 58, pp. 2049–2056, July 2011. © 2011 IEEE).

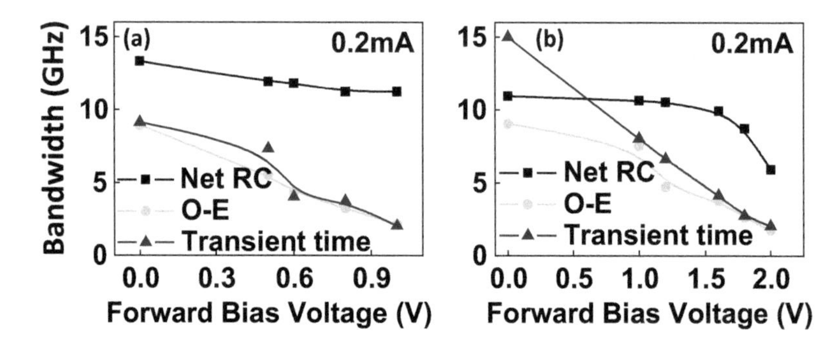

FIGURE 4.15
Measured O-E, extracted RC, and transient time–limited 3 dB bandwidths versus forward-bias voltages for (a) the single device and (b) the cascade device. (J.-W. Shi et al., "Dynamic Analysis of Cascade Laser Power Converters for Simultaneous High-Speed Data Detection and Optical-to-Electrical DC Power Generation," *IEEE Transactions on Electron Devices*, vol. 58, pp. 2049–2056, July 2011. © 2011 IEEE).

the forward operation voltage increases further, the 3 dB bandwidth performance seriously degrades due to the significant increase in carrier transient time. This phenomenon can be attributed to the current-blocking effect in the GaAs/AlGaAs hetero-junction, which can be minimized by using type II hetero-structures at the interface between the absorption and collection layers [29, 30]. Furthermore, the cascaded device has a superior speed performance to that of single device. This result can be attributed to an imbalance in the optical power injected into the two cascaded devices (1:1.2), which causes the device with the lower injected optical power to suffer from a smaller effective DC or instantaneous AC forward voltage and a larger built-in field compared with the device with a higher injected power. The larger built-in field should avoid the slow-down of the over-shoot electron drift velocity and a higher 3 dB bandwidth.

Figure 4.16 shows the measured bit error rate (BER) versus the forward-bias voltage under different output photocurrents for the single device at 5 Gbit/s and the linear-cascade device at 10 Gbit/s (pseudo-random bit sequence [PRBS]: $2^{15}-1$). We can clearly see that the maximum operation speed of the single device is only around 5 Gbit/s, but by using the cascade structure, the operation speed can be further improved to 10 Gbit/s. The faster speed performance of the cascaded device can mainly be attributed to the shorter internal carrier transient time, as previously discussed.

Although the cascaded device can attain 10 Gbit/s operation, a certain ratio of optically injected power in the cascaded two-LPC device is necessary, which limits its practical application in OI systems. The key to further improving the speed performance of LPCs is to minimize the current-blocking effect.

In order to attain this goal, a GaAs/$In_{0.5}Ga_{0.5}P$ LPC was tested [28]. The inset in Figure 4.17a,b shows a top view and conceptual cross-sectional view of the fabricated device, respectively. The main difference between this structure and the GaAs/$Al_{0.15}Ga_{0.85}As$-based LPC is that the AlGaAs-based collector is replaced by a 730 nm $In_{0.5}Ga_{0.5}P$ n-type collector layer with a graded doping profile (1×10^{16} cm^{-3} [top] to 5×10^{18} cm^{-3} [bottom]). The graded doping profile induced built-in E-field in the collector layer can accelerate the electron diffusion/drift process, which will significantly benefit the high-speed performance of our LPC operated under forward bias with a very small net E-field inside. Although the use of graded n-type doping in the collector will increase the C_J and degrade the RC-limited bandwidth, compared with those of devices with an undoped collector [18,19],

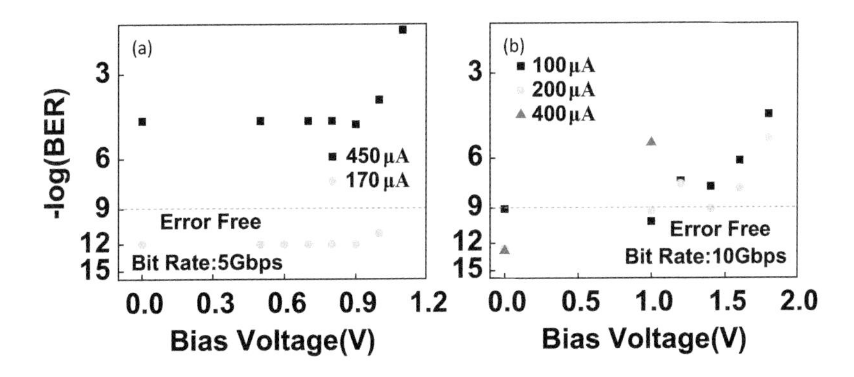

FIGURE 4.16
Measured BER versus forward-bias voltages of (a) the single device at 5 Gbit/s and (b) the linear-cascade device at 10 Gbit/s under different output photocurrents (J.-W. Shi et al., "Dynamic Analysis of Cascade Laser Power Converters for Simultaneous High-Speed Data Detection and Optical-to-Electrical DC Power Generation," *IEEE Transactions on Electron Devices*, vol. 58, pp. 2049–2056, July 2011. © 2011 IEEE).

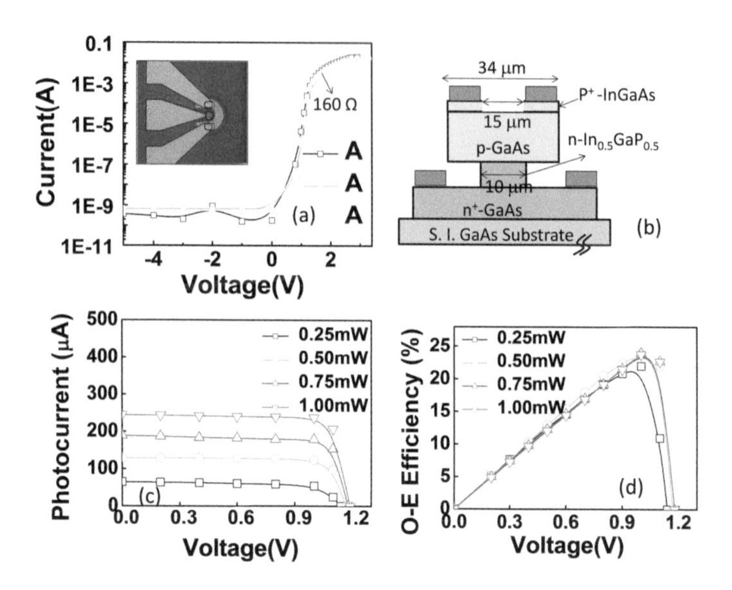

FIGURE 4.17
(a) The dark I-V curves, (b) the conceptual cross-sectional view of the device with an undercut mesa, (c) photocurrents measured under different optical pumping powers and forward voltages for device A (open symbols), and (d) corresponding DC O-E power conversion efficiency. The inset to (a) shows the top view of the demonstrated LPC (J.-W. Shi et al., "GaAs/$In_{0.5}Ga_{0.5}P$ Laser Power Converter with Undercut Mesa for Simultaneous High-Speed Data Detection and DC Electrical Power Generation," *IEEE Electron Device Letters*, vol. 33, pp. 561–563, April 2012. © 2012 IEEE).

such problems can be minimized by using an $In_{0.5}Ga_{0.5}P$ mesa with an undercut (mushroom) structure, as will be discussed later. Furthermore, the $In_{0.5}Ga_{0.5}P$ collector can greatly minimize the conduction band offsets between the GaAs-based absorption and collection layers, which will block the electron current and seriously limit the speed performance of the LPC [27], as previously discussed.

In order to further reduce C_J without seriously reducing the device's active area and increasing the differential resistance, an undercut mesa profile (as shown in Figure 4.17b) was realized in our device [28]. By properly controlling the wet-etching time, the active diameter of the final fabricated device is around 10 μm, as specified in Figure 4.17b.

Figure 4.17a shows the measured I-V curves of the device with an undercut mesa profile. As can be seen, the differential resistance is ~160 Ω, which is close to the reference device without an undercut profile. The measured DC photocurrent of our LPC versus the forward voltage under different optical pumping powers with the wavelength at 830 nm is shown in Figure 4.17c, and the corresponding O-E power conversion efficiency is shown in Figure 4.17d. As can be seen, the maximum O-E conversion efficiency happens at the bias ~+1.0 V and the corresponding maximum conversion efficiency is ~23%. A higher conversion efficiency can be achieved by inserting a DBR with central wavelength at around ~830 nm below the active layers of the device (as shown in Figure 4.12) to enhance its photo-absorption process.

Figure 4.18a,b shows the measured O-E response (f_{O-E}) of a device with an undercut (device A) and a reference device without an undercut (device B), respectively, under a fixed output photocurrent (~90 μA) and two different operating voltages (−5 V and +0.8 V). Under +0.8 V operation, the corresponding maximum conversion efficiency is up to ~ 20%. As can be seen for device A, even when the operating voltage is pushed near to the turn-on point, the degradation in speed is not so serious (10 to 8 GHz). On the other hand, device B exhibits a more serious degradation in speed performance (9 to 2 GHz) due to its larger C_J than that of device A.

The influence of a forward operation voltage on the measured high-speed eye patterns is a key issue for the practical application of the demonstration LPC. Figure 4.19a,b shows the measured BER of device A versus a forward operation voltage under 8 and 10 Gbit/s operations, respectively. The insets show the corresponding error-free ($BER < 10^{-9}$) eye patterns. As can be seen, under 8 Gbit/s operations, the error-free operation can be sustained from 0 to +0.8 V, which corresponds to 20% DC O-E power generation efficiency. As shown in Figure 4.19b, even at data rates up to 10 Gbit/s, error-free ($BER < 10^{-9}$) performances can still be maintained when the operation voltage is pushed to +0.8 V.

4.5 Type II InP-Based UTC-PDs

Under zero-bias operation, the major limiting factor of the saturation current and speed in UTC-PDs is usually the current-blocking effect between the interface of the InGaAs absorption and InP collection layers, where an

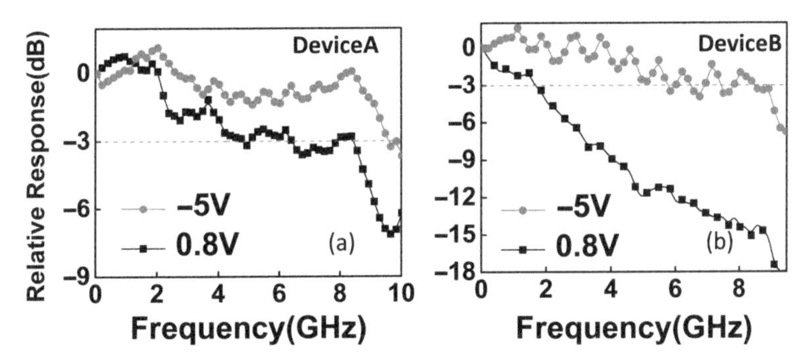

FIGURE 4.18
Measured O-E frequency responses of (a) device A and (b) device B under +0.8 V and −5 V under a fixed photocurrent (90 μA). (J.-W. Shi et al., "GaAs/$In_{0.5}Ga_{0.5}P$ Laser Power Converter with Undercut Mesa for Simultaneous High-Speed Data Detection and DC Electrical Power Generation," *IEEE Electron Device Letters*, vol. 33, pp. 561–563, April 2012. © 2012 IEEE).

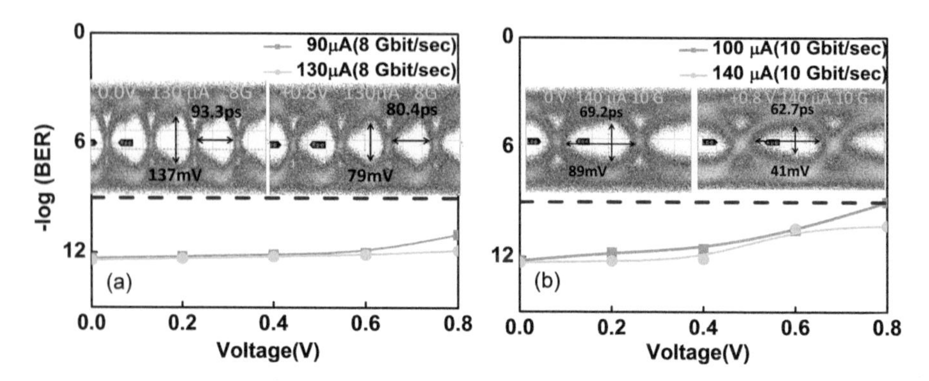

FIGURE 4.19
Measured BER versus forward operation voltages of device A with different output photocurrents at (a) 8 Gbit/s and (b) 10 Gbit/s operations. The insets show the corresponding error-free eye patterns.

electron barrier exists due to the discontinuity of bandgaps [20], as discussed in the previous section on LPCs. In order to further improve the speed performance of UTC-PD under zero-bias operation, using a type II material system ($GaSb_{0.5}As_{0.5}$/InP) in an absorption–collection (A-C) junction is one possible solution [29, 30] and has been successfully demonstrated by our group [29].

Figure 4.20 shows a simulated band diagram for the demonstrated device structure under zero-bias operation. The position of the Fermi level is also specified (see the dotted line in the figure). Here, we adopted the $GaAs_{0.5}Sb_{0.5}$/InP type II A-C interface [29, 30]. The p-type $GaSb_{0.5}As_{0.5}$ absorption layer with a thickness of 160 nm and a linearly graded doping profile

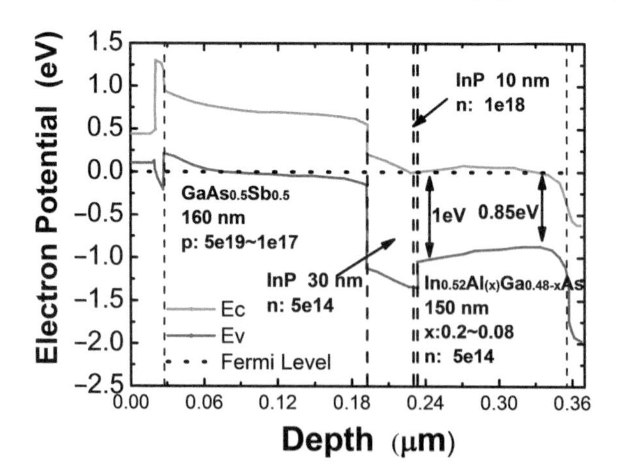

FIGURE 4.20
Simulated band diagram of demonstrated UTC-PDs under zero-bias operation. The unit of doping density is cm^{-3}. (J.-M. Wun et al., "Type-II $GaAs_{0.5}Sb_{0.5}$/InP Uni-Traveling Carrier Photodiodes with Sub-THz Bandwidth and High-Power Performance under Zero-Bias Operation," *IEEE/OSA Journal of Lightwave Technology*, vol. 35, pp. 711–716, February 2017. © 2017 IEEE).

(5×10^{19} cm^{-3} [top] to 1×10^{17} cm^{-3} [bottom]) accelerates the electron diffusion process. This type of band alignment can provide the injected electrons from the p-type $GaAs_{0.5}Sb_{0.5}$ layer to the InP collector with high excess energy (~0.22 eV) and minimize the current-blocking effect, which is the major bandwidth-limiting factor of UTC-PDs with type I A-C ($In_{0.53}Ga_{0.47}As$/InP) junctions under zero-bias operation [15]. In our collector, we adopted intrinsic InP and $In_{0.52}Al_xGa_{0.48-x}As$ layers with an ultra-low background doping density in order to relax the increase in C_J and RC-limited bandwidth under zero-bias (nearly forward-bias) operation [18, 19]. However, the position of the simulated Fermi level suggests that in the proposed device structure, the undoped layers should become n-type doped layers under zero bias due to the type II band alignment in the A-C junction, as previously discussed. This will lead to an increase in the C_J and allow the net O-E frequency responses to be RC limited. This issue will be discussed in detail later using the device modeling technique. The insertion of a thin (10 nm) n-type charge layer among these intrinsic layers at around the type II A-C junction results in a drop in the electron potential and further suppresses the current-blocking effect [16]. In addition, a graded bandgap layer (GBL) comprised of three $In_{0.52}Al_xGa_{0.48-x}As$ layers with different mole fractions (x: 0.2 [top] to 0.08 [bottom]) is used as the collector to create a built-in E-field and accelerate the electron drift process. Among our GBLs, the longest cutoff wavelength for the $In_{0.52}Al_xGa_{0.48-x}As$ alloy ($In_{0.52}Al_{0.08}Ga_{0.4}As$), 1.46 μm (0.85 eV), is chosen in order to avoid any undesired absorption process under 1.55 μm wavelength excitation.

Figure 4.21 shows the simulated E-field distribution inside the absorption and collection layers for our device under zero-bias operation. The blue and red lines in this figure represent the operation of the device under dark conditions and with an output photocurrent of 5 mA, respectively. The electron-induced space-charge field inside this device is calculated assuming a reasonable electron drift velocity (4×10^7 cm/s) inside the collector and a 6 μm active diameter, which is the actual size of the fabricated PD. In addition, with a 5 mA output photocurrent there is a shift in the effective bias voltage applied in the device simulation from 0 to +0.25 V due to the fact that a load of 50 Ω is used during real measurement. As shown in Figure 4.21, the proposed GBL structure can produce a built-in E-field, which will not be screened by the high output photocurrent under zero-bias operation. As can be seen, under a 5 mA output photocurrent, the net E-field in the A-C junction is nearly zero, which implies that a saturation phenomenon happens at this output current. This value is very close to the measured saturation current for our device, which will be discussed in detail later. Overall, the type II band alignment and GBL in our proposed device structure induce a more significant built-in electron potential drop from anode to cathode than with the traditional UTC-PD. This is of great benefit to the speed and output power performance of the UTC-PD when there is no externally applied bias voltage (E-field).

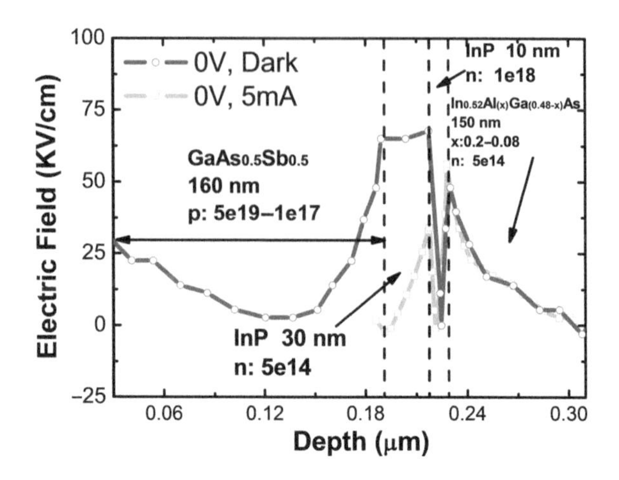

FIGURE 4.21
The simulated E-field distributions inside the absorption and collection layers of our demonstrated device structure under zero-bias operations. Dark gray line with circles: E-field distribution without light illumination (i.e., dark). Light gray line with squares: E-field distribution with 5 mA output photocurrent under a 50 Ω load. (J.-M. Wun et al., "Type-II $GaAs_{0.5}Sb_{0.5}$/InP Uni-Traveling Carrier Photodiodes with Sub-THz Bandwidth and High-Power Performance under Zero-Bias Operation," *IEEE/OSA Journal of Lightwave Technology*, vol. 35, pp. 711–716, February 2017. © 2017 IEEE).

Figure 4.22 shows top views of the active PD chip, the coplanar waveguide (CPW) pad on the AlN substrate (for flip-chip bonding, which can provide a 3 dB bandwidth of nearly ~400 GHz [31]), and the PD chip after flip-chip bonding. The inset to Figure 4.3a shows an enlarged image of the fabricated PD, which has a diameter of 6 μm and measured DC responsivity at ~0.09 A/W.

Figures 4.23 shows the measured bias-dependent O-E frequency responses of the device under a 2 mA output photocurrent. The values of the data points in each O-E trace have been normalized to the output power of the PD at near

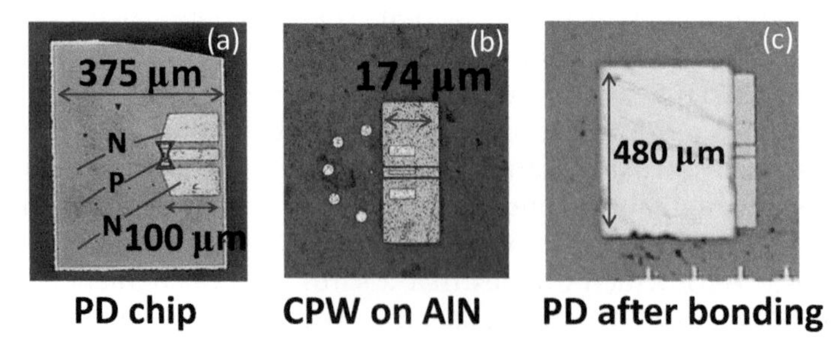

FIGURE 4.22
(a) Top views of the active PD chip, (b) bonding pads on the AlN substrate, and (c) PD chip after flip-chip bonding. The inset in (a) shows a zoomed-in picture of the active PD. (J.-M. Wun et al., "Type-II $GaAs_{0.5}Sb_{0.5}$/InP Uni-Traveling Carrier Photodiodes with Sub-THz Bandwidth and High-Power Performance under Zero-Bias Operation," *IEEE/OSA Journal of Lightwave Technology*, vol. 35, pp. 711–716, February 2017. © 2017 IEEE).

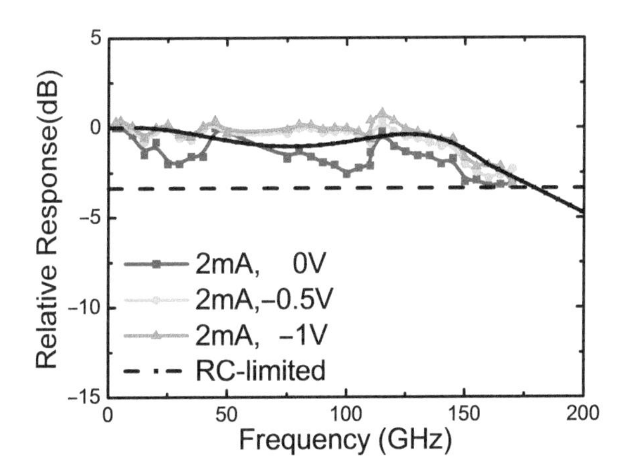

FIGURE 4.23
Measured bias-dependent (0, −0.5, and −1V) O-E frequency responses with a 6 μm active diameter PD measured at an output photocurrent of 2 mA. The black trace is the extracted RC-limited frequency response under zero-bias operation. (J.-M. Wun et al., "Type-II $GaAs_{0.5}Sb_{0.5}$/InP Uni-Traveling Carrier Photodiodes with Sub-THz Bandwidth and High-Power Performance under Zero-Bias Operation," *IEEE/OSA Journal of Lightwave Technology*, vol. 35, pp. 711–716, February 2017. © 2017 IEEE).

DC and under zero-bias operation. The black trace represents the extracted RC-limited frequency response of our demonstration device, which will be discussed in detail later. We can clearly see that even under zero-bias operation with a moderate output photocurrent value (2 mA), an extremely wide 3 dB O-E bandwidth (170 GHz) can be achieved. To the best of our knowledge, this is the best high-speed performance reported for any zero-bias PD [19–21]. Furthermore, the corresponding output photocurrent (>2 mA) for the wide O-E bandwidth is high enough for most high-speed (100 Gbit/s) photoreceiver circuit applications [18, 19]. In addition, we can clearly see that under a low output photocurrent (2 mA), the measured bandwidth is insensitive to the reverse-bias voltage (0 to −1 V). This implies that the built-in E-field inside the active layers is strong enough to overcome the space-charge-screening (SCS) effect induced by the output photocurrent (~2 mA).

The net O-E 3 dB bandwidth f_{3dB} of a PD is determined by the carrier transit time ($1/f_t$) and the RC time constant ($1/f_{RC}$). We used the *equivalent circuit modeling* technique to investigate whether the internal carrier transit or RC-limited bandwidth dominates the measured net O-E bandwidth of our device. With this approach, the RC-limited bandwidth (f_{RC}) can be extracted by the use of the measured scattering parameters of the microwave reflection coefficients (S_{11}) [31] of the PD. Figure 4.24a shows the adopted equivalent circuit models for the fitting of the S_{11} parameters and the fitted values of each circuit element, except for R_T and C_T, which are shown in the table inserted into Figure 4.24b. Here, C_J, R_J, and R_C represent the junction capacitance, junction resistance, and differential resistance of the active p-n diode,

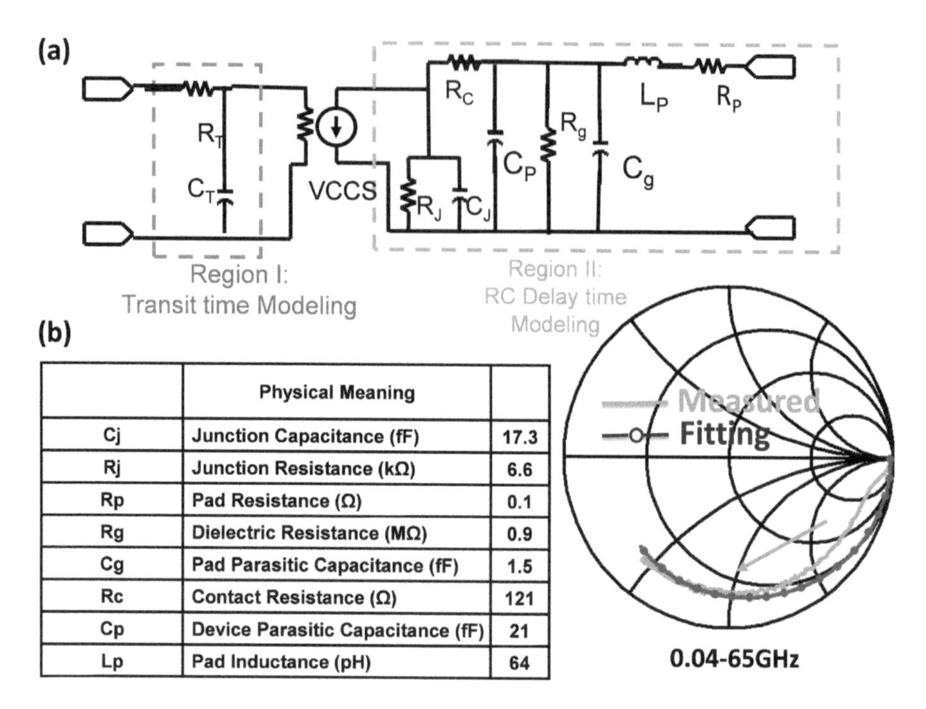

	Physical Meaning	
Cj	Junction Capacitance (fF)	17.3
Rj	Junction Resistance (kΩ)	6.6
Rp	Pad Resistance (Ω)	0.1
Rg	Dielectric Resistance (MΩ)	0.9
Cg	Pad Parasitic Capacitance (fF)	1.5
Rc	Contact Resistance (Ω)	121
Cp	Device Parasitic Capacitance (fF)	21
Lp	Pad Inductance (pH)	64

FIGURE 4.24
(a) Equivalent circuit model (VCCS: voltage controlled current source). (b) Measured (continuous line) and fitted (open circles) S_{11} parameters from near DC to 65 GHz under a fixed DC bias (0 V). The head of the light gray arrow indicates the increase in the sweep frequency. The inserted table shows the values of the circuit elements used in the modeling process. (J.-M. Wun et al., "Type-II $GaAs_{0.5}Sb_{0.5}$/InP Uni-Traveling Carrier Photodiodes with Sub-THz Bandwidth and High-Power Performance under Zero-Bias Operation," *IEEE/OSA Journal of Lightwave Technology*, vol. 35, pp. 711–716, February 2017. © 2017 IEEE).

respectively; C_P represents the parasitic capacitance induced by the interconnected metal lines between the passive CPW and the active diode; R_P and L_P indicate the ohmic loss and inductance of the metallic CPW pads, respectively; and R_g and C_g represent the dielectric loss and capacitance caused by the buried dielectric layer (polymethylglutarimide [PMGI]) below the metal pads, respectively.

During the device modeling process for the extraction of the extrinsic f_{RC} for the PD chips, two artificial circuit elements (R_T and C_T) are removed, due to the fact that they are used to mimic the low-pass frequency response of the internal carrier transit time [31]. Figure 4.24b shows the fitted and measured S_{11} parameters as a Smith chart, while the extracted RC-limited frequency responses are given in Figure 4.23, as previously discussed. We can clearly see that the fitted S_{11} trace matches the measured ones well, from nearly DC to 65 GHz (on the Smith chart), and that the resonant frequency (~120 GHz) of the extracted RC-limited frequency response is very close to the net-measured O-E one. These results are indicative of the accuracy of the proposed equivalent circuit model. From the extracted RC-limited 3 dB bandwidth at around 175 GHz, we can conclude that the measured net O-E bandwidth (~170 GHz) is mainly determined by the RC time constant. The root cause of our large RC time constant might be the additional C_J induced by the non-depleted carriers, which originates from the type II band alignment, as shown in Figure 4.20. One feasible way to further relax the RC bandwidth and improve the net speed performance of the demonstrated device is to appropriately thicken the intrinsic layer. In addition, directly downscaling the device's active area can also enhance the RC bandwidth of the device; however, such an approach will usually lead to an increase in the differential resistance and a significant degradation in the speed performance of the PD under zero-bias operation [15, 28]. As in our previous discussion on the GaAs/$In_{0.5}Ga_{0.5}P$ high-speed LPC [28], an undercut mesa structure may be used in our demonstration type II $Ga_{0.5}As_{0.5}Sb$/InP UTC-PD to release the trade-off between the C_J and the differential resistance of the UTC-PD for zero- and forward-bias operation and further improve its dynamic performance.

Figures 4.25 and 4.26 show the measured transfer curves of the reverse-bias voltages versus the output photocurrent and normalized sub-THz power at 170 GHz, respectively. Here, two types of devices are measured for comparison. One is the proposed PD structure with a $Ga_{0.5}As_{0.5}Sb$/InP type II A-C junction and the other is the traditional $In_{0.53}Ga_{0.47}As$/InP UTC-PD structure with an additional n-type charge layer near the A-C interface [31]. Both devices share the same flip-chip bonding packaged structure and absorption/collection (~160/~160 nm) layer thicknesses.

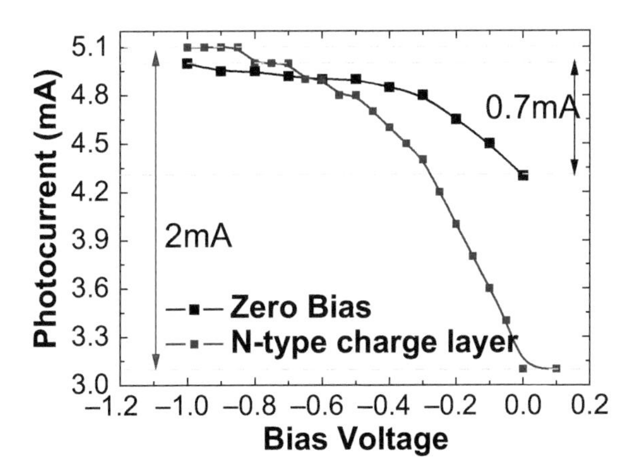

FIGURE 4.25
Measured transfer curve (bias voltage versus photocurrent) at 170 GHz for the demonstrated type II UTC-PD structure and reference UTC-PD structure. (J.-M. Wun et al., "Type-II $GaAs_{0.5}Sb_{0.5}$/InP Uni-Traveling Carrier Photodiodes with Sub-THz Bandwidth and High-Power Performance under Zero-Bias Operation," *IEEE/OSA Journal of Lightwave Technology*, vol. 35, pp. 711–716, February 2017. © 2017 IEEE).

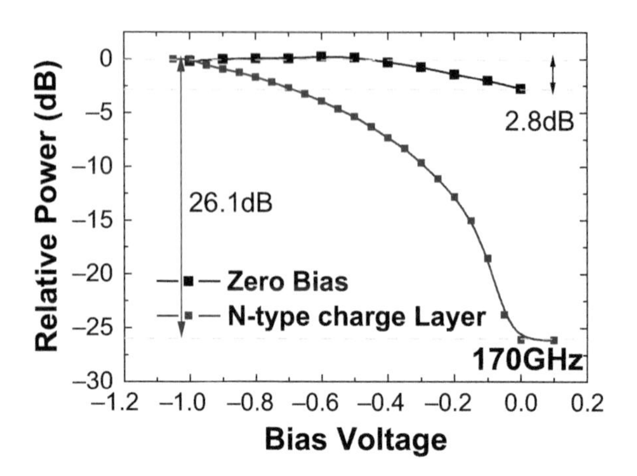

FIGURE 4.26
Measured transfer curve (bias voltage versus output MMW power) at 170 GHz for the demonstrated type II UTC-PD structure and reference UTC-PD structure. (J.-M. Wun et al., "Type-II $GaAs_{0.5}Sb_{0.5}$/InP Uni-Traveling Carrier Photodiodes with Sub-THz Bandwidth and High-Power Performance under Zero-Bias Operation," *IEEE/OSA Journal of Lightwave Technology*, vol. 35, pp. 711–716, February 2017. © 2017 IEEE).

We can clearly see that compared with the traditional UTC-PD, our proposed device structure shows a much smaller variation in both the photocurrent (0.7 versus 2 mA) and sub-THz output power (2.8 versus 26.1 dB) when the bias voltage swings from reverse to nearly forward (zero bias). The measurement results strongly indicate the superior speed and power performance of our device compared with the traditional UTC-PD under zero-bias operation.

Figure 4.27 shows the photo-generated MMW power versus the output photocurrent obtained with our PD under sinusoidal signal excitation at the 170 GHz operating frequency. The output power was measured by a thermal THz power meter (PM4, VDI-Erickson), and the value of the THz power shown here has been carefully de-embedded, taking into account an insertion loss of around 3 dB for the D-band WR-6 waveguide probe, 0.33 dB for the waveguide taper (WR-6 to WR-10), and 0.185 dB for the WR-10 waveguide section at the 170 GHz operating frequency. As can be seen, the optimum bias for maximum output power should be around −0.5 V, and there is around a 3 dB enhancement in saturation output power compared with that obtained under zero-bias operation.

We can clearly see that under zero-bias operation, there is significant saturation in the output power when the photocurrent is over 5 mA, which is consistent with our calculation results, as illustrated in Figure 4.21. Furthermore, compared with the results reported for the traditional UTC-PD with the $In_{0.53}Ga_{0.47}As$/InP A-C junctions and the excellent performance in terms of speed and power under zero-bias operation [18,19], our device has an output power ~7.3 dB higher (−11.3 versus −18.6 dBm), a higher saturation current (8 versus 2 mA), and at a higher operation frequency (170 versus 100 GHz).

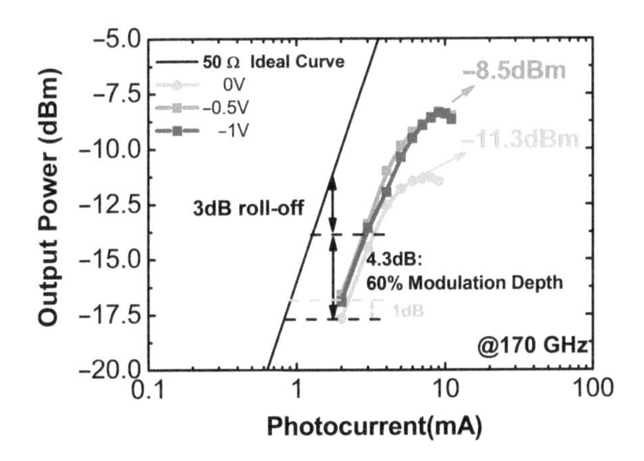

FIGURE 4.27
Measured photo-generated MMW power versus photocurrent under sinusoidal signal excitation and different reverse-bias voltages (0, −0.5, and −1 V) at an operating frequency of 170 GHz. The solid line shows the ideal trace for a 100% modulation depth and 50 Ω load. (J.-M. Wun et al., "Type-II $GaAs_{0.5}Sb_{0.5}$/InP Uni-Traveling Carrier Photodiodes with Sub-THz Bandwidth and High-Power Performance under Zero-Bias Operation," *IEEE/OSA Journal of Lightwave Technology*, vol. 35, pp. 711–716, February 2017. © 2017 IEEE).

To the best of our knowledge, this is the highest continuous wave output power ever reported for the photonic generation of sub-THz waves from a PD under zero-bias (bias-free) operation [19, 21]. Besides, as specified in figure 4.27, there is an approximately 7.3 dB difference in power between the ideal (100% optical modulation depth) and measured traces. Such a discrepancy can be attributed to the 3 dB high-frequency roll-off of the device itself at 170 GHz operation and the ~60% optical modulation depth in our optical system during power measurement, which corresponds to the other 4.3 dB loss of power.

Overall, through the use of a type II A-C junction and graded bandgap collector, the UTC-PD, with an advanced flip-chip bonding package, exhibits an extremely wide O-E bandwidth (170 GHz), reasonable responsivity (0.09 A/W), and a moderate value of output power (−11.3 dBm) at the sub-THz regime (170 GHz) under zero-bias operation. The proposed device structure may find applications in next-generation OI or coherent fiber communication systems, where a densely packaged high-speed (100 Gbit/s) photo-receiver array with extremely small thermal–electrical cross-talk and compact module size is highly desired.

4.6 High-Speed Ge-on-Si-Based PDs

One of the major trends in the development of PDs for application in Si-photonic OI systems is Ge-on-Si-based PDs [32–37]. They can be monolithically integrated with Si-based CMOS ICs and the further reduction in both system size and cost of package can be expected. By using a Ge-based photo-absorption layer in the waveguide PD structure, excellent high-speed performance for 40 Gbit/s operation with reasonable responsivity (including waveguide coupling loss) under zero-bias operation has been successfully demonstrated [36, 37]. The capabilities for high-speed and zero-bias operations of the CMOS-compatible Ge-based PD are mainly due to the photo-generated hole in the Ge layer having a much faster drift velocity than that of the hole in the III-V semiconductor–based photo-absorption layer. Such excellent performance indicates its strong potential for green OI applications.

The important advantage of the Ge layer directly grown on the Si substrate is not only for the issue of PD integration with CMOS ICs but also its unique strain characteristics, which will enhance the photo-absorption process. Ge-on-Si should be compressively strained when grown coherently because of the larger lattice constant of the Ge layer. However, the thick Ge epi-layer on the Si substrate shows tensile strain in the epi-layers of the PDs. This is because the linear-lattice thermal expansion coefficient of Ge is larger than that of Si, provided the epi-layer is not relaxed during growth cooling [38]. This tensile-strained Ge epi-layer can extend the absorption cut-off wavelength of the Ge layer to around 1.55 μm. Figure 4.28 shows the measured absorption coefficient versus the wavelengths for the Ge layer, with and without tensile strain [39]. As can be seen, the area of selective growth in the Ge layer will have residual tensile strain (0.141%) and enhanced photo-absorption on the long wavelength side compared with that of the fully relaxed Ge layer. The value of the absorption constant for the strained Ge layer is comparable with that of the III-V–based $In_{0.53}Ga_{0.47}As$ layer. Very high-speed performance (>50 GHz) has been demonstrated by

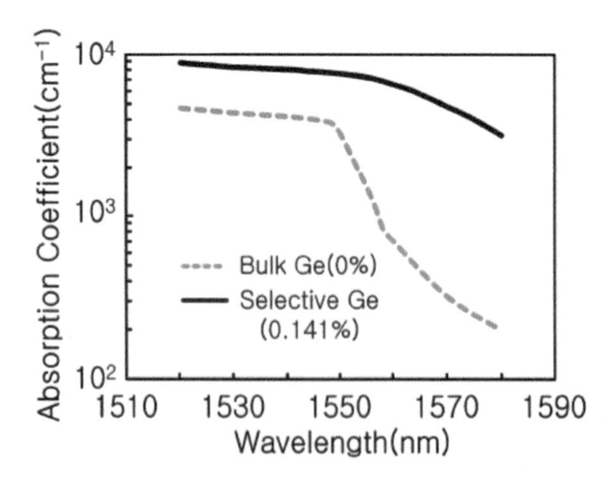

FIGURE 4.28
Absorption constants of tensile-stained Ge film and full-relaxed Ge film on Si substrate (H.-Y. Yu, "High-Efficiency p-i-n Photodetectors on Selective-Area-Grown Ge for Monolithic Integration," *IEEE Electron Device Letters*, vol. 30, pp. 1161–1163, November 2009. © 2009 IEEE).

the selective-area Ge PD growth on a silicon-on-insulator (SOI) substrate, which has a thick buried insulator (oxide) layer (~1 μm), under 1.55 μm wavelength excitations [32–37].

Figure 4.29 shows a conceptual cross-sectional view of high-speed Ge PDs with a vertically illuminated structure on the SOI substrates [33]. This kind of substrate plays an important role in the realization of chip-level optical interconnects and Si-based photonic integrated circuits (PICs). This is because there is a huge difference in the index between the buried oxide layer and the active Si/Ge-based epi-layers, and a strong index-guided ridge waveguide on an SOI substrate can thus be realized. Figure 4.30a,b shows the conceptual and real cross-sectional views of a Ge waveguide PD with a vertical n-i-p junction on an SOI substrate for chip-level optical interconnect

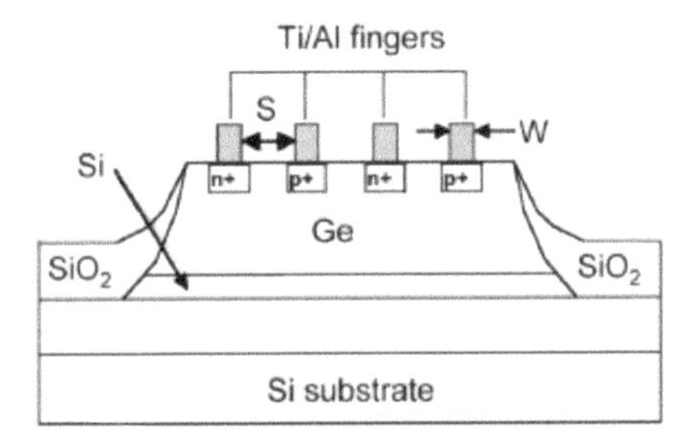

FIGURE 4.29
Conceptual cross-sectional view of a Ge-on-SOI PD with the vertically illuminated structure (C. L. Schow et al., "A 15-Gb/s 2.4-V Optical Receiver Using a Ge-on-SOI Photodiode and a CMOS IC," *IEEE Photonics Technology Letters*, vol. 18, pp. 1981–1983, October 2006. © 2006 IEEE)

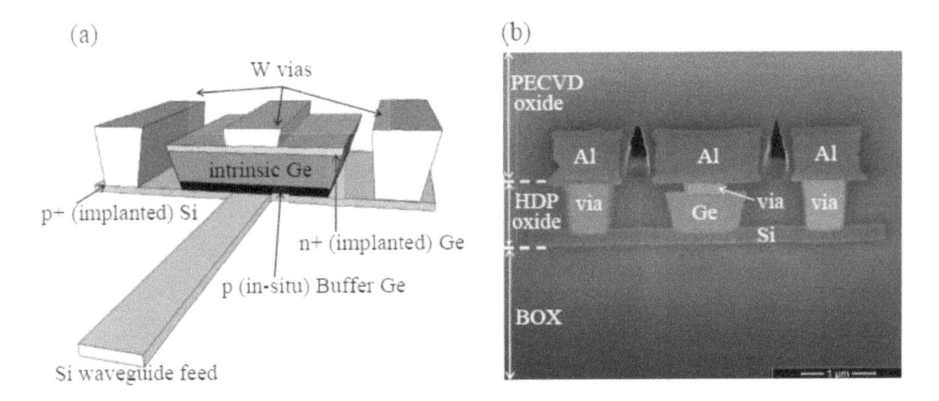

FIGURE 4.30
(a) Conceptual and (b) real cross-sectional views of a GOI-based waveguide PD with a vertical n-i-p junction (C. T. DeRose et al., "Ultra Compact 45 GHz CMOS Compatible Germanium Waveguide Photodiode with Low Dark Current," *Optics Express*, vol. 19, no. 25, pp. 24897–24904, December 2011. © 2011 OSA)

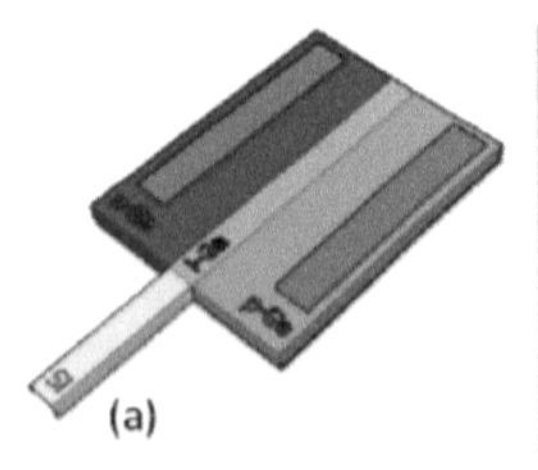

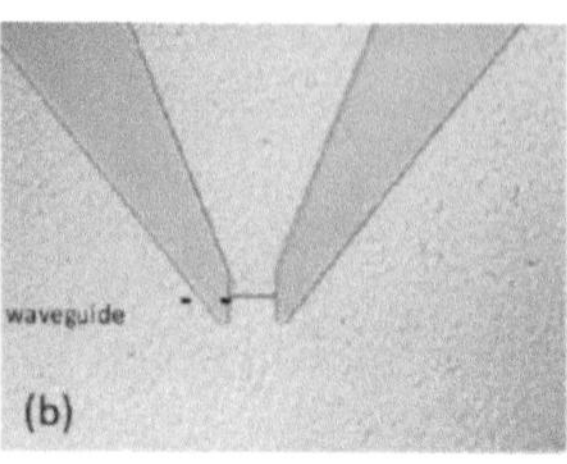

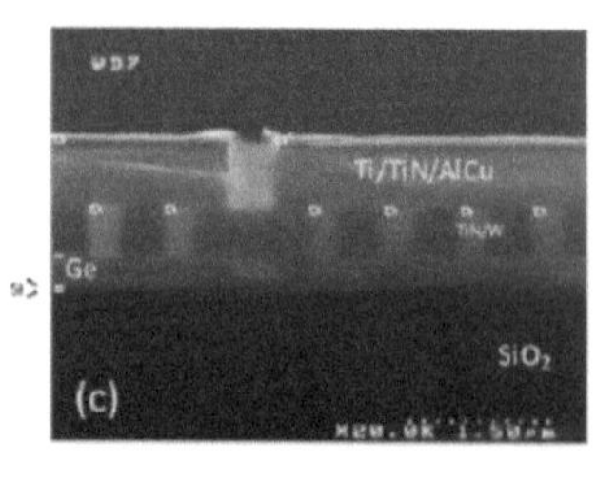

FIGURE 4.31
(a) Conceptual and (b) real top views of a GOI-based waveguide PD with a lateral n-i-p junction; (c) real cross-sectional view of a fabricated GOI PD. (L. Vivien et al., "Zero-Bias 40 Gbit/s Germanium Waveguide Photodetector on Silicon," *Optics Express*, vol. 20, no. 2, pp. 1096–1101, January 2012. © 2012 OSA).

applications. As can be seen, the topmost Ge layer serves as a photo-absorption region, and the 1.55 μm-wavelength photons propagating in the bottom SOI waveguide can be absorbed through the use of an evanescent coupling process [36]. An alternative structure of the Ge waveguide PD with a transverse (lateral) p-i-n junction is shown in Figure 4.31 [37]. The main difference between the vertical- and lateral-junction Ge PDs is in the optical coupling process. As can be seen, in the case of the lateral-junction PD, the Si core of the SOI waveguide is etched away and replaced by a Ge-on-insulator (GOI) ridge waveguide by use of the selective area growth (SAG) technique. This butt joint–coupling scheme between the SOI and GOI waveguides on the same plane can eliminate the extra insertion loss of the evanescent coupling process in the vertical n-i-p junction Ge waveguide PD, as previously discussed. Both of the aforementioned Ge PD structures have demonstrated ultra-high-speed (>40 Gbit/s) performance with reasonable responsivity (~0.8 A/W) and low to reasonable dark currents (nA to μA) under 1.55 μm wavelength excitation and a low bias voltage (–1 V). Furthermore, in the case of the lateral Ge waveguide PD, a clear eye opening at 40 Gbit/s under zero-bias operations has been successfully demonstrated. Figure 4.32 shows the measured eye patterns of a lateral-junction GOI PD [37] at different data rates and under 0 and –1 V bias operations. The excellent zero-bias speed performance of the Ge-based p-i-n waveguide PD can be attributed to the fast drift velocity of the hole in the Ge photo-absorption layer. This performance is even comparable with the high-performance III-V–based UTC-PDs under zero-bias operation, as previously discussed [23].

4.7 Conclusion

In this chapter, we have reviewed several kinds of high-speed PDs that can sustain ultra-fast performance under zero- or even forward-bias operations for green OI applications. Traditional p-i-n PDs suffer from significant

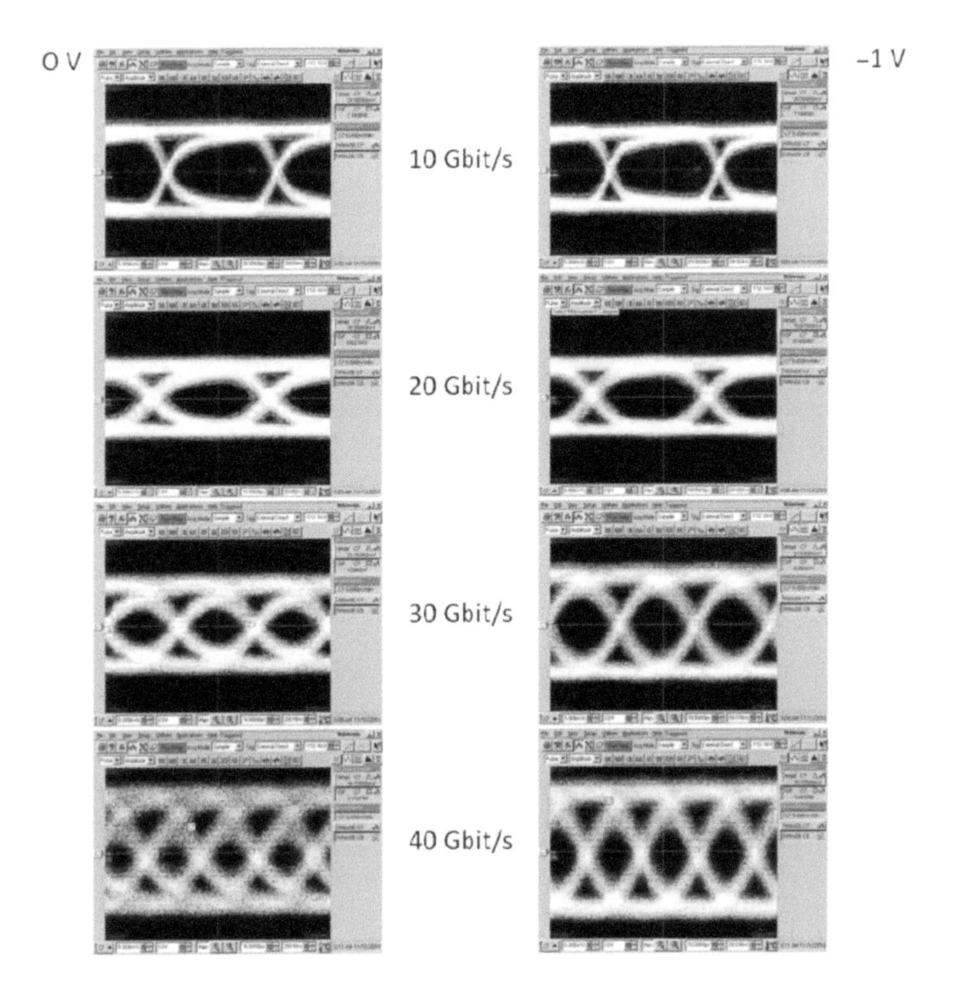

FIGURE 4.32
Measured eye patterns at different data rates of a GOI-based waveguide PD with a lateral n-i-p junction under zero bias and −1 V bias. (L. Vivien et al., "Zero-Bias 40 Gbit/s Germanium Waveguide Photodetector on Silicon," *Optics Express*, vol. 20, no. 2, pp. 1096–1101, January 2012. © 2012 OSA)

bandwidth degradation when the bias voltage reaches zero, due to the increase in C_J and the slow-down of the photo-generated hole under such a small externally applied E-field. In order to overcome the two aforementioned speed-limiting factors, UTC-PD structures, which can eliminate the hole transport, have successfully demonstrated excellent speed and power performance under zero- or forward-bias operations. By using the GaAs and $In_{0.5}Ga_{0.5}P$ as the absorption and collection layers, respectively, in the UTC-PD structure, problems with the hole transport and the current-blocking effect can be minimized. The demonstrated PD can achieve invariable high-speed performance at forward-bias voltages of 0 to +0.8 V, and 10 Gbit/s error-free operation with a 20% power conversion efficiency. This result challenges the common belief that high-speed PDs must be power-consuming devices under reverse bias. Furthermore, by combining a UTC-PD that has a type II $GaAs_{0.5}Sb_{0.5}$/InP A-C junction with an advanced flip-chip bonding structure, sub-THz bandwidth and high output power under zero- (forward-) bias operation can be achieved.

Compared with III-V–based PDs, the major advantages of Ge-on-Si (or GOI) PDs is that they can be integrated with mature CMOS ICs, and the mobility of the hole in an active Ge absorber is much higher than that in an III-V material–based absorber. This can reduce the required bias voltage for high-speed performance in traditional III-V–based p-i-n PDs and benefit by reducing the power consumption of the Si photonics–based OI system. The GOI-based waveguide PD with a lateral n-i-p junction has successfully demonstrated 40 Gbit/s operation under zero-bias operation.

References

1. G. Cook, D. Pomerantz, K. Rohrbach, B. Johnson, and J. Smyth, *Clicking Clean: A Guide to Building the Green Internet*, Washington, DC, Greenpeace, 2015, ch. 1.
2. C. Lange, D. Kosiankowski, R. Weidmann, and A. Gladishch, "Energy Consumption of Telecommunication Networks and Related Improvement Options," *IEEE Journal of Selected Topics in Quantum Electronics*, vol. 17, pp. 285–295, March/April 2011.
3. B. Kellogg, V. Talla, S. Gollakota, and J. R. Smith, "Passive Wi-Fi: Bringing Low Power to Wi-Fi Transmissions," *Proc. NSDI'16*, Santa Clara, CA, pp. 151–164, March 2016.
4. B. E. Lemoff, M. E. Ali, G. Panotopoulos, G. M. Flower, B. Madhavan, A. F. J. Levi, and D. W. Dolfi, "MAUI: Enabling Fiber-to-Processor with Parallel Multiwavelength Optical Interconnects," *IEEE/OSA Journal of Lightwave Technology*, vol. 22, pp. 2043–2054, September 2004.
5. L. Schares, J. A. Kash, F. E. Doany, C. L. Schow, C. Schuster, D. M. Kuchta, P. K. Pepeljugoski, J. M. Trewhella, C. W. Baks, R. A. John, L. Shan, et al., "Terabus: Terabit/Second-Class Card-Level Optical Interconnect Technologies," *IEEE Journal of Selected Topics in Quantum Electronics*, vol. 12, pp. 1032–1044, May 2006.
6. K. Kurata, "High-Speed Optical Transceiver and Systems for Optical Interconnects," *Proc. OFC 2010*, San Diego, CA, pp. OThS3, March 2010.
7. J. A. Kash, A. F. Benner, F. E. Doany, D. M. Kuchta, B. G. Lee, P. K. Pepeljugoski, L. Schares, C. L. Schow, and M. Taubenblat, "Optical Interconnects in Exascale Supercomputers," *IEEE Photonic Society Meeting 2010*, Denver, CO, pp. WR1, November 2010.
8. R. S. Tucker, "A Green Internet," *IEEE Lasers and Electro-Optics Society 2008 (LEOS/2008) Annual Meeting*, Newport Beach, CA, pp. PLE3, November 2008.
9. D. Bimberg, "Green Data and Computer Communication," *IEEE Photonic Society Meeting 2011*, Arlington, VA, October, TuN3.

10. M. A. Taubenblatt, "Optical Interconnects for High-Performance Computing," *IEEE/OSA Journal of Lightwave Technology*, vol. 30, no. 4, pp. 448–458, February 2012.
11. Clint L. Schow, "Power-Efficient Transceivers for High-Bandwidth, Short-Reach Interconnects," *Proceedings of OFC 2012*, Los Angeles, CA, pp. OTh1E4, March 2012.
12. H. Chen A. Beling, H. Pan, and J. C. Campbell, "A Method to Estimate the Junction Temperature of Photodetectors Operating at High Photocurrent," *IEEE Journal of Quantum Electronics*, vol. 45, pp. 1537–1541, December 2009.
13. D. M. Kuchta, A. V. Rylyakov, C. L. Schow, J. E. Proesel, C. W. Baks, P. Westbergh, J.S. Gustavsson, and A. Larsson, "A 50 Gb/s NRZ Modulated 850 nm VCSEL Transmitter Operating Error Free to 90 °C," *IEEE/OSA Journal of Lightwave Technology*, vol. 33, no. 4, pp. 802–810, February 2015.
14. J. Schubert E. Oliva, F. Dimroth, W. Guter, R. Loeckenhoff, and A. W. Bett, "High-Voltage GaAs Photovoltaic Laser Power Converters," *IEEE Transactions on Electron Devices*, vol. 56, pp. 170–175, February 2009.
15. J.-W. Shi, F.-M. Kuo, Chan-Shan Yang, S.-S. Lo, and Ci-Ling Pan, "Dynamic Analysis of Cascade Laser Power Converters for Simultaneous High-Speed Data Detection and Optical-to-Electrical DC Power Generation," *IEEE Transactions on Electron Devices*, vol. 58, pp. 2049–2056, July 2011.
16. J.-W. Shi, J.-M. Wun, C.-Y. Tsai, and J. E. Bowers, "Mushroom-Mesa GaAs/ $In_{0.5}Ga_{0.5}P$ Based Laser Power Converter for Simultaneous 10 Gbit/s Data Detection and DC Electrical Power Generation," *IEEE Photonic Society Meeting 2012*, San Francisco, CA, pp. TuL4, September 2012.
17. J.-M. Wun, J.-W. Shi, C.-Y. Tsai, and Y.-M. Hsin, "Undercut GaAs/$In_{0.5}Ga_{0.5}P$ High-Speed Laser Power Converter for Simultaneous 10 Gbit/s Data Detection and Efficient DC Electrical Power Generation," *2012 International Conference on Solid State Devices and Materials*, Kyoto, Japan, A-6-5, 2012.
18. T. Umezawa, K. Akahane, N. Yamamoto, A. Kanno, K. Inagaki, and T. Kawanishi, "Zero-Bias Operational Ultra-Broadband UTC-PD above 110 GHz for High Symbol Rate PD-Array in High-Density Photonic Integration," *OFC 2015*, Los Angeles, CA, pp. M3C.7, March 2015.
19. T. Umezawa, A. Kanno, K. Kashima, A. Matsumoto, K. Akahane, N. Yamamoto, and T. Kawanishi, "Bias-Free Operational UTC-PD above 110 GHz and Its Application to High Baud Rate Fixed-Fiber Communication and W-Band Photonic Wireless Communication," *IEEE/OSA Journal of Lightwave Technology*, vol. 34, no. 13, pp. 3138–3147, July 2016.
20. K. Kato, "Ultrawide-Band/High-Frequency Photodetectors," *IEEE Transactions on Microwave Theory and Techniques*, vol. 47, pp. 1265–1281, July 1999.
21. H. Ito, S. Kodama, Y. Muramoto, T. Furuta, T. Nagatsuma, and T. Ishibashi, "High-Speed and High-Output InP-InGaAs Unitraveling-Carrier Photodiodes," *IEEE Journal of Selected Topics in Quantum Electronics*, vol. 10, pp. 709–727, July/ August 2004.
22. N. Shimizu, N. Watanabe, T. Furuta, and T. Ishibashi, "InP-InGaAs Uni-Traveling-Carrier Photodiode with Improved 3-dB Bandwidth of over 150GHz," *IEEE Photonics Technology Letters*, vol. 10, pp. 412–414, March 1998.
23. H. Ito, T. Furuta, S. Kodama and T. Ishibashi, "Zero-Bias High-Speed and High Output Voltage Operation of Cascade-Twin Unitravelling-Carrier Photodiode," *Electronics Letters*, vol. 36, no. 24, November 2000.

24. F.-M. Kuo, M.-Z. Chou, and J.-W. Shi, "Linear-Cascade Near-Ballistic Uni-Traveling-Carrier Photodiodes with an Extremely High Saturation-Current-Bandwidth Product," *IEEE/OSA Journal of Lightwave Technology*, vol. 29, no. 4, pp. 432–438, February 2011.
25. Y. Muramoto, H. Fukano, and T. Furuta, "A Polarization-Independent Refracting-Facet Uni-Traveling-Carrier Photodiode with High Efficiency and Large Bandwidth," *IEEE/OSA Journal of Lightwave Technology*, vol. 24, pp. 3830–3834, October 2006.
26. J.-W. Shi, Y.-T. Li, C.-L. Pan, M. L. Lin, Y. S. Wu, W. S. Liu, and J.-I. Chyi, "Bandwidth Enhancement Phenomenon of a High-Speed GaAs-AlGaAs Based Unitraveling Carrier Photodiode with an Optimally Designed Absorption Layer at an 830nm Wavelength," *Applied Physics Letters*, vol. 89, p. 053512, 2006.
27. J.-W. Shi, F.-M. Kuo, C.-S. Yang, S.-S. Lo, and C.-L. Pan, "Dynamic Analysis of Cascade Laser Power Converters for Simultaneous High-Speed Data Detection and Optical-to-Electrical DC Power Generation," *IEEE Transactions on Electron Devices*, vol. 58, pp. 2049–2056, July 2011.
28. J.-W. Shi, C.-Y. Tsai, C.-S. Yang, F.-M. Kuo, Y.-M. Hsin, J. E. Bowers, and C.-L. Pan, "GaAs/$In_{0.5}Ga_{0.5}P$ Laser Power Converter with Undercut Mesa for Simultaneous High-Speed Data Detection and DC Electrical Power Generation," *IEEE Electron Device Letters*, vol. 33, pp. 561–563, April 2012.
29. J.-M. Wun, R.-L. Chao, Y.-W. Wang, Y.-H. Chen, and J.-W. Shi, "Type-II $GaAs_{0.5}Sb_{0.5}$/InP Uni-Traveling Carrier Photodiodes with Sub-THz Bandwidth and High-Power Performance under Zero-Bias Operation," *IEEE/OSA Journal of Lightwave Technology*, vol. 35, pp. 711–716, February 2017.
30. L. Zheng, X. Zhang, Y. Zeng, S. R. Tatavarti, S. P. Watkins, C. R. Bolognesi, S. Demiguel, and J. C. Campbell "Demonstration of High-Speed Staggered Lineup GaAsSb–InP Unitraveling Carrier Photodiodes" *IEEE Photonics Technology Letters*, vol. 17, pp. 651–653, March 2005.
31. J.-M. Wun, H.-Y. Liu, Y.-L. Zeng , S.-D. Yang, C.-L. Pan, C.-B. Huang, and J.-W. Shi, "Photonic High-Power CW THz-Wave Generation by Using Flip-Chip Packaged Uni-Traveling Carrier Photodiode and Femtosecond Optical Pulse Generator," *IEEE/OSA Journal of Lightwave Technology*, vol. 34, pp. 1387–1397, February 2016.
32. M. Jutzi, M. Berroth, G. Wohl, M. Oehme, and E. Kasper, "Ge-on-Si Vertical Incidence Photodiodes with 39-GHz Bandwidth," *IEEE Photonics Technology Letters*, vol. 17, pp. 1510–1512, July 2005.
33. C. L. Schow, L. Schares, S. J. Koester, G. Dehlinger, R. John, and F. E. Doany, "A 15-Gb/s 2.4-V Optical Receiver Using a Ge-on-SOI Photodiode and a CMOS IC," *IEEE Photonics Technology Letters*, vol. 18, pp. 1981–1983, October 2006.
34. G. Masini, G. Capellini, J. Witzens, and C. Gunn, "A Four-Channel 10Gbps Monolithic Optical Receiver in 130nm CMOS with Integrated Ge Waveguide Photodetectors," postdeadline paper, *OFC* 2007, Anaheim, CA, pp. PDP31, March 2007.
35. T. Yin, R. Cohen, M. M. Morse, G. Sarid, Y. Chetrit, D. Rubin, and M. J. Paniccia, "31GHz Ge n-i-p Waveguide Photodetectors on Silicon-on-Insulator Substrate," *Optics Express*, vol. 15, no. 21, pp. 13965–13971, October 2007.

36. C. T. DeRose, D. C. Trotter, W. A. Zortman. A. L. Starbuck, M. Fisher, M. R. Watts, and P. S. Davids, "Ultra Compact 45 GHz CMOS Compatible Germanium Waveguide Photodiode with Low Dark Current," *Optics Express*, vol. 19, no. 25, pp. 24897–24904, December 2011.
37. L. Vivien, A. Polzer, D. Marris-Morin, J. Osmond, J. M. Hartmann, P. Crozat, E. Cassan, C. Kopp, H. Zimmermann, and J. M. Fedeli, "Zero-Bias 40 Gbit/s Germanium Waveguide Photodetector on Silicon," *Optics Express*, vol. 20, no. 2, pp. 1096–1101, January 2012.
38. K. Wada, S. Park, and Y. Ishikawa, "Si Photonics and Fiber to the Home," *IEEE Proceedings*, vol. 97, no. 7, pp. 1329–1336, July 2009.
39. H.-Y. Yu, S. Ren, W. S. Jung, A. K. Okyay, D. A. B. Miller, and K. C. Sarawat, "High-Efficiency p-i-n Photodetectors on Selective-Area-Grown Ge for Monolithic Integration," *IEEE Electron Device Letters*, vol. 30, pp. 1161–1163, November 2009.

5

Semiconductor Sensors for Direct X-Ray Conversion

Kris Iniewski and Toby Astill

CONTENTS

5.1 Scintillators versus Direct-Conversion Sensors 98
5.1.1 Introduction 98
5.1.2 Scintillator Technology 100
5.1.2.1 Principle of Operation 100
5.1.2.2 Inorganic Scintillators 101
5.1.2.3 Organic Scintillators 101
5.1.2.4 Fundamental Limitations 101
5.1.3 Photodiode Technologies 102
5.1.3.1 Photomultiplier Tubes 102
5.1.3.2 Avalanche Photodiodes 103
5.1.3.3 Solid-State Photomultipliers 104
5.1.4 Direct-Conversion Detectors 105
5.1.4.1 Case Study: CZT Direct Detection in Baggage Scanning 107
5.1.4.2 Case Study: CZT Direct Detection in Computed Tomography 108
5.2 Sensor Selection for Direct-Conversion X-Ray Detection 109
5.2.1 Introduction to High-Z Materials 109
5.2.2 Germanium Sensors 110
5.2.3 Gallium Arsenide Sensors 111
5.2.4 Cadmium Telluride Sensors 111
5.2.5 Cadmium Zinc Telluride Sensors 112
5.2.6 Other Semiconductor Materials 114
5.2.7 Photon Energy Range 115
5.2.7.1 Minimum Photon Energy 115
5.2.7.2 Maximum Photon Energy 115
5.2.7.3 Extensions to Higher Energies 116
5.3 ASIC Readout Electronics 117
5.3.1 Introduction 117
5.3.2 ASIC to Sensor Attachment 117
5.3.3 ASICs for Photon Counting 118

5.3.4 Photon-Counting ASICs 120
5.3.4.1 Timepix 120
5.3.4.2 Medipix3 120
5.3.4.3 ChromAIX 121
5.3.5 Spectroscopic ASICs 122
5.3.5.1 IDeF-X 122
5.3.5.2 VAS UM/TAT4 123
5.3.5.3 HEXITEC 124
5.4 Conclusions 125
References 126

5.1 Scintillators versus Direct-Conversion Sensors

5.1.1 Introduction

There are two ways to detect X-ray radiation today: indirectly using scintillator materials and directly using semiconductor sensors. The main difference between an indirect-conversion scintillator detector and a direct-conversion semiconductor detector can be summarized as follows. The scintillation-based indirect detector is an optical device using light photons as intermittent information carriers. A direct-conversion detector omits the need for the conversion of radiation to light and directly generates charge carriers (electrons) from the energy of the absorbed X-ray photon. It is an electrical device that employs electrons and holes to transfer the event information to the electrodes. The difference between the two types of detection is schematically summarized in Figure 5.1. Details of that comparison will be discussed in this section.

In general, radiation detectors can be operated in integrating or counting modes. In integrating mode, the charge information is sampled over an integration time and converted to a digital signal. In counting mode, the process in direct-conversion detectors, the total number of events is measured by counting the charge pulses, as illustrated in Figure 5.2. In addition to this, the energy of each absorbed photon can be obtained by measuring the total charge or pulse amplitude of each photon. Direct-conversion detectors thus offer a spectral resolution of the incoming radiation. The additional spectral information leads to the well-recognized benefits of photon-counting technology: lower noise, higher efficiency, and better spatial resolution. X-ray imaging applications benefit from direct conversion as it requires less image filtering to obtain the same image resolution or the same amount of image filtering to obtain better image resolution.

Currently, most digital radiation detectors are based on integrating the X-ray photons emitted from the X-ray tube for each frame. This technique

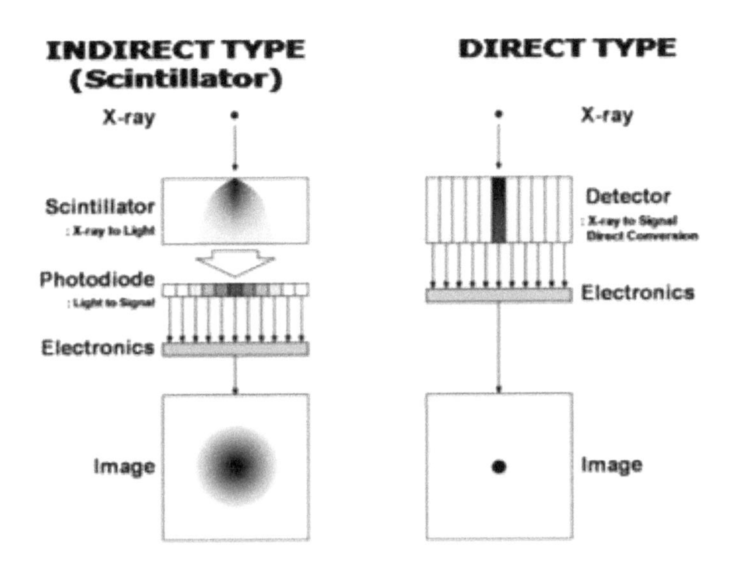

FIGURE 5.1
Schematic comparison between conventional detector using a scintillator (left) and a direct-conversion detector (right) [1].

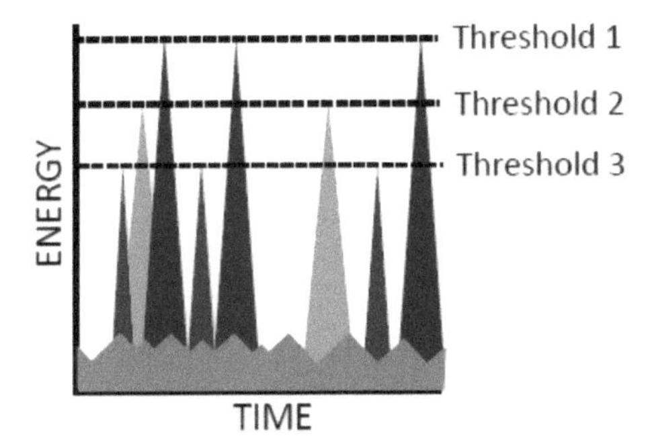

FIGURE 5.2
Schematic illustration of the PC principle. Multiple thresholds can be used to classify the incoming photon energy.

is vulnerable to noise due to variations in the magnitude of the electric charge generated per X-ray photon. Higher-energy photons deposit more charge in the detector than lower-energy photons, so that the higher-energy photons receive greater weight in the integrating detector. This effect is undesirable, as the higher part of the energy spectrum provides lower differential attenuation between various materials or tissues, and hence these energies yield low-contrast images. X-ray-counting detectors solve the noise problem associated with photon weighting by providing better weighting of information from X-ray photons with different energies. In an X-ray quantum-counting system, all photons detected with energies above a certain predetermined noise threshold are assigned as a signal. Adding energy ranges, referred to as *energy binning*, to the system (i.e., counting photons within a specified energy range) theoretically eliminates the noise associated with photon weighting and decreases the required X-ray dosage by up to 50% compared with the integrating systems.

5.1.2 Scintillator Technology

5.1.2.1 Principle of Operation

Scintillator materials are used to detect and convert incoming X-rays into visible light. Upon receiving X-ray photons, scintillators capture their energy and release it as photons of lower energy. The material should capture as much of the energy of each photon striking it as possible in order to build an accurate photon energy spectrum and thus identify the material emitting the photons. Ideally, a photon should deposit its full energy in the scintillator material, a so-called full-energy interaction. It is also important that the deposited energy be efficiently converted into photons of visible light, which are then counted to determine the original X-ray photon energy.

An example of the detection process using scintillator material is shown in Figure 5.3. The X-ray photon of energy (E) might be absorbed by the scintillator, which is the desired process in the X-ray detection. However, the X-ray photon can also scatter and be detected in a neighboring piece of the scintillator detector or escape the detector volume completely. After absorption, multiple visible photons are generated and propagate in all directions, irrespective of the direction of the original photon. This leads to significant loss

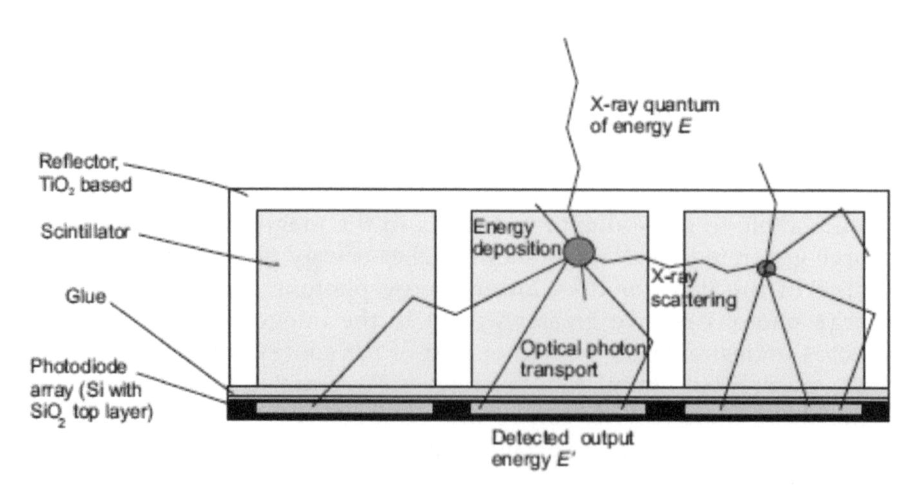

FIGURE 5.3
Illustration of the two-step detection process using a scintillator and a Si photodiode [2].

of instrument efficiency. Finally, the visible photons are typically detected by standard silicon (Si) photodiodes.

5.1.2.2 Inorganic Scintillators

There are two main types of scintillators: inorganic and organic. Inorganic scintillators (those not containing carbon) are typically single crystals, such as sodium iodide (NaI), with a small amount of thallium (TI) added. The probability of full-energy interaction increases sharply with the atomic number (Z) of the scintillator material and is high for inorganic crystals. The more energy from each photon a scintillator absorbs and then gives off, the better the correlation between energy input and output and the more precise the spectrum that can be constructed.

Inorganic scintillators are quite inexpensive but have several disadvantages. The materials are typically fragile. Many inorganic crystals absorb water and are sensitive to light, so they must be protected from environmental conditions. NaI crystals are easy and inexpensive to grow, but other higher-resolution scintillators are harder to grow, more costly, and even sensitive to moisture. For NaI, light output varies strongly with temperature, so the temperature must be stabilized or the data corrected.

5.1.2.3 Organic Scintillators

Organic scintillators have the opposite set of properties. They can be made of plastic, such as polyvinyltoluene (PVT). As such, they are easy and cheap to make and are much less fragile than crystals. They can be produced in bulk, making them suitable for deployment in large sheets, such as for radiation portal monitors. On the other hand, since they are composed mostly of hydrogen and carbon, both very low-Z elements, they are very inefficient at absorbing the total energy of gamma rays.

5.1.2.4 Fundamental Limitations

Regardless of whether they are inorganic or organic, all scintillators suffer from a major fundamental limitation called *afterglow*, since scintillation crystals contain a number of luminescent components. The main component corresponds to decay time; however, less intense and slower components also exist. Commonly, the strength of these components is estimated by using the intensity of a scintillator's glow, measured a specified period after the decay time. Afterglow is the ratio of the intensity measured at this specified time to the intensity of the main component measured at the decay time. Afterglow, which is typically in the order of milliseconds, limits the speed at which the X-ray signals can be read using conventional scintillator technology. As discussed in the next section, only direct-conversion detectors offer high speeds of operation due to their potential glow-free properties.

5.1.3 Photodiode Technologies

Conventional X-ray detection systems are be made of three parts:

1. A scintillator crystal
2. Photodetectors
3. A multichannel digital acquisition system

As discussed previously, the role of the scintillator crystal is to absorb X-ray photons and emit visible light while the photodetection system converts the visible light signal into an electric charge or current. When these X-rays interact in the scintillator via the photoelectric effect, electrons are released and travel short distances to luminescent centers in the crystal, where subsequent energy transitions lead to the emission of multiple low-energy photons. The energy of these emitted photons varies with the scintillator material, but these photons are typically in the visible light spectrum. While the signal produced in the scintillator is proportional to the energy of the detected photon, it is usually quite weak and needs to be amplified with a sensitive photodetector coupled to the crystal. The readout process is summarized in Figure 5.4.

5.1.3.1 Photomultiplier Tubes

The device most commonly used to detect the low intensity light produced in the scintillator used to be the photomultiplier tube (PMT). A PMT is a vacuum tube containing a light-sensitive photocathode at one end and a series of metal plates called *dynodes* along the length of the tube. At the cathode, visible light from the scintillator is converted into electrons, which are then

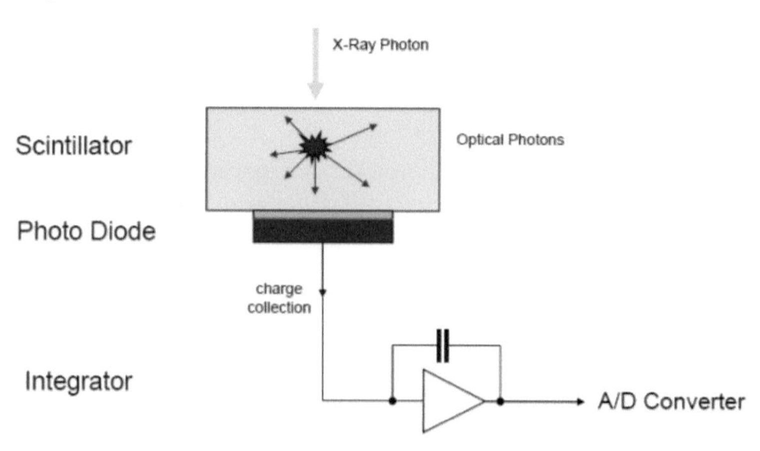

FIGURE 5.4
Schematic representation of indirect readout process, which uses a scintillator, photodiodes, and integrating electronics.

accelerated due to the applied potential difference to the first dynode. For each electron that collides with the dynode, several electrons are released and accelerate toward the next dynode, where the interacting electrons further generate more electrons. This process of amplification repeats itself through all dynode stages until eventually the electrons reach the anode. At each dynode stage, for each electron that collides, it is common for between three and six electrons to be released. As most PMTs consist of between eight and ten dynodes, it is common to achieve an amplification of around a factor of a million through the course of the dynode amplification process.

At the anode, the large number of electrons present create a measurable electrical signal on the order of several microamps. When passed through an appropriate resistor, voltage drops on the order of several volts can be obtained. While PMTs have been around for many years and are used extensively in many instruments, they suffer from many disadvantages. For example, they do not operate properly in the vicinity of a magnetic field, which precludes their use in MRI applications. In fact, even the earth's relatively weak magnetic field greatly affects the quality of images obtained with conventional PET and SPECT cameras. Care must be taken when orienting conventional PMT-based radiation detectors and proper calibrations are required. Most SPECT and PET cameras are PMT based and rotate. There is always a metal shield around each tube to prevent influence of magnetic field, which increase the unit costs. Additionally, PMTs are known to be extremely sensitive to temperature variations and their high operating voltages make them sensitive to noise fluctuations.

5.1.3.2 Avalanche Photodiodes

As PMTs suffer from the various limitations outlined previously, alternative photodetectors are being used. The workhorse of the photodiode technology is avalanche photodiodes (APDs). APDs essentially consist of a pn-junction semiconductor that, when exposed to light photons from a scintillator crystal, generates electron–hole pairs. When a high reverse bias is applied across the APD diode, an electric field is produced between the anode and cathode and the electron–hole pairs migrate across the depletion layer, with the holes moving to the cathode and the electrons moving to the anode. If the applied bias is high enough, the electrons will be accelerated such that they collide with the surrounding crystal lattice of the semiconductor, thereby creating additional ionization effects that lead to higher number of additional electron–hole pairs, hence the term *avalanche effect*. As the electrons reach the anode, there is a voltage drop in the high voltage bias that is proportional to the original number of electron–hole pairs produced. The signal gain of an APD is proportional to the voltage bias applied across it. There comes a point, however, where further increases to the bias will not cause a proportional voltage drop across the diode. At this voltage, commonly known as the *breakdown voltage*, the APD will conduct incident light.

As the voltage drop across an APD tends to be so small, it is beneficial to utilize APDs with large active areas in order to collect as many light photons as possible and thereby maximize the output voltage signal. When used for radiation detection, most APD detectors typically use a 1:1 coupling of scintillator crystal to APD. This is primarily done because the APD signal is so small that any light sharing among multiple APD detectors as a result of using a monolithic detector will result in a subsequent reduction in the APD signal. This type of arrangement tends to limit the physical size of the detector due to

- The high cost of pixilated scintillators
- Difficulty in fabrication
- The need for many thousands of digitization channels

For these reasons, APDs are still not necessarily the best choice of photodetectors for many applications. The primary downside is their relatively low signal gain and high noise levels compared with conventional PMTs. While PMTs have typical amplifications of 10E+6, APDs typically only have gains of about 50–200. This low gain, when combined with a relatively large background noise component, reduces spatial resolution and limits the range of photon energies that can be imaged with such a system. Since the magnitude of the APD signal depends on the number of light photons produced in the scintillator, APDs have typically been limited to high-energy applications. In addition, due to variations in manufacturing, it is important that each APD has its own voltage bias control in order to operate at optimal signal gain. Confounding the issue even further, APDs exhibit significant gain variations with temperature, thus leading to the requirement for a temperature-sensitive operating bias.

5.1.3.3 Solid-State Photomultipliers

In the last few years, a new technology based on APDs has become available. These so-called solid-state photomultipliers (SSPMs), are similar in many respects to APDs but provide signal amplifications much closer to that of conventional PMTs (up to 10^6). This is achieved by operating the APD in *Geiger mode*. As APDs can be made very small (several μm across) and many of these single-element APDs can be placed in an array on a single substrate and all operated in Geiger mode, then by simply counting the number of elements that cascade into Geiger mode, an accurate measure of the original light intensity can be made. Thus, even though each APD element produces a relatively small signal (several mV), the summation of all the elements following a scintillation event yields rather large, easily detectable signals through the diode, thus causing a much larger than expected voltage drop. At this point, the APD is said to be operating in Geiger mode as the device no longer outputs a voltage proportional to the amount of incident light.

5.1.4 Direct-Conversion Detectors

Direct-conversion radiation detectors offer new capabilities for medical imaging, non-destructive testing (NDT), and security applications due to their superior energy resolution (ER), detector quantum efficiency (DQE), variable energy weighting, noise reduction, and increased spatial resolution. These advantages in turn enable new applications such as material decomposition as well as improved image quality in low-dose screening applications and high spatial resolution imaging. In direct-conversion detectors, absorbed X-ray-imaging photons generate individually detectable and measurable electric current pulses without the production and detection of light photons as an intermediary step. Since the signals are immediately available on the back of the detectors, they are sensed with close-fitting miniature electronics that significantly reduce sources of noise and expensive infrastructure overhead. Schematic illustration of the direct detection architecture is shown in Figure 5.5.

Various semiconductor materials such as silicon (Si), germanium (Ge), amorphous selenium (a-Se), gallium arsenide (GaAs), cadmium telluride (CdTe), and cadmium zinc telluride (CZT) can be used in direct detection. The

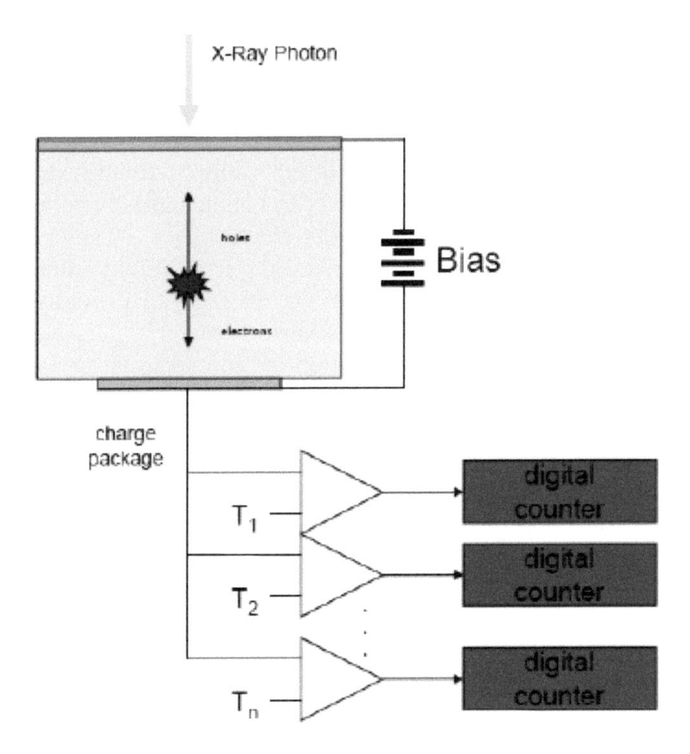

FIGURE 5.5
Schematic representation of the direct readout process, which uses a semiconductor detector and photon-counting electronics.

comparison of different detector systems shows that semiconductor materials such as CdTe/CZT are excellent material choices for high-performance photon-counting X-ray detectors for high X-ray energies. ER expressed as a full-width half-max (FWHM) value is superior to common scintillators such as NaI and CsI, as shown in Figure 5.6.

In addition, CdTe/CZT carrier trapping (expressed as carrier mobility lifetime product) is superior to that of a-Se, allowing for fast detectors free from the image ghosting (residual images left over from previous scans) and lag that are typical of scintillator materials. Finally, photon absorption efficiency is very high.

High levels of energy discrimination and photon-counting binning enable material determination in CdTe/CZT systems, in contrast to conventional integrating detectors that do not offer such possibilities. The dynamic range measurements on the CdTe/CZT detectors illustrate one of the key benefits of our photon-counting technology: the detector is free from electronic noise and has excellent dynamic range, from zero up to hundreds of millions of photons/pixels. In conclusion, photon-counting CdTe/CZT detectors offer unique performance with excellent spatial resolution, no electronic noise, high quantum efficiency, high frame rates, and extremely wide dynamic range. This, in combination with the electronic energy separation, has the potential to improve present applications and enable new ones in medical imaging, security, NDT, and related fields.

Despite intense efforts on the part of established scintillator suppliers, none of the materials available in the marketplace can compete with CZT sensors. As discussed previously, CZT material offers a simple direct-conversion process that does not require additional X-ray to visible light conversion and the use of PMTs or APDs. CZT produces 20,000 electrons for 100 keV photons versus only about 1000 for Tl-doped NaI, resulting in a 20-fold enhancement in detection efficiency. Finally, as shown in Figure 5.6, CZT provides better ER, which enables the clear identification of various materials, tissues, or radioisotopes, a critical feature in NDT and medical and security applications.

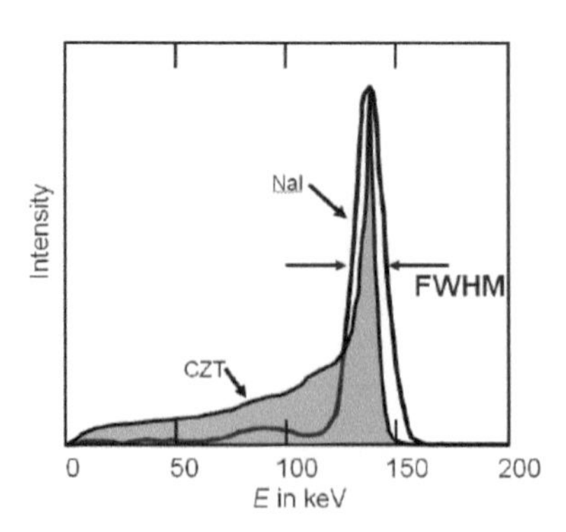

FIGURE 5.6
Comparison between CZT and NaI scintillators illustrating better ER [2].

5.1.4.1 Case Study: CZT Direct Detection in Baggage Scanning

Baggage-scanning equipment needs to detect explosives in bulk, sheet, liquid, and slurry form. Traditional radiation detection systems use scintillator technology, as shown in Figure 5.7. X-rays produced by the X-ray tube are being used to scan the object. The ceramic scintillator detects those photons that have passed through the object and produce a visible light signal. The received light signal is in turn converted by the photodiode to produce analog electrical signals that can be used to produce an image.

Most detection systems take linear projection through the luggage traveling on a conveyor belt in a so-called line scan mode. The detectors used are efficiently collimated linear arrays. For dual-energy capabilities, two solutions have been used. In the first technique, a linear scan is performed twice, the second time with an additional X-ray filter in front of the beam. This elementary technique allows low- and high-energy photon separation to be obtained.

In the second technique, a linear scan is performed only once but sandwich detectors have been optimized for dual-energy scanning. They consist of two layers of the scintillator–photodiode type, separated by a metal filter. The first layer absorbs low-energy photons and the second layer absorbs high-energy photons.

In both cases, due to the poor energy separation of those acquisition systems and a significant noise level resulting from the acquisition speed, the obtained accuracy at best only allows materials to be classified into broad bands such inorganic and organic.

Modern baggage-scanning equipment uses a more efficient, direct method of radiation detection, as shown in Figure 5.8. The two-step process involving scintillator and photodiode is replaced by a semiconductor detector that converts X-rays directly into an electric charge. CZT detectors fit perfectly into these applications due their high stopping power, reasonable cost, high stability, and reliability.

In addition, through the use of CZT, multiple energy analysis and multiple energy sources, the equipment might be able to detect thin sections and subtle differences in atomic number. The equipment might also provide full volumetric reconstruction and analysis of the object being imaged. The result of

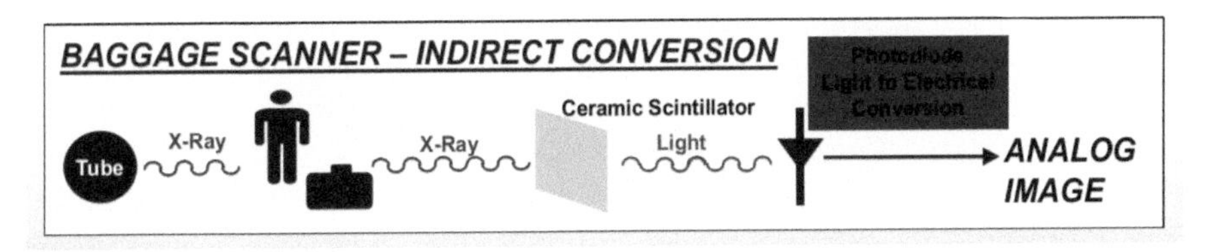

FIGURE 5.7
Scintillator-based baggage-scanning technology as typically used in standard systems.

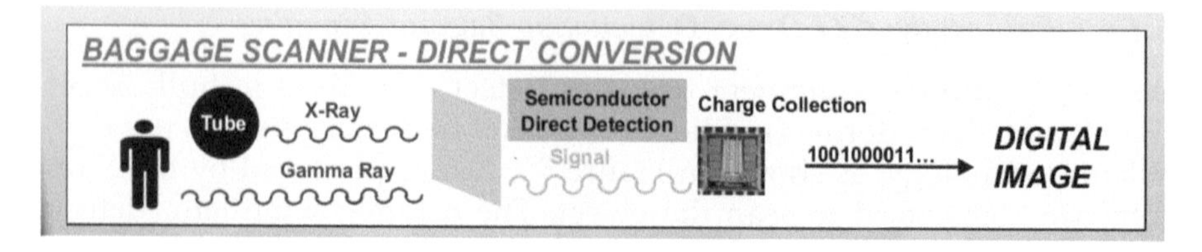

FIGURE 5.8
Semiconductor-based baggage-scanning technology being introduced today.

the advanced CZT technology could be the exceptional ability to detect and discriminate a wide variety of threat materials, providing enhanced overall security. One of the key motivations in technological developments is to eliminate the liquid carry-on ban currently in place. Once that ban is eliminated (already legislated in Europe), airport operators will be forced to use better technology for liquid/security detection.

5.1.4.2 Case Study: CZT Direct Detection in Computed Tomography

In computed tomography (CT), traditional scanners use an indirect-conversion process with scintillation detectors such as gadolinium oxysulfide (GOS) or cesium iodide (CsI), where each pixel is enclosed by an epoxy compound filled with backscattering particles. Typical pixel dimensions of around 1 mm are being used. A registered photosensor detects the secondary light photons at the bottom surface of each pixel. The primary interaction in a detector pixel is given by an absorption of an incoming X-ray quantum by a gadolinium atom (in case of GOS). The X-ray energy is converted into light photons. The energy conversion rate is around 12%. Secondary light photon transport takes place, whereby photons that reach the photosensor contribute to the output energy signal.

Radiography and mammography detectors follow similar designs. CsI is usually employed as a scintillator. Due to its vertical needle structure, it has the advantage of providing intrinsic light-guiding properties; thus, no back scattering septa are required. This allows for an improved detector resolution at the expense of reduced stopping power and signal speed.

Two main physical effects influence the spatial and spectral resolution in pixelized scintillator detectors. First, the primary energy deposition is not perfectly localized. For the high-Z atom gadolinium, absorption is governed by the photoelectric effect. This generates fluorescence escape photons with mean free path lengths in the order of several hundreds of microns. They might be reabsorbed in the pixel volume, become registered in a neighboring pixel, or leave the detector volume completely. Secondly, light transport is affected by optical cross-talk. Septa walls are designed with a limited thickness to optimize overall dose usage and light yield. As a consequence, a significant portion of the light is transferred to adjacent "false" pixels, creating a blurry image in the process.

The direct-conversion CZT scheme does not have these disadvantages. A common-cathode design with pixelized anodes on the bottom surface of the semiconductor bulk is typically used. Imaging voxels are established with electrical fields of the order of 1 kV/cm. The physics of the primary energy deposition are comparable to the indirect-conversion detector. However, the deposited energy is converted to charges instead of optical photons. The holes and electrons are separated and accelerated by the electrical field. Electrical pulse signals are induced on the electrodes. The main signal pulse is generated when the electrons follow the stronger curved electrical field in the bulk region close to the anodes.

The main signal degradation mechanisms are comparable to indirect-conversion scintillator detectors: First, fluorescence scattering takes place. Due to the lower K-edge energy, the mean free path lengths of fluorescence quanta in CZT are about 100 μm. The smaller the pixel size, the more fluorescent cross-talk will affect the behavior of the detector. Secondly, the charge signal transport is affected by charge sharing. The moving charge cloud also induces electrical pulses in neighboring pixels—again, mostly at the bottom part of the pixel field configuration.

CZT-based CT readout systems utilize multiple-bind photon counting (PC), where the radiologist can set various thresholds (energy bins). In PC, only pulses whose heights are greater than the thresholds are counted and the basic operation is noise free, as the lowest threshold can be set above the noise level. The number of energy bins varies from two to six, with four being typical. The ability to count photons within separate energy bins enables the use of some innovating energy weighting scenarios in image reconstruction.

5.2 Sensor Selection for Direct-Conversion X-Ray Detection

5.2.1 Introduction to High-Z Materials

Si is the most commonly used semiconductor sensor material, due to its various advantages. Si wafer–processing technology is mature as it relies on advancements in an annual $300 billion CMOS industry. Si wafers are available with near-perfect crystal quality; low-leakage currents can be achieved by using a photodiode structure; and Si sensors are relatively robust, both mechanically and chemically.

However, the X-ray absorption efficiency of Si is very limited at higher energies. Si sensors with a typical thickness of 500 μm provide 90% photoelectric absorption efficiency at 12 keV, but their efficiency falls quickly as the photon energy is increased beyond this value, as shown in Figure 5.9. For many applications in X-ray detection, such a low efficiency is not acceptable as it limits the throughput of the system.

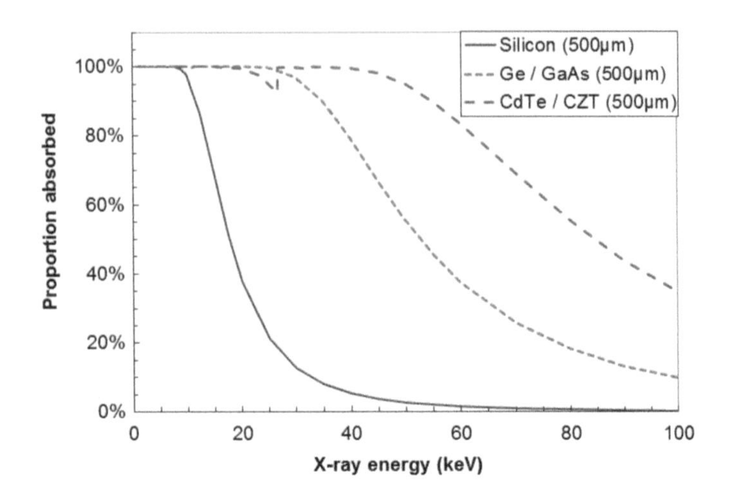

FIGURE 5.9
Photoelectric absorption efficiency of Si, Ge/GaAs, and CdTe/CZT.

Efficient direct detection of hard X-rays can be achieved using so-called high-Z semiconductors such as Ge, GaAs, CdTe, and CZT. Each of these high-Z materials has its own strengths and weaknesses. A common issue for these materials (with the exception of Ge) is that they cannot be produced with the same level of crystal perfection and uniformity as Si. This is largely due to these materials being compound semiconductors, meaning they contain more than one component, which makes their crystal growth more challenging.

5.2.2 Germanium Sensors

Unlike most high-Z semiconductors, Ge is an elemental semiconductor, which is a great advantage for producing high-quality crystals. Single high-purity Ge (HPGe) crystals are grown using the Czochralski process. Net impurity concentrations can be extremely low (10E+9 cm^{-3}) in both p- and n-type Ge over large volumes. However, because of the relatively low bandgap (0.7 eV), HPGe detectors cannot be operated at room temperature due to the large intrinsic carrier concentration (10E+13 cm^{-3}) leakage current from the thermally generated charge carriers. As a result, cooling down to 77 K, the temperature of liquid nitrogen (LN2), is required to reduce the leakage currents to the acceptable level (a few pA per pixel). The cooling requirements make this solution quite expensive and difficult to use in commercial applications.

The detection principle of an HPGe semiconductor sensor consists of developing a p-i-n diode structure in which the intrinsic region has high resistivity and is sensitive to ionizing radiation when a reverse-bias voltage is applied. These properties are due to the excellent mobility of electrons and holes, the relatively high Z (32) and density (5.32 g/cm^3), and the availability of large, pure, and high-quality single crystals. For example, an ER of 0.5 keV can be achieved at 60 keV photon energy. Compared with other semiconductors, HPGe presents the best properties for high-resolution and high-efficiency spectroscopy, but very few applications can afford its steep price and cooling requirements.

5.2.3 Gallium Arsenide Sensors

GaAs is widely used in applications such as high-speed and high-power electronics (GaAs MESFETs) and optoelectronics. The material is available in large wafers (6 in.), which makes it an appealing option for radiation detection. However, detecting X-rays requires a combination of good carrier transport, high resistivity, and large sensor thickness (more than 1 mm), which is impossible to achieve today using commercial bulk GaAs produced for other uses in electronics.

In recent years, it has been shown that a moderate level of radiation performance can be achieved by taking standard n-type GaAs (which does not have high resistivity) and compensating for excess electrons with chromium (Cr) doping. It has been experimentally shown that a material with resistivity of about 10E+9 Ω cm, which is close to the theoretical limit, can be produced using this technology with reasonable charge transport properties. With a strong electric field applied to the detector, the mean carrier path for electronics is at best of the order of 5–10 mm for electrons but only 50–100 μm for holes; this is acceptable for low-flux applications but not for high-flux applications as NDT. The best ER obtained to date is about 2–3 keV at 60 keV. Poor hole transport precludes the use of GaAs sensors in most commercial applications, although its low price can be appealing in some cases where the poor hole transport can be tolerated.

5.2.4 Cadmium Telluride Sensors

For applications in NDT, CdTe might be an appealing option because the higher atomic numbers of Cd (48) and Te (52) result in significantly shorter X-ray attenuation lengths compared with Ge and GaAs. For example, 500 μm of CdTe will absorb 55% of 80 keV photons. CdTe is mostly used for applications such as thin-film solar cells, but the detector grade production requires special manufacturing steps. Typically, polycrystalline CdTe ingots are grown from a Te-rich melt, and then the material is progressively recrystallized by the traveling heater method (THM).

The carrier mean free path in CdTe is about 1–10 cm for electrons and 0.1–1 mm for holes, better than GaAs but worse than CZT. Defects in CdTe crystals cause a range of problems, and due to the fragility of the material, some defects such as dislocations can be introduced during processing; for example, mechanical force can cause atomic planes to slip with respect to each other.

CdTe sensor technology has been extensively studied over the last several years [4–15]. One severe issue with CdTe is material polarization, where the electric field within the sensor changes with time. The changes in electric field are due to carrier trapping (mostly holes trapped by deep-level acceptors in CdTe). Changing electric fields lead to changes in measured radiation spectra, making the use of these sensors challenging even in low-flux applications. Under high-flux conditions, the polarization effect is much stronger in CdTe due to the amount of holes generated. The trapped charge can build up to the point where the electric field collapses, resulting in a complete loss of sensor operation. However, even before the catastrophic loss of device operation, serious degradation occurs if the electric field changes with time.

The effects of polarization also depend on the contact technology being used. By using different metals for electrodes, it is possible to create Schottky contacts with the material (giving diode-like behavior) or conductive (ohmic) contacts. Schottky contacts reduce current flow in the sensor, resulting in better spectroscopic performance, but polarization effects increase. Significant signal loss after operating a Schottky contact sensor even for several minutes has been widely reported in the literature. However, this can be combated by temporarily switching off the high voltage in the sensor and reversing its polarity. Another possibly is to illuminate the sensor with light to create additional hole–electron pairs that would help the carrier recombination process. Needless to say, both solutions lead to major complications in system design. For this reason, Ohmic contacts are more commonly used for high-count rate applications, but even with the use of these contacts, polarization remains a major challenge, typically limiting the thickness of CdTe to 1 mm (in some demanding cases, such as CT scanners, 1.4 mm thickness was used).

5.2.5 Cadmium Zinc Telluride Sensors

Replacing some Cd atoms with Zn produces a $Cd_{1-x}Zn_xTe$ (CZT) compound that increases the bandgap of the material from 1.44 to approximately 1.6 eV for typical Zn concentrations of 10%. The wider bandgap leads to higher sensor resistivity, giving CZT a 10-fold advantage over CdTe. Higher resistivity offers the benefit of a lower leakage current, which in turn can improve spectroscopic behavior. Finally, a lower leakage current improves detector ER and the resulting contrast-to-noise ratio (CNR) in the phantom studies, as shown in Figure 5.10.

Superior CZT sensor performance is due to dramatic advances in crystal growth and sensor manufacturing. Different growth methods have been used to grow CZT. The high-pressure Bridgman technique produces large, polycrystalline CZT ingots, which can then be diced to obtain single crystals, typically up to a few cubic centimeters. This can provide CZT sensors for spectroscopic detectors, but PC sensors typically require larger single-crystal areas. Recent advances in THM growth has made high-flux applications feasible.

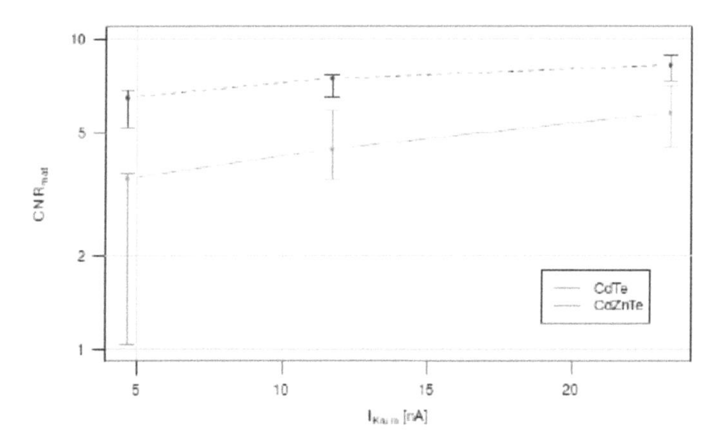

FIGURE 5.10
Measured phantom CNR for CdTe and CZT sensors [3].

As previously discussed for CdTe, until recently, the high-flux operation of CZT sensors presented well-known challenges due to poor hole transport and material defects that led to polarization and other instability effects. Improvements in THM growth and device fabrication that require additional processing steps have enabled CZT to show dramatically improved hole transport properties and reduced polarization effects. As a result, the high-flux operation of CZT sensors at rates in excess of 200 million counts/s/mm^2 is now possible. The availability of fast, stable, and reliable CZT sensors has enabled multiple medical imaging equipment manufacturers to start building cameras for CT, baggage scanning, and NDT.

The main advantage of the CZT sensor over its CdTe equivalent is the flexibility in the choice of sensor thickness and the resulting working energy range. As CZT does not suffer from polarization, CZT sensors can be made significantly thicker. For example, a comparison between 5 mm-thick CZT and 1 mm-thick CdTe measured at an independent national laboratory in the United Kingdom illustrated a dramatic difference in detector efficiency, as shown in Figure 5.11.

Both sensors have been exposed to a Co-57 source for 10 min at a distance of 25 cm. During that exposure, the CZT sensor detected 7145 peak events, while the CdTe sensor detected only 795. The difference of almost an order of magnitude was mostly due to the sensor thickness, which was 5 mm for CZT and 1 mm for CdTe. The additional loss of efficiency for CdTe was due to smaller pixel pitch and the resulting number of charge-shared events. The excellent ER of the CZT sensor (better than 2 keV) is clearly visible due to clean separation between two Co-57 peaks at 122 and 136 keV.

In order to prepare for high-volume commercial production, the CZT industry is moving from individual tile processing to whole-wafer processing using Si-like methodologies. Parametric-level screening is being developed at the wafer stage to ensure high wafer quality before detector fabrication in

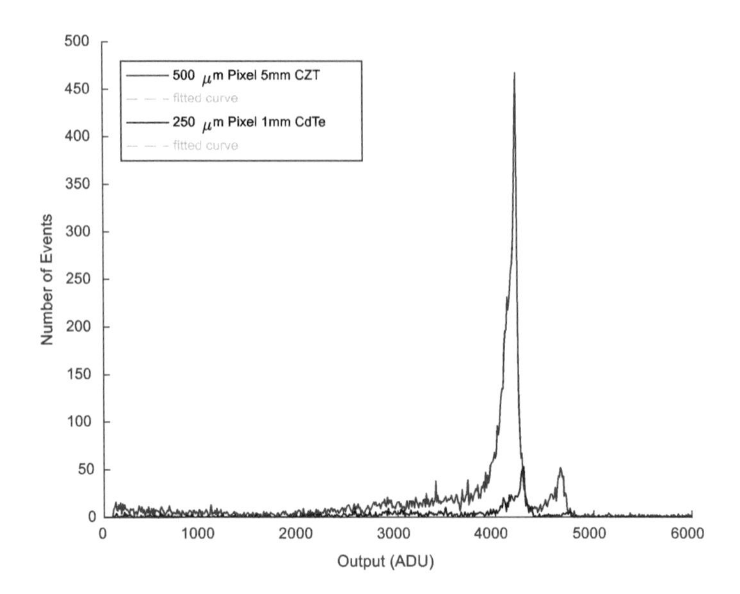

FIGURE 5.11
Comparison between 5 mm-thick CZT sensor (dark gray) and 1 mm-thick CdTe sensor (light gray) exposed to Co-57 radiation. Courtesy: Rutherford Appleton Laboratory.

order to maximize production yields. These process improvements enable CZT manufacturers to provide high-volume production for photon-counting applications in an economically feasible manner.

CZT sensors are capable of delivering both high count rates and high-resolution spectroscopic performance, although it is challenging to achieve both of these attributes simultaneously. It is possible to achieve 1 keV ER in spectroscopy at 60 keV, not as good as HPGe but better than GaAs, and a 300 Mcps/mm^2 flux rate in PC, significantly better than CdTe. In fact, CZT sensors provide the highest flux rate response from high-Z materials; therefore, several large medical imaging OEMs are building next-generation CT scanners using this technology. Recent publications discuss material challenges, detector design trade-offs, and the hybrid pixel readout chip architectures required to build cost-effective CZT-based detection systems [16–21]

5.2.6 Other Semiconductor Materials

There has been some effort in the research community to produce new semiconductor materials that can compete with CZT, such as thallium bromide (TIBr) and mercury iodide (HgI2). A basic summary of semiconductor material properties is shown in Table 5.1. At present, none of these materials are anywhere near the production readiness required for volume manufacturing, so they are not discussed any further here.

TABLE 5.1

Basic Properties of Selected Semiconductor Materials Used for Radiation Detection

Material	Ge	Si	CZT	TlBr	HgI2
Average Z	32	14	49	80	80
Density (g/cm³)	5.32	2.33	5.78	7.56	6.4
Resistivity (Ohm-cm)	50	1E5	1E10	1E12	1E12

5.2.7 Photon Energy Range

5.2.7.1 Minimum Photon Energy

Each sensor technology presents its own limitation on the energy range of incoming photons that can be detected. The limitation on the low end typically comes from the readouts from application-specific integrated circuit (ASIC) electronics, not from sensor properties. More detailed discussion on this topic and a description of PC ASIC electronics is provided later in this chapter, so at this point we will only make some estimates on the minimum photon energy (E_{min}) that can be detected.

A useful rule of thumb is that E_{min} is three times the ER of the system. This condition ensures that no false trigger events are detected due to noise events. The ER is a function of both the sensor and the ASIC electronics. Typical values for high-Z materials are given in Table 5.2. As shown, there are no relevant differences between all sensor materials (with the exception of GaAs) unless very low energies are of interest.

5.2.7.2 Maximum Photon Energy

Maximum photon energy is in practice related to the maximum thickness of the sensor. If the sensor is not thick enough, a significant portion of the incoming radiation will escape the sensor material. This has severe undesired consequences. Firstly, sensor efficiency suffers, requiring a long exposure time in order to collect a sufficient number of photons to ensure good signal-to-noise ratio (SNR) and CNR values. Second, escaping radiation will cause radiation damage in the underlying electronics. The degree of the

TABLE 5.2

Typical Values for Energy Resolution, Minimum Photon Energy, Maximum Sensor Thickness, and Maximum Photon Energy for Selecting High-Z Sensors

Material	Ge	Si	GaAs	CdTe	CZT
Energy Resolution (keV)	0.5	0.5	3	1	1
Minimum Energy (keV)	1.5	1.5	9	3	3
Maximum Thickness (mm)	0.5	0.5	0.5	1	5
Maximum Energy (keV)	35	12	35	60	100

radiation damage depends on the radiation hardness of the electronics, but it will lead to high implementation costs (if radiation-hard technologies are used) and/or high servicing costs (if frequent replacement costs have to be accepted).

As discussed earlier, Si, Ge, and GaAs are limited to the standard 500 µm thickness typical of the microelectronics industry. As a result, to obtain 90% detector quantum efficiency (DQE), their maximum photon energy is severely limited, as shown in Table 5.2, being as little as 12 keV for Si and only 35 keV for Ge and GaAs. CdTe sensors can be thicker, but in practice the polarization effect limits their thickness to 1 mm, which corresponds to a maximum energy of 60 keV. CZT sensors do not exhibit polarization effects if properly designed and fabricated; therefore, even 5 mm-thick material might be suitable for NDT applications, extending the energy range significantly to 1000 keV. CT scanners that require very high count rates typically use 2–3 mm-thick CZT.

The values calculated in Table 5.2 are for a 90% absorption level in the sensor. If lower absorption levels are acceptable, the energy range can obviously be extended. CZT sensors are used in CT scanners up to 160 keV and can easily operate at high count rates for energies up to 200 keV.

5.2.7.3 Extensions to Higher Energies

It is possible to extend the range of detected energies even further than shown in Table 5.2. The major challenge is the various ways CZT sensors interact with X-rays. As shown in Figure 5.12, the photoelectric effect is dominant in the typical energy range of interest: 10–300 keV. At an energy of 122 keV, the photoelectric effect has an 82% probability of happening, while Rayleigh scattering has a probability of 7% and Compton scattering 11%. At higher energies, Compton scattering becomes more probable. It is possible to build Compton cameras using CZT sensors and detect photon energies up

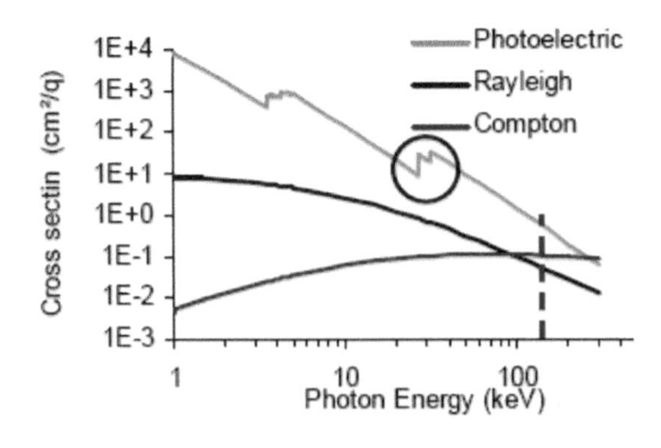

FIGURE 5.12
Effective photon cross-sections for the photoelectric effect and Rayleigh and Compton scattering. The dashed line indicates 122 keV cobalt.

to 9 MeV, but this topic will be discussed separately, as it requires the use of different physical concepts and specialized electronics.

5.3 ASIC Readout Electronics

5.3.1 Introduction

CZT pixelated sensors have a high level of segmented multichannel readout requirements. Several decades ago, the only way to achieve this was via massive fan-out schemes to route signals to discrete low-density electronics. At the present time, CMOS technology allows one to build very dense low-power electronics with many channels that can be bonded directly or indirectly (through a common carrier PCB) to the detector [21–37].

There are different requirements for the CMOS technology used for the analog front-end signal processing, as opposed to that for the digital signal processing. For the analog part of the electronics, there is a requirement for a robust technology that has low electronic noise and high dynamic range, which typically requires large power supply voltages. Digital signal processing in turn requires very high speed and high density, which is more compatible with modern low-voltage supply, deep submicron processes.

There seems to be a technology optimum at around 0.35–0.18 μm minimum feature size for the analogue requirements. The small feature size limits the complexity of the circuitry that can be integrated in a pixel, but even at 0.35 μm, it is possible to place millions of transistors on a reasonably sized Si die. In comparison, digital signal processing can benefit from the rapid development of deep submicron processes. Research studies are now taking place using 90, 65, and 45 nm processing nodes. These technologies are well suited to high-speed ADC architectures and to fast data manipulation and compression. The deep submicron technologies have their own limitations in terms of gate oxide thickness, noise, and extremely high non-recurring engineering (NRE) costs.

5.3.2 ASIC to Sensor Attachment

CZT pixel detectors require a connection from the pads on the detector material to the bond pads on the ASICs. In some cases, the pixel pitch on the readout ASIC is the same as the detector pixel pitch. It is also possible to fan out the connections on the detector with multilevel metal routing or with the use of the interposer board. This fanned-out routing has to be done carefully as there is great danger of signal cross-talk. With integrated readout and synchronous input signals where the signal is totally removed from the detector, this might not be a problem, but transients can still upset thresholds in these systems.

The pixel pitch of X-ray imaging systems currently falls between about 100 μm to 1 mm. For small pixel pitches, bump bonding is used to connect the detector pixels to the ASICs. There are many different technologies to do this, depending on the requirements of the detectors and environmental constraints. The industry-standard bump-bonding method is to deposit solder onto the under-bump metallization on the pads of the detector and ASIC and then to align the two and heat them to reflow the solder, as shown schematically in Figure 5.13. Various solders are used, including lead-tin, bismuth-tin, indium alloys, and silver alloys, depending on the temperature to reflow and the operating temperature required. Typically, these materials require 140–240 °C to reflow. Indium is used in either lower-temperature reflow processes or straight compression bonding and gives good results, but it cannot be used if high operating temperatures will ever be experienced.

With all these processes, either fluxes or special gas environments including nitrogen, hydrogen, or formic acid have to be used to ensure good contact between parts. These processes can allow 10–20 μm bond diameters to be used, which allows 50–100 μm pixels to be connected. Another method used particularly for larger-pixel (above 150 μm) CZT detectors relies on gold studs and silver-loaded epoxy dots. The gold studs are either bonded or deposited on the ASIC part and the silver-loaded epoxy is screen printed on the detector.

5.3.3 ASICs for Photon Counting

One of the major advantages of PC detectors is electronics noise rejection. A well-designed PC detector has an ASIC electronics threshold high enough to reject noise pulses while still counting useful signals. Therefore, quantum-limited operation of the PC detector can be achieved, as image noise is

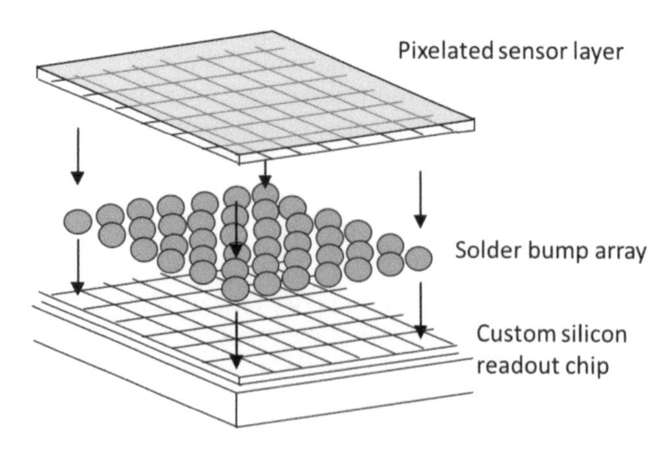

FIGURE 5.13
Schematic illustration of the bump-bonding process that connects the sensor and ASIC readout chip.

determined only by statistical variations of X-ray photons. On the other hand, energy-integrating detectors suffer from electronics noise that is mixed with useful photon signals, and separating it from statistical noise is not possible. Electronics noise rejection is important because its magnitude for currently used digital X-ray detectors is not negligible, and a high SNR is essential for forward-scattering X-ray systems.

After converting the CZT-generated charge to voltage via the charge-sensitive amplifier (CSA) and subsequently shaping the signal, given enough gain in the system, it should be ready for digitization. Figure 5.14 shows a typical block diagram of the ASIC electronics used for CZT sensors; other semiconductor sensors use very similar architectures. Typically, the signal is compared with a user-selected threshold voltage (quantizer box in Figure 5.14) to produce a 1 bit trigger signal indicating the detection of the pulse. In parallel, the value of the shaped signal is sent to the ADC converter (or time-over-threshold [ToT] processor) with n-bit accuracy. The conversion resolution n is typically between 8 and 16 bits depending on the system accuracy, noise levels, and degree of signal precision achieved.

One important consideration in the practical system is CSA reset. As the feedback capacitor C_f is charged by the input signal, there must be some means of discharging this capacitor in order for the CSA to be ready for the next signal. This circuitry is schematically shown as a reset block in Figure 5.14. There are two possible configurations for the reset block: digital and analog. The digital one involves using a switch that will discharge the feedback capacitor quickly. Unfortunately, this process typically creates too much disturbance for the sensitive CSA. The analog solution involves using a resistor (or MOSFET operating in a triode regime) and provides continuous discharging during the entire process. The discharge can't be too slow, in which case the capacitor will not be fully discharged before the next event, or too fast, as that will affect signal formation.

On a final note, let us point out that while the principles of CSA signal amplifying, pulse shaping, and ADC conversions outlined in this section are fairly

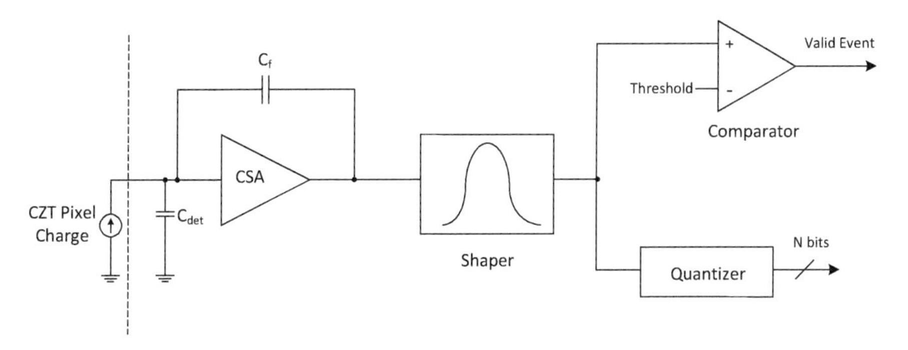

FIGURE 5.14
CZT readout signal chain.

simple, practical implementations can be very challenging due to the very small input signals involved (below 1 mV). One has to worry about system noise, power supplies decoupling, EMI radiation, and op-amp stability issues.

5.3.4 Photon-Counting ASICs

Fast-readout electronic circuits have been developed to reach count rates of several million counts per second. These systems provide coarse ER given by a limited number of discriminators and counters. This section provides information about some selected PC devices.

5.3.4.1 Timepix

Timepix is an ASIC developed in the framework of the Medipix2 collaboration [14]. The pixel matrix consists of 256×256 pixels with a pitch of 55 μm, which gives a sensitive area of about 14×14 mm. Timepix has been designed in a 0.25 μm CMOS process and has about 500 transistors per pixel.

The chip has one threshold and can be operated in PC, ToT, or time-of-arrival (ToA) modes. The principles of the different operating modes are described in detail in [14]. In PC mode, the counter is incremented once for each pulse that is over the threshold; in ToT mode, the counter is incremented as long as the pulse is over the threshold; and in ToA mode, the counter is incremented when the signal crosses the threshold and keeps counting until the shutter is closed.

5.3.4.2 Medipix3

While Timepix is a general-purpose chip, the Medipix3 is aimed specifically at X-ray imaging [27]. It can be configured with up to eight thresholds per pixel and features analog charge summing over dynamically allocated 2×2 pixel clusters. The intrinsic pixel pitch of the ASIC is 55 μm, as in Timepix. Si die can be bump bonded in this mode, called *fine pitch mode*; the chip can then be run with either four thresholds per pixel in single-pixel mode or with two thresholds per pixel in charge-summing mode. Optionally, the chip can be bump bonded to have a 110 μm pitch, which combines counters and thresholds from four pixels. Operation is then possible in single-pixel mode with eight thresholds per pixel or in charge-summing mode with four thresholds and a 220×220 μm^2 area.

Being a very versatile and configurable chip, it is also possible to utilize the two counters per pixel and run in continuous read/write mode, where one counter counts while the other one is being read out. This eliminates the readout dead-time but comes at the cost of losing one threshold, since both counters need to be used for the same threshold. Finally, charge-summing mode is a very important feature, as it combats contrast degradation by charge sharing in CZT detectors with small pixels.

5.3.4.3 ChromAIX

A proprietary multi-energy-resolving ASIC called ChromAIX has been designed by Philips to support the study of spectral CT applications [24]. In order to enable K-edge imaging, at least three spectrally distinct measurements are necessary; for a photon-counting detector, the simplest choice is to have at least the same number of different energy windows. As the number of energy windows increases, the spectrum of incident X-ray photons is sampled more accurately, thus improving the separation capabilities.

The ChromAIX ASIC accommodates a sufficient number of discriminators to enable K-edge imaging applications. Post-processing allows one to separate the photo effect, the Compton effect, and one or possibly two contrast agents with their corresponding quantifications. The ChromAIX ASIC is a pixelated integrated circuit that has been devised for direct flip-chip connection to a direct-conversion crystal such as CZT. The design target in terms of observed count rate performance is 10 Mcps/pixel (Poissonian), which corresponds to approximately 27.2 MHz/pixel periodic pulses, assuming a paralyzable dead-time model. Although the pixel area in CT is typically slightly larger than 1 mm^2, both the ASIC and the direct converter feature a significantly smaller pixel or *subpixel*. In this way, significantly higher rates can be achieved at an equivalent CT pixel size, while further improving the spectral response of the detector via the exploitation of the so-called small pixel effect. The subpixel should not be made too small, since charge sharing effects then start to deteriorate the spectral performance. Very small pixels need countermeasures as implemented in Medipix3, the effectiveness of which at higher rates remains doubtful, as previously discussed in this chapter.

The ChromAIX ASIC consists of a CSA and a pulse shaper stage, as with any other PC device. The CSA integrates the fast transient current pulses generated by the direct converter, providing a voltage step-like function with a long exponential decay time. The Shaper stage represents a band-pass filter that transforms the aforementioned step-like function into voltage pulses of a defined height. The height of such pulses is directly proportional to the charge of the incoming X-ray photon. A number of discriminator stages are then used to compare a predefined value (i.e., energy threshold) with the height of the produced pulse. When the amplitude of the pulse exceeds the threshold of any given discriminator, the associated counter will increment its value by one count.

In order to achieve 10 Mcps Poisson rates, which would typically correspond to incoming rates exceeding 27 Mcps, a very high bandwidth is required. The two-stage approach using a CSA and a shaper achieves such high rates while relaxing the specification of its components. The design specification in terms of noise is 400 e^-, which corresponds to approximately 4.7 keV FWHM. Simulations of the analogue front-end have been carried out by the Philips group to evaluate the noise performance of the channel. According to these simulations, the complete analogue front-end electronic noise (i.e., of the CSA, shaper, and discriminator input stages together) amounts to approximately 2.51 mVRMS, which in terms of ER corresponds to approximately 4.0 keV FWHM for a given input equivalent capacitance.

As the ChromAIX ASIC has been designed to be able to cope with CZT for both Ohmic and Schottky contacts, some sort of detector leakage current compensation is required. Schottky-type CZT detectors of the geometry considered here deliver sufficiently low-leakage currents to regard their effect on the dynamic range negligible. Ohmic contacts, however, exhibit significantly larger leakage currents that might reduce the dynamic range of the CSA stage, compromising the count rate performance and reducing the spectral performance due to additional electronic noise. To overcome this, a passive detector leakage compensation circuit has been included in each pixel. The leakage current compensation circuit consists of a 6 bit digital-to-analog converter (DAC) and a comparator. The comparator is used to sense positive voltages at the output node of the CSA. This information is then utilized to set a current source controlled by the 6 bit DAC to a level that makes the comparator trip. It has been designed to compensate leakage currents of up to 20 nA/pixel. All ChromAIX components are integrated into a pixel area below 300×300 μm^2, including the corresponding asynchronous counters and readout latches that enable continuous acquisition; that is, any given frame is read out while the next frame is being acquired. In CT applications, continuous readout is a mandatory requirement in order to minimize X-ray doses. The counters are 12 bits wide.

5.3.5 Spectroscopic ASICs

5.3.5.1 IDeF-X

IDeF-X HD is the latest generation of low-noise radiation-hard front-end ASICs designed by CEA/Leti for spectroscopy with CZT detectors. The chip includes 32 analog channels to convert the impinging charge into an amplified pulse-shaped signal and a common part for slow control and readout communication with a controller. The first stage of the analog channel is a CSA based on a folded-cascade amplifier with an inverter input amplifier. It integrates the incoming charge on a feedback capacitor and converts it into a voltage; the feedback capacitor is discharged by a continuous reset system realized with a PMOS transistor. The increase of drain current in this transistor during the reset phase is responsible for a non-stationary noise; to reduce the impact of this noise on the equivalent noise charge, a so-called non-stationary noise suppressor was implemented for the first time in this chip version using a low-pass filter between the CSA output and the source of the reset transistor to delay the noise.

The second stage is a variable-gain stage to select the input dynamic range from 10 fC (250 keV) to 40 fC (1 MeV). The third stage is pole zero (PZ) cancellation, implemented to avoid long-duration undershoots at the output and to perform a first integration. The next stage of the analog channel is a second-order low-pass filter (RC^2) with variable shaping time. To minimize the influence of the leakage current on the signal baseline, a so-called baseline holder (BLH) was implemented by inserting a low-pass filter in the feedback loop between the output of the RC^2 filter and the input of the PZ stage. The DC level at the output is stabilized for leakage currents up to 7 nA per channel. The output of each analog channel feeds a discriminator and a stretcher. The discriminator compares the amplitude with an in-pixel reference low-level threshold to detect events. The stretcher consists of a peak detector and a storage capacitor to sample and hold the amplitude of the signal, which is proportional to the integrated charge and hence to the incident energy. In addition, each channel can be switched off by slow control programming to reduce the total power consumption of the ASIC when using only a few channels of the whole chip.

The slow control interface was designed to minimize the number of signals and to connect up to eight ASICs and address them individually. This optimization has reduced the electrical interface from 49 pins in Caliste 256 to 16 pins in Caliste-HD for the same number of channels and using low-voltage differential signals (LVDSs). When an event is detected by at least one channel, a global trigger signal (TRIG) is sent out of the chip. The controller starts a readout communication with three digital signals (DIN, STROBE, and DOUT) to get the address of the hit ASIC and then the hit channels. Then the amplitudes stored in the peak detectors of the hit channels are multiplexed and output using a differential output buffer (AOUT). The whole readout sequence lasts between 5 and 20 μs, according to the set delays and clock frequencies and the number of channels to read out.

5.3.5.2 VAS UM/TAT4

The VAS UM/TAT4 ASIC chip is used to read out both the amplitude of the charge induction and the electron drift time independently for each anode pixel. The ASIC has 128 channels, each with a charge-sensitive preamp and two CR–RC unipolar shapers with different shaping times. The slow shaper has a 1 μs peaking time and is coupled to a peak-hold stage to record the pulse amplitude. The fast shaper has a 100 ns shaping time and is coupled to simple level discriminators for timing pickoff.

Of the 128 channels, 121 are connected to the pixels, 1 is connected to the grid, and 1 is connected to the cathode. Compared to the anodes, the polarity of the signals is reversed for the cathode and the grid. The peak-hold properties, signal shaping, ASIC noise, and triggering procedures are included in the ASIC readout system model. The fast shaper can trigger pulses as small as 30 keV for the anode and 50 keV for the cathode. Only the pixels with slow-shaped signals greater than a noise discrimination threshold of 25 keV are typically used in operation.

VAS UM/TAT4 is particularly well suited for three-dimensional imaging and detection using thick CZT detectors (>10 mm) with high-energy photons (>1 MeV). Three-dimensional position-sensing techniques enable multiple-pixel events of pixelated CZT detectors to be used for 4π Compton imaging. Multiple-pixel events occur by either multiple gamma ray interactions or charge sharing from a single electron cloud between adjacent pixels. To perform successful Compton imaging, one has to correct for charge sharing. There is a large research effort at the University of Michigan under the direction of Professor Zhing He to resolve these complicated signal processing issues and to reconstruct the trajectory of incoming photons for dirty bomb detection.

5.3.5.3 HEXITEC

HEXITEC was a collaborative project between the Universities of Manchester, Durham, Surrey, Birkbeck, and the Science and Technology Facilities Council (STFC). The objective of the program was to develop a new range of detectors, such as CZT for high-energy X-ray imaging applications. The project was funded by the Engineering and Physical Sciences Research Council (EPSRC) on behalf of Research Councils UK (RCUK) under the Basic Technology program.

The HEXITEC ASIC consists of an 80 × 80 pixel array on a pitch of 0.25 mm. Each pixel contains a 52 μm bond pad that can be gold stud bonded to a CZT detector. Figure 5.8 shows a block diagram of the electronics contained in each HEXITEC ASIC pixel. A charge is read from each of the CZT detector pixels using a charge amplifier, which has a selectable range and a feedback circuit that compensates for detector leakage currents up to 50 pA.

The output from the each charge amplifier is filtered by a 2 μs peaking circuit comprising a CR-RC shaper followed by a second-order low-pass filter, as shown schematically in Figure 5.11. A peak-hold circuit maintains the voltage at the peak of the shaped signal until it can be read out. Three track-and-hold buffers are used to sample the shaper and peak-hold voltages sequentially prior to the pixel being read.

HEXITEC is read out using a rolling shutter technique. A row select register is used to select the row that is to be read out. The data from each pixel becomes available on all column outputs at the same time, and at this point the peak-hold circuits in that row can be reset to accept new data. The data being held on the column output is read out through a column multiplexer. The column readout rate is up to 25 MHz, and the total frame rate depends on the number of pixels being read out. The main limitation of HEXITEC is its maximum count rate due to the 10 kHz frame readout scheme.

5.4 Conclusions

The purpose of this chapter was to describe semiconductor sensors used for direct X-ray detection. In the first section, we described typical X-ray detection applications and their resulting sensor requirements. In the second section, we contrasted traditional scintillator detectors with modern direct-conversion detector technology using semiconductor detectors. In the third section, we analyzed direct-conversion sensors and contrasted Si, Ge, GaAs, CdTe, and CZT. We have shown why semiconductor sensors are typically the best choice for high-count and/or high-energy applications in particular. In the fourth section, we discussed the requirements of the sensor electronics required to read signals generated by direct-conversion sensors. In the fifth section, we demonstrated some examples of the usage of semiconductor sensors in NDT, medical imaging (including CT), and baggage scanning. Finally, we concluded by highlighting future opportunities in product development that are offered by semiconductor sensor technology.

Based on the review of the aforementioned available sensor technologies, CZT seem like the most suitable choice for high-count and/or high-energy applications. Recent advances in THM growth and device fabrication have dramatically improved hole transport properties and reduced the polarization effects in CZT material. As a result, high-flux operation of CZT sensors at rates in excess of 200 Mcps/mm^2 is now possible and has enabled multiple medical imaging companies to start building prototype CT scanners and high-efficiency linear-array detectors. Due economies of scale, CZT sensors are now finding new commercial applications in NDT and baggage scanning. CZT sensors are capable of delivering both high count rates and high-resolution spectroscopic performance, although it is challenging to achieve both of these attributes simultaneously. We have been discussing here the material challenges, detector design trade-offs, and ASIC architectures required to build cost-effective CZT-based detection systems for various markets.

In addition to CZT sensors, photon-counting ASICs are an essential part of the integrated module platforms, as charge-sensitive electronics needs to deal with charge sharing and pile-up effects. Using CZT sensor and suitable ASIC electronics, one can generate X-ray spectral information from multiple (typically between two and four) energy bins. This type of acquisition enables a level of material discrimination analysis that is not possible using integrating detectors that contain no spectral information. The first examples of the application of spectral PC are in medical imaging (CT) and baggage scanning. CZT-based spectral imaging in CT is bringing new clinical information to the diagnosis and characterization of many diseases. Spectral functionality brings additional contrast to the image, which makes subtle diseases more conspicuous and brings chemical composition to the diagnosis to better characterize diseases. This makes CZT-based CT more cost-effective by reducing the need for downstream tests, especially in the diagnosis of cancer, vascular disease, and kidney stones. The first spectral CT scanners are currently being tested in phantom trials as well as in first clinical trials. Some technical challenges related to charge sharing between pixels and pile-up are currently being solved. Intense research efforts are also being directed toward applying K-edge detection, a concept that can be exploited in NDT applications as well.

Traditional baggage-scanning systems use energy-integrating scintillator detectors to measure the X-rays transmitted directly through an object. While these systems do an adequate job of detecting traditional threats based on macroscopic shape (e.g., guns and knives) and can achieve a high throughput, they require significant operator involvement, have high false-alarm rates in identifying known explosives, and are not effective in detecting new threats posed by liquids, aerosols, and gels (LAGs) and home-made explosives (HMEs). Successful stand-off detection of these new threats requires the capacity to perform spatially resolved material identification. Achieving the necessary material discrimination capabilities relies on the development of cost-effective multipixel, energy-sensitive, semiconductor sensor–based systems.

References

1. i3system, Digital X-ray. Source: http://www.i3system.com/eng/n_tech/tech2.html. Accessed March 27, 2018.
2. B. J. Heismann, D. Henseler, D. Niederloehner, P. Hackenschmied, M. Strassburg, S. Janssen, and S. Wirth, Spectral and spatial resolution of semiconductor detectors in medical X- and gamma ray imaging, in K. Iniewski and T. Farncombe, (Eds.), *Medical Imaging: Technology and Applications*, Boca Raton: CRC Press, 2013. eBook ISBN 9781466582637.
3. K. Zuber, A comparison of the temporal instabilities found in a comparison of the temporal instabilities found in state-of-the-art CdTe and CZT sensors used in spectral CT measurements with the Medipix3RX detector, *IEEE NSS-MICS*, San Diego, CA, 2015.
4. R. Redus, *Charge Trapping in XR-100-CdTe and -CZT Detectors*, Amptek, 2007, www.amptek.com. Accessed July 13, 2017.
5. R. Macdonald, Design and implementation of a dual-energy X-ray imaging system for organic material detection in an airport security application, *Proceedings of SPIE*, 4301, San Jose, CA, 2001.
6. I. Farella, Study on instability phenomena in CdTe diode-like detectors, *IEEE Transactions on Nuclear Science*, 56(4), 1736–1742, September 2009.
7. H. Toyama, A. Higa, I. Owan, S. Yamanoha, M. Yamazato, T. Maehama, R. Ohno, and M. Toguchi, Analysis of polarization phenomenon and deep acceptor in CdTe radiation detectors, IEEE NSS-MICS, 2006.
8. A. Arodzero, A system for the characterization and testing of CdZnTe/CdTePixel detectors for X-ray and gamma-ray imaging, *IEEE NSS-MICS*, San Diego, CA, 2006.

9. T. Seino, S. Kominami, Y. Ueno, and K. Amemiya, Pulsed bias voltage shutdown to suppress the polarization effect for a CdTe radiation detector, *IEEE Transactions on Nuclear Science*, 55(5), October 2008.
10. L. Verger, E. Gros d'Aillon, O. Monnet, G. Montemont, and B. Pelliciari, New trends in γ-ray imaging with CdZnTe/CdTe at CEA-Leti, *Nuclear Instruments and Methods in Physics Research Section A*, 571, 33–43, Feb. 2007.
11. X. Wang, D. Meier, B. Sundal, B. Oya, P. Maehlum, G. Wagenaar, D. Bradley, E. Patt, B. Tsui, and E. Frey, A digital line-camera for energy resolved X-ray photon counting, *Proceedings of IEEE* Nuclear Science Symposium *Conference Record*, 3453–3457, 2009
12. A. Brambilla, C. Boudou, P. Ouvrier-Buffet, F. Mougel, G. Gonon, J. Rinkel, and L. Verger, Spectrometric performances of CdTe and CdZnTe semiconductor detector arrays at high X-ray flux, *Proceedings of IEEE Nuclear Science Symposium Conference Record*, 1753–1757, 2009.
13. J. Rinkel, G. Beldjoudi, V. Rebuffel, C. Boudou, P. Ouvrier-Buffet, G. Gonon, L. Verger, and A. Brambilla, Experimental evaluation of material identification methods with CdTe X-ray spectrometric detector, *IEEE Transactions on Nuclear Science*, 58(2), 2371–2377, October 2011.
14. L. Tlustos, Spectroscopic X-ray imaging with photon counting pixel detectors, *Nuclear Instruments and Methods A*, 623(2), 823–828, November 2010.
15. J. Iwanczyk, E. Nygard, O. Meirav, J. Arenson, W. Barber, N. Hartsiugh, N. Malakhov, and J. Wessel, Photon counting energy dispersive detector arrays for X-ray imaging, *Proceedings of IEEE Nuclear Science Symposium Record*, 2741–2748, 2007.
16. C. Szeles, S. Soldner, S. Vydrin, J. Graves, and D. Bale, CdZnTe semiconductor detectors for spectrometric X-ray imaging, *IEEE Transactions on Nuclear Science*, 55(1), 572–582, February 2008.
17. E. Kraft and I. Peric, Circuits for digital X-ray imaging: Counting and integration, in K. Iniewski (Ed.), *Medical Imaging Electronics*, New York: Wiley, 2008.
18. S. Mikkelsen, D. Meier, G. Maehlum, P. Oya, B. Sundal, and J. Talebi, An ASIC for multi-energy X-ray counting, *Proceedings of IEEE Nuclear Science Symposium Record*, 294–299, 2008.
19. S. Awadalla, Solid-state radiation detectors: Technology and applications, in *Devices, Circuits, and Systems*, Boca Raton, FL: CRC Press, 2015.
20. G. Prekas, The effect of crystal quality on the behavior of semi-insulating CdZnTe detectors for X-ray spectroscopic and high flux applications, *Proceedings of IEEE Nuclear Science Symposium and Medical Imaging Conference*, Seattle, WA, November 8–15, 2014.
21. K. Iniewski, CZT growth, characterization, fabrication and electronics for operation at 100 Mcps/mm^2, *Proceedings of Workshop on Medical Applications of Spectroscopic X-ray Detectors*, CERN, Geneva, Switzerland, April 20–23, 2015.
22. J. Iwanczyk, Radiation detectors for medical imaging, in *Devices, Circuits, and Systems*, Boca Raton, FL: CRC Press, 2015.
23. R. Turchetta, Analog electronics for radiation detection, in *Devices, Circuits, and Systems*, Boca Raton, FL: CRC Press, 2016.
24. R. Steadman, H. Christoph, O. Mülhens, D. G. Maeding, J. Colley, T. Firlit, R. Luhta, M. Chappo, B. Harwood, and D. Kosty, ChromAIX, A high-rate energy resolving photon-counting ASIC for spectral computed tomography, *Proceedings of SPIE*, 7622, 762220, 2010.

25. W. Barber, E. Nygard, J. C. Wessel, N. Malakhov, N. E. Hartsough, T. Gandhi, G. Wawrzyniak, and J. S. Iwanczyk, Photon counting energy resolving CdTe detectors for high-flux X-ray imaging, *Proceedings of IEEE Nuclear Science Symposium and Medical Imaging Conference*, Knoxville, TN, October 30–November 6, 3953, 2010.
26. C. Ullberg, M. Urech, N. Weber, A. Engman, A. Redz, and F. Henckel, Measurements of a dual-energy fast photon counting CdTe detector with integrated charge sharing correction, *Proceedings of SPIE*, 8668, 86680P, 2013.
27. R. Ballabriga J. Alozy, G. Blaj, M. Campbell, M. Fiederle, E. Frojdh, E.H.M. Heijne, et al., The Medipix3RX: A high resolution, zero dead-time pixel detector readout chip allowing spectroscopic imaging, *JINST*, 8, C02016, 2013.
28. P. Maj, P. Grybos, R. Szczygiel, P. Kmon, R. Kłeczek, A. Drozd, P. Otfinowski, and G. Deptuch, Measurements of matching and noise performance of a prototype readout chip in 40 nm CMOS process for hybrid pixel detectors, *IEEE Transactions on Nuclear Science*, 62, 359, 2015.
29. G. Deptuch, G. Carini, T. Collier, P. Grybos, P. Kmon, R. Lipton, P. Maj, D. Siddons, R. Szczygiel, and R. Yarema, Results of tests of three-dimensionally integrated chips bonded to sensors, *IEEE Transactions on Nuclear Science*, 62, 349, 2015.
30. R. Ballabriga, J. Alozy, M. Campbell, E. Frojdh, E.H.M. Heijne, T. Koenig, X. Llopart, et al., Review of hybrid pixel detector readout ASICs for spectroscopic X-ray imaging, *JINST*, 11, P01007, 2016.
31. P. Kraft, A. Bergamaschi, Ch. Brönnimann, R. Dinapoli, E. F. Eikenberry, H. Graafsma, B. Henrich, et al., Characterization and calibration of PILATUS detectors, *IEEE Transactions on Nuclear Science*, 56, 758, 2009.
32. T. Koenig, T. Koenig, M. Zuber, E. Hamann, A. Cecilia, R. Ballabriga, M. Campbell, M. Ruat, L. Tlustos, A. Fauler, and M. Fiederle, How spectroscopic X-ray imaging benefits from inter-pixel communication, *Physics in Medicine & Biology*, 59, 6195, 2014.
33. K. Iniewski et al., High precision medium flux rate CZT spectroscopy, *Proceedings of IEEE Nuclear Science Symposium and Medical Imaging Conference*, San Diego, CA, October 31–November 7, 2015.
34. K. Taguchi and J.S. Iwanczyk, Vision 20-20: Single photon counting X-ray detectors in medical imaging, *Medical Physics*, 40, 100901, 2013.
35. S. Kappler, F. Glasser, S. Janssen, E. Kraft, and M. Reinwand, A research prototype system for quantum-counting clinical CT, *Proceedings of SPIE*, 7622, 76221Z, 2010.
36. A. Altman, State of the art and future trends in radiation detection for computed tomography, AAPM Virtual Library, 2013. Source: http://www.aapm.org/education/VL/vl.asp?id=2325.
37. K. Mahnken, S. Stanzel, and B. Heismann, Spectral RhoZ-projection method for characterization of body fluids in computed tomography: ex-vivo experiments, *Academic Radiology*, 16, 2009.
38. K. Iniewski and J. Jakubek, Small pixel CZT characterization for non-destructive testing, in preparation.

6

Organic Imagers

Dario Natali

CONTENTS

6.1 Introduction 129
6.2 Imager Structure 130
6.3 Passive Imagers 131
6.4 Active Imagers 133
6.5 Unconventional Architectures 140
6.6 Conclusions 144
References 145

6.1 Introduction

Organic semiconductors[1] (OSCs) have attracted much attention in recent years. OSCs refer to carbon-based compounds, small molecules, oligomers, or polymers where carbon is present in its sp^2 hybridization form; this gives rise, in addition to highly energetic and localized σ-bonds, to delocalized π-bonds. The latter constitute the frontier orbitals and are responsible for (macro)molecule optoelectronic properties. The physical, chemical, and electronic properties of OSCs can be tuned by means of chemical tailoring, giving an unprecedented freedom in the adaption of semiconductor properties to the application. Most OSCs can be processed by solution-based processes with a limited (below 200°C) thermal budget: this makes the process far more eco-friendly than inorganic semiconductors and allows for direct deposition onto lightweight, flexible substrates. On the other hand, solid-state organic semiconductors give rise to van der Waals–bonded molecular solids with a limited degree of long-range order, spanning from nano- to micro-crystalline structures otherwise embedded into an amorphous matrix (with the notable exception of single crystals[2], which require dedicated growing techniques to be obtained). The consequences of the more or less amorphous nature of OSCs are twofold:

1. Charge carrier mobilities are relatively low, up to a few tens of $cm^2 V^{-1} s^{-1}$, but only at high carrier densities such as those achieved in the transistor-accumulated channel (about 10^{19} cm^{-3}), otherwise they lie well below 1 $cm^2 V^{-1} s^{-1}$.

2. Light–matter interaction results in relatively deeply bound excitons that hardly spontaneously dissociate at room temperature, also because of the low (≅3) relative dielectric constant[3,4].

To circumvent the last issue, binary mixtures of OSCs called donor/acceptor (DA) blends have been devised where the presence of suitable heterojunctions forces exciton dissociation to occur at room temperature.

The possibility of solution processing offered by OSCs opens the way to the adoption of deposition techniques borrowed from the printing arts, such as gravure printing, flexography, inkjet printing, screen printing, and so on[5]. This requires the formulation of the dissolved OSC as a proper functional ink, in order to match the rheological requisites of the chosen deposition technique. In principle, large areas at high throughput and hence at low cost can be addressed by means of printing. In addition, printing techniques are scalable: a demonstration at the lab scale can be easily extended to the industrial scale (in contrast, for instance, to non-scalable techniques such as spin coating). Finally, inks can be deposited selectively onto the substrate, so that a complex optoelectronic system requiring many materials and layers can be developed additively without recurring to lithography.

All the basic building blocks of electronics can be realized with OSCs: diodes[6], organic light-emitting diodes (OLEDs)[7], thin-film transistors (TFTs)[8–12] and circuits[13–15], memories[16], solar cells[17], and, of course, organic photodetectors (OPDs)[3,18–23]. While OLEDs have already been on the market for few years, the other devices are still subject to intense research, even though some startups have begun their commercialization.

At a glance, organic electronics is suited to address applications where large areas, light weight, and flexibility represent a premium feature. As such, it is by no means in competition with inorganic, silicon (Si)-dominated traditional electronics, which on the contrary is mainly focused on miniaturization and on adding more and more functionalities to areas as small as possible. Imagers[24], the array of light detectors able to transduce an optical image into a digital one, represent a paradigmatic application for OSCs and are the focus of this chapter.

6.2 Imager Structure

Arrays of light detectors can be roughly divided according to their structure as *passive* or *active matrices*. The former are simply a crossbar arrangement of rows and columns: their crossings define the detector area. Albeit having a very simple structure that makes them very attractive and relatively easy to develop, passive matrices suffer from the *sneak path* problem[25]. Let's say we want to address the (i,j) detector by measuring the current flowing from the *i*-th row to the *j*-th column. Indeed, current is not restricted to flow across the *(i,j)* detector but has access to a number of alternative paths involving other detectors; such parasitic paths, which are in parallel to the *(i,j)* detector, are actually called sneak paths. If the detectors involved in the sneak path are not illuminated, then the result is a deterioration of the dark current of the *(i,j)* pixel; if the pixels involved in the sneak path are illuminated, then cross-talk phenomena and image blurring can occur. The larger the array, the worse the situation, so that the passive matrix topology is suitable when the imager comprises a limited number of elements.

To solve this issue, it is necessary to complicate the structure by adding a selector in series to the light detector; the subsystem comprised of the detector and the selector is termed the *passive pixel*, whereas an array thereof is termed the *active matrix*. The selector can be either a diode or a TFT. In the former case, the pixel is a two-terminal device, which helps to keep the system topology simple. From a processing point of view, the detector and the selector diode are very similar; therefore, their integration is expected to be relatively easy. The suppression of the sneak path issue exploits the fact that along a sneak path at least one reverse-biased diode is present, thus making the parasitic current very low. On the other hand, the inherent non-linearity of the diode makes the reading of the pixel somewhat more complicated than in the case of TFT addressing. In a TFT-based pixel, the TFT is used as a switch that decouples the detector from the rest of the array, unless the TFT is switched on by a suitable voltage at its gate terminal. The light detector is connected to the ground on one side and to the TFT source terminal on the other side; the gates and drains of TFTs are connected to rows and columns. While very effective, a TFT-based pixel is a three-terminal device (ground, gate, and drain), thus making the topology more elaborated; in addition, the integration of a light detector and of a TFT on the same substrate may prove complicated. When signal amplification occurs at the pixel level, the latter is termed an *active pixel*. Such a configuration requires at least three transistors and a trade-off between the pixel fill factor (viz., the fraction of pixel area occupied by the light detector), and the signal-to-noise ratio has to be taken into account. The various possible pixel and matrix topologies are shown in Figure 6.1; the sneak path problem and its solution in active matrix topology are shown in Figure 6.2.

6.3 Passive Imagers

Only two realizations of passive imagers based on OSCs can be found in the literature (Table 6.1). Xu et al., adopting the simplified passive matrix topology, developed a direct-transfer method to develop an imager array on a planar substrate, which is easier to handle and process, and then to transfer

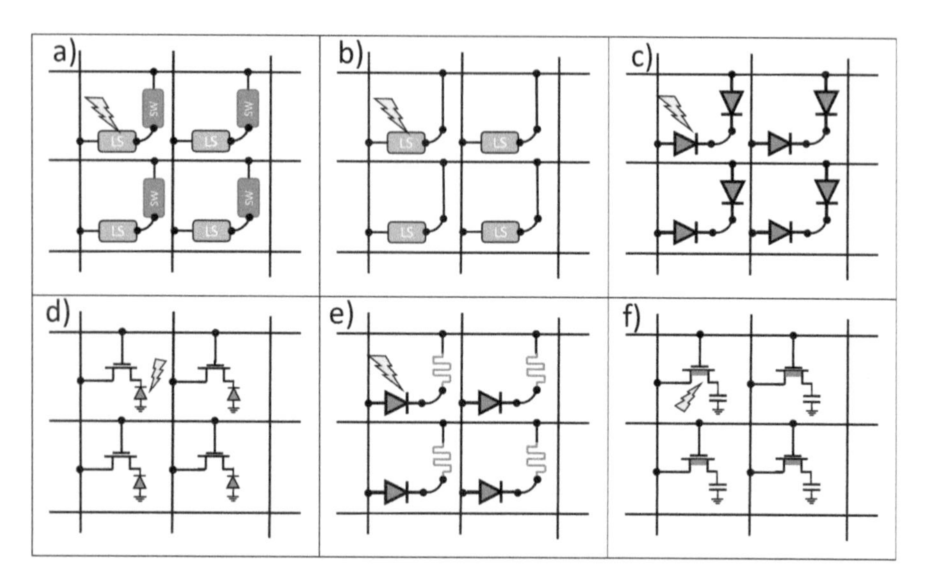

FIGURE 6.1
Various possible imager architectures. a) A generic Light Sensitive element (LS) is shown in series with a switch (SW) element in a passive pixel (PP) configuration. The 2 x 2 array shown is an active matrix (AM). b) Without the switch, a simple passive matrix, crossbar configuration is obtained. c) PP constituted by a photodiode and a switching diode. AM configuration. d) PP constituted by a photodiode and a switching transistor. AM configuration. e) PP constituted by a photodiode and a resistive switching element (memristor). AM configuration. f) PP constituted by a phototransistor in series with biasing capacitor to suppress dark current [56]. AM configuration.

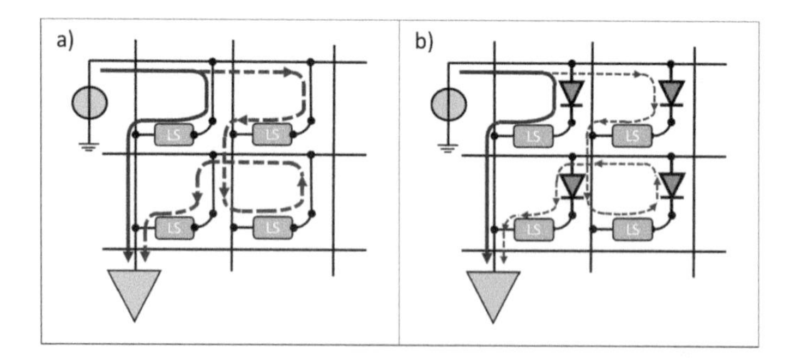

FIGURE 6.2
The sneak path issue. Panel a) shows in a passive, 2×2 matrix that when the element (1,1) is addressed, current flows also in elements (2,1), (2,2), (2,1) and finally adds to the current coming from element (1,1). In panel b) an active matrix with a switching diode is shown: in the sneak path made up of pixels (2,1), (2,2), (2,1), the switching diode of pixel (2,2) is reverse biased thus largely suppressing the sneak path current.

TABLE 6.1

Imagers based on Photodetectors Arranged into a Passive Matrix Topology

Pixel Number	Pixel Size	Fabrication	Notes	Ref.
100×100	40 μm^2	Direct-transfer patterning on hemispherical surface	Semitransparent Au anode	(26)
100	500 μm^2	Direct-transfer patterning on hemispherical surface	Semitransparent ITO anode	(27)

Note: Realizations are chronologically sorted.

it onto a 3D deformed surface by means of elastomeric stamps and cold welding[(26)]. A vacuum-deposited array of 100×100 photodiodes with a 40 μm^2 area was developed and tested. The possibility of creating non-planar photodiode arrays opens the way to imaging systems with greatly simplified optics. The process was later extended in order to address indium tin oxide (ITO) electrodes in spite of semitransparent Au electrodes: the higher transparency of ITO in the visible spectrum led to an increase in external quantum efficiency (EQE) up to 65%, as demonstrated in a 100-pixel array[(27)].

6.4 Active Imagers

The vast majority of organic imagers with active matrix architecture rely on passive pixel topology (with the exception of Ref.[(28)]). Details are summarized in Table 6.2. The goal of ongoing research is the realization of an entirely organic optoelectronic system developed by means of scalable printing techniques in order to easily address large-area applications. Indeed, the mere integration of an organic TFT with an organic photodetector into an array poses serious processing problems, and the additional constraint of exclusively adopting scalable techniques poses further processing issues. To date, no such realization has been reported in the literature. Few publications have dealt with *single all-organic pixels*, exploring both coplanar[(29), (30), (31)] and vertically stacked architecture[(32)]. In the former the photodetector and the TFT lie on the same plane, whereas in the latter they lie one on top of the other, with obvious advantages in terms of fill factor but possibly with more integration issues (e.g., the development of means to connect the two elements). Very few realizations have succeeded in the development of *all-organic imagers* (Figure 6.3)—that is, where both the photodetector and the TFT are based on OSCs—but they have at least partially relied on non-scalable (e.g., lithographic) techniques. This is the case in Refs.[(33), (34,35), (36), (37)], with array dimensions as large as 72×72 in the case of Ref.[(33)]. Very often, instead of developing a single photodetector for each pixel, the photoactive medium is deposited as a continuous, unpatterned layer on the switching

TABLE 6.2

Imagers Based on the Active Matrix Topology

Type[a]	Transistor Active Material	Pixel Number	Pixel Size	Resolution	Fabrication	Notes	Ref.
H	α-Si	512×512	100×100 μm²		OPD vacuum deposition	Unpatterned photoactive materials	(38,39)
A	Organic (pentacene)	72×72	700×700 μm²	36 ppi	Vacuum deposition	TFTs and OPDs manufactured on separate plastic sheets and laminated	(33)
A	Organic (pentacene)	4×4	700×700 μm²		Partially printed; scalable process	OPD is a lateral photoconductor; active materials patterned	(34,35)
H	α-Si	180×180	340×340 μm²	75 ppi	α-Si inkjet printed	Unpatterned photoactive material; flexible	(41)
H	α-Si	256×256	–	–	OPD solution processed; top contact evaporated	NIR sensitive; unpatterned photoactive material	(40)
H	ZnO	128×96	100×100 μm²		Vacuum deposition	Color imaging by means of vertically stacked RGB sensors; unpatterned photoactive material	(42)
H	Si CMOS	30×30	15×15 μm²		Photoactive material spray coated	Unpatterned photoactive material; NIR sensitivity	(43)
A	Organic (pentacene from soluble precursor)	32×32	200×200 μm²	127 ppi	Solution processed apart from metals	Unpatterned photoactive material; X-ray imaging; flexible substrate	(36)

(Continued)

TABLE 6.2 (CONTINUED)

Imagers Based on the Active Matrix Topology

Type[a]	Transistor Active Material	Pixel Number	Pixel Size	Resolution	Fabrication	Notes	Ref.
H	Carbon nanotube	18×18	1×1 mm^2		Active materials solution processed	Unpatterned photoactive material; X-ray imaging; flexible substrate	(45)
A	Organic (proprietary molecule)	32×32		25–125 ppi	Evaporated OPD; solution-processed TFT	Cross-linkable dielectric; unpatterned photoactive material	(37)
H	CMOS 90 nm	2592×1944	≅2.7 μm^2		Evaporated OPD	Unpatterned photoactive material	(44)
H	α-IGZO	120×160	126 μm^2	200 ppi		Unpatterned photoactive material; X-ray imaging; flexible substrate	(46)
H	Si CMOS (65nm)	970×550	6 μm^2			Unpatterned photoactive material; dual-sensitivity pixel	(28)

Note: Unless specified, pixels are passive. All realizations are transistor addressed. All photodetectors are based on organic semiconductors. Realizations are chronologically sorted.[a] Hybrid (H) imagers adopt a non-organic switching matrix; all-organic (A) imagers employ an organic-based switching matrix.

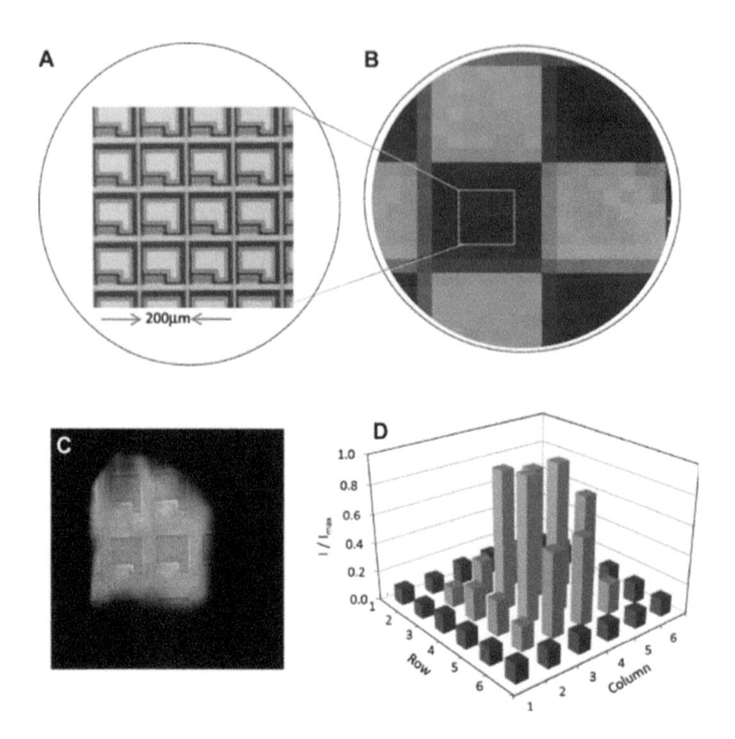

FIGURE 6.3
Example of an all-organic imager based on a passive pixel, active matrix topology. The resolution is 127 ppi. Panel a) shows the top view micrograph of 200 × 200 mm² sensor pixels. Panel b) shows a photo-image recorded by illuminating the detector array through a metal mask with square holes of 1.6 mm in a checkerboard pattern. Pixels in the lighter parts are exposed, pixels in the darker parts are not exposed. Panel c) shows the micrograph of ~10 pixels in array that are fully or partially exposed through a pinhole in a metal. Panel d) shows the related photo-image, obtained by biasing the transistors with a gate voltage of −20 V or +10 V during selection and non-selection, respectively, and the common cathode at −2 V. Reproduced with permission from [36]. Copyright Elsevier (2013).

matrix. Analogously, the photodetector contact that is held at fixed potential, being in common to all photodetectors, is deposited as the topmost contact and is a continuous film as well. The photodetector contact connected to the TFT is patterned and defines the photodetector area. This solution simplifies the process, since it avoids patterning the photoactive medium, but for the same reason it can give rise to cross-talk between adjacent pixels. All-organic imagers developed according to this scheme are those of Refs.(36) and (37). In Ref.(33), a stacked configuration is obtained by separately realizing the switching matrix and the photodetector arrays and then laminating one on top of the other. Only in Refs.(34,35) are the TFT and the photodetector (actually a lateral photoconductor) patterned, yet a limited (4×4) number of pixels was developed. Details for each realization are given in the following.

Someya et al.[33] developed a 72×72, 36 dpi imager with a 700×700 μm^2 pixel area addressed with organic TFTs. The realization exploited vacuum evaporation. To circumvent the integration of OPDs and TFTs the two device arrays were realized on separate plastic sheets and then laminated one on top of the other.

The first realization of an imager with organic active materials patterned on the same substrate and employing a low thermal budget process was by Nausieda et al.[34,35]. Despite the limited number of pixels (the array being 4×4), this realization was obtained by combining photolithographic steps together with inkjet printing.

Gelinck et al. developed an entirely organic imager where all layers are solution processed apart from the metals, which are evaporated. The 32×32 pixels of 200 μm^2 were capable of a 127 ppi resolution.[36] The imager, developed on plastic substrate, was able to acquire X-ray images at dose levels as low as 0.27 mGys^{-1} when coupled with a suitable phosphor.

Malinowski et al. realized a 32×32 imager with a resolution as high as 125 ppi where both TFT and OPD are based on OSCs (Figure 6.3); the former is solution processed and the latter is vacuum evaporated.[37] The production process took advantage of the cross-linkable dielectrics, which helped to decouple and integrate the TFT process and the OPD process.

Another difficulty in realizing an all-organic imager lies in the relative immaturity of organic TFT technology, in terms of mobility, supply voltage, and reliability. Henceforth, in many cases, *hybrid imagers* have been developed, with a non-organic switching matrix covered by a non-patterned photoactive medium and with a continuous topmost fixed-potential contact (Figure 6.4). α-Si[38,39] [40][41], ZnO[42], standard Si CMOS[43] [44] [28], carbon nanotubes[45], and α-IGZO[46] have been used. Thanks to the higher maturity of the adopted switching matrix technology, hybrid imagers with very large numbers of pixels (up to 2592×1944, as in Ref.[44]) have been developed. Details for each realization are given in the following.

Starting in early 2001, Street et al. demonstrated a 512×512 imager based on a vacuum-evaporated organic photodiode coupled with an α-Si switching matrix[38,39]. The pixel size was 100 μm×100 μm. Coupled with GdO_2S_2:Tb

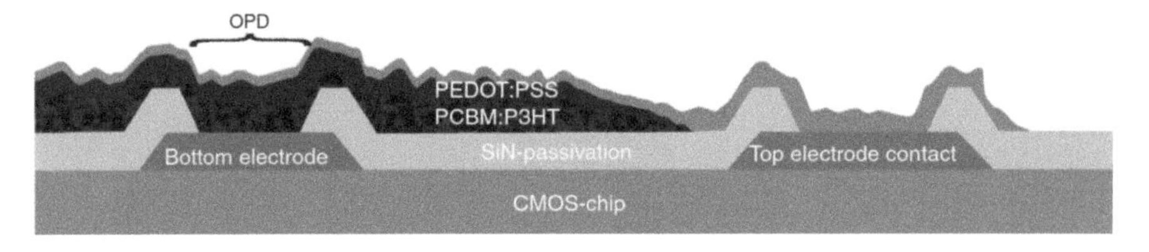

FIGURE 6.4
The structure of a hybrid imager where the photoactive semiconductor is unpatterned and deposited on top of an inorganic-based chip. The photodetector area is defined by the bottom contact, whereas the top contact is unpatterned. Reproduced with permission from [43]. Copyright MacMillan Publisher Limited (2012).

phosphor, the system was capable of recording X-ray images. In this case, the photoactive medium was unpatterned and deposited on top of the addressing matrix; this simplified the process, even though a certain degree of pixel cross-talk is inevitably introduced. Later, Steet and coworkers extended the approach and developed a flexible imager based on an inkjet-printed α-Si switching matrix deposited on a plastic substrate. The imager is comprised of 180×180 pixels, each with an area of 340 μm^2, achieving a resolution of 75 ppi. The photosensitive layer was solution processed and its thickness tailored in order to suppress dark currents.[41]

The absorption spectrum of organic semiconductors usually lies in the visible range. Rauch et al. succeeded in extending the response into the NIR by adding PbS quantum dots to the organic blend[40]. A 256×256 imager with an unpatterned photoactive medium was developed by adopting a α-Si switching matrix.

Seo et al.[42] produced a 128×96 color imager by stacking three active matrices (passive pixels addressed by means of ZnO transistors) with 100×100 μm^2 photodiodes with absorption and EQE peaked in the blue, green, and red regions of the spectrum. Care to ensure electronics transparency (gate and signal lines) was paid in order to not deteriorate light transmission from one stack to another; the large (3.3 eV) ZnO gap was fundamental to achieve this goal.

Baierl et al.[43] were the first to demonstrate an hybrid organic/CMOS imager comprised of 30×30 pixels with a 15 μm^2 area (Figure 6.4). The photoactive medium is unpatterned and spray coated on top of an ad hoc CMOS addressing circuit. By adopting CMOS instead of α-Si, a faster and less noisy readout can be achieved.

Takashi et al. used solution-processed carbon nanotubes to develop a transistor switching matrix (Figure 6.5).[45] A 18×18 imager with 1×1 mm^2 pixels was fabricated onto a plastic sheet. Visible and X-ray imaging (by means of coupling with a scintillator at a dose rate of 100 $mGys^{-1}$) was demonstrated. The photoactive material was unpatterned.

FIGURE 6.5
Hybrid imager (the switching matrix is based on carbon nanotubes) with active matrix, passive pixel topology. Panel a) shows a pixel cross-section. Coupled to a Gd_2O_2S/Tb (GOS) scintillator as shown in panel b), this flexible imager is able to record X-ray images: panel c) shows the spatial mapping when a circle-shaped (diameter of 4 mm) X-ray source is projected with a dose rate of 100 mGy/s. Reproduced with permission from [45]. Copyright American Chemical Society (2013).

Lim et al. exploited evaporated OPDs to actually double the functional surface area of a CMOS imager in the framework of a stacked architecture[44]. The CMOS imager was developed to be sensitive to red and blue only (in contrast to the conventional RGB Bayer pixel arrangement), whereas an array of OPDs was placed on top of the imager to record the green part of the spectrum. Thanks to the large variety of available molecules, the photoactive materials for the OPD were chosen to mainly absorb the green part of the spectrum. A 5 Mpixel prototype with a 2.7 μm^2 pixel area has been developed and tested.

Gelinck et al.[46] developed a hybrid 120×160 imager with a 126 μm^2 pixel area, employing a switching matrix based on α-IGZO, thanks to its superior noise performances with respect to α-Si. The photoactive blend, unpatterned, is deposited on top of the α-IGZO by means of solution processing. Thanks to the overall limited thermal processing budget, the imager was placed onto a 25 μm-thick plastic sheet. Coupled to a scintillator, this imager was capable of recording X-ray images at doses as low as 3 μGy/frame.

Nishimura et al.[28], taking advantage of a 65 nm Si CMOS backplane, developed an imager of 970×550 pixels with a 6 μm^2 area. The organic photoactive layer is unpatterned and laminated on top of the inorganic chip. Each pixel has double circuitry to effectively address both dark subjects and ultrabright subjects. In doing so, a very high dynamic range is achieved.

Among the possible applications of large-area imaging, X-ray imaging is highly appealing[47–49], because the lack of a convenient focusing medium forces the imager to have the same dimensions as the target object. For instance, in the medical radiography field this implies relatively large imager areas (in excess of 1000 cm^2). Nowadays, modern radiographic apparatus are digital and employ inorganic technology, but this results in brittle and ponderous systems. There is a great need for lightweight and portable radiographic systems that can shift the point of care closer to the place where the patient was injured. In addition, the possibility of adopting flexible substrates can open the way to non-planar, conformable imagers. Given this high interest, some realizations were built on a flexible substrate and tested the imager capability of recording X-ray images upon coupling with a suitable phosphor layer, as in Refs.[36], [45], [46]. The clear advantage here would be to develop an organic medium capable of directly converting X-ray photons into electron–hole pairs without recurring to the intermediation of the scintillator. Organic compounds, being mainly carbon based, are usually made of low-Z molecules, so their inherent capability of absorbing X-rays is limited. However, the inclusion of suitable high-Z heteroatoms could improve the situation. Interested readers are referred to Ref.[50] for a review on the development of X-ray-sensitive organic photodetectors, to Ref.[51] for hybrid OPDs containing PbS nanoparticles for the direct conversion of X-ray photons into charge carriers, and to Ref.[52], where single OSC crystals are shown to have high X-ray sensitivity despite their low Z thanks to a photoconductive gain mechanism.

6.5 Unconventional Architectures

Recently, unconventional imager architectures have been introduced that are not based on the concept of a photodetector coupled to a switch. Details are given in Table 6.3.

Unconventional imagers can be divided into two categories.

The first one is based on *organic phototransistors*—that is, transistors where the drain current can be controlled both by the gate voltage and by the impinging light[53], [54], [55], [56]. Thanks to suitably engineered trapping states, located either at the semiconductor/dielectric interface or buried in the dielectric (e.g., a floating gate), one of the two photo-generated species gets trapped and results in a threshold voltage shift. Since photoconversion, data retention, and addressing are all embedded into a single device, the integration issues of standard imagers are avoided. On the other hand, the requisites for efficient photogeneration often contrast with those for attaining high mobility and good transistor properties, requiring either a trade-off or the adoption of ad hoc multilayers, as in Ref.[56]. Details of the most notable realizations are given in the following.

Zhang et al.[53] developed light-charge organic memories based on photo-writeable organic transistors. Each pixel is comprised of a transistor with a suitably engineered electron trapping layer at the semiconductor–dielectric interface. Upon illumination, photo-generated electrons are collected in the charge-trapping layer (provided the gate is positively biased); this results in a threshold voltage shift in the TFT proportional to the amount of absorbed light. Since electrons are deeply trapped, a memory functionality is realized with data retention up to 20,000 s. Suitable gate voltage pulses can empty the traps and erase the pixel. A 12×12 array developed on a rigid Si substrate and a 3×24 array on a plastic substrate have been demonstrated.

Chen et al. exploited the memory effect in organic TFTs with a suitable bilayer gate dielectric to develop a 30×30 imager prototype[54]. Since addressing, light sensitivity, and memory are embedded into the transistor, a very compact pixel structure is obtained. The potential for information storage is related to the bilayer gate, which is comprised of a self-assembled monolayer of a photochromic molecule and a sol-gel-prepared HfO_2 thin film. At the photochromic–HfO_2 interface, trapping states are present; once charged, they induce a threshold voltage shift in the TFT. Access to these trapping sites (analogous to the floating gate of a flash inorganic memory) is controlled by the photochromic SAM, which is in between the TFT semiconductor (pentacene) and the HfO_2 dielectric. Generally speaking, photochromic molecules undergo reversible photo-isomerization between two isomers upon irradiation with light of different wavelengths. The two isomers may possess markedly different photophysical and electronic properties. In Chen et al.'s paper, photochromic molecules were chosen so that in one state the energy gap is so large that electrons cannot be effectively transferred from pentacene

TABLE 6.3

Unconventional Imagers

Pixel Number	Pixel Size	Resolution	Fabrication	Operating Principle	Ref.
12×12 on Si (3×24 on plastic)			Solution processing and vacuum evaporation	Organic Photo-TFT with non-volatile memory	(53)
32×32			OPD solution processed; switch evaporated	PVD based on OPD coupled to non-volatile resistive switch	(57)
30×30	120×800 μm^2	15 ppi	Evaporation, solution processing	Organic Photo-TFT with non-volatile memory	(54)
12×13		9 ppi	Evaporation, solution processing, peeling	PVD coupled to organic TFT (passive matrix). Bendable	(58)
10×10		10 ppi	Evaporation, solution processing,	Organic Photo-TFT. flexible	(55)
10×10		220 devices/cm^2	Evaporation, solution processing, DUV patterning	PVD based on organic phototransistor and IGZO TFT. Skin compatible.	(60)
5×5	650×450 μm^2	37 ppi	Solution processing and evaporation	Crossbar matrix based on TiO_2. UV exposure results in permanent photodoping	(61)
4×4			Scalable techniques	Organic photo-TFT	(56)

to the floating gate, whereas in the other state the gap is sufficiently low to ensure efficient electron transfer and trap charging. Once the floating gate is charged, since the energy of the trapping level lies within the gap of both pentacene and the photochromic molecule, it remains permanently charged unless a suitable gate voltage pulse or white light pulse is applied. Image recording occurs as follows: firstly, a preconditioning UV light pulse brings photochromic molecules into the low-gap state; then a visible light pulse photogenerates electron–hole couples within the pentacene; photo-generated electrons then charge the floating gate, producing a threshold voltage shift proportional to the amount of photons impinging on the pixel.

Liu et al. developed a 10×10 phototransistor array on flexible substrate(55). The transistor's photoconductivity was enhanced by means of a suitable Ru-based solution-processable molecule; the latter acts as an efficient light absorber and, in addition, the interface with the transistor semiconductor is effective in splitting photo-generated excitons.

Pierre at al. developed a 4×4 imager based on organic phototransistors, notably developed by means of scalable techniques(56). Organic phototransistors have a layer for charge photogeneration and a high-mobility layer for hole transport. Upon illumination (applying a positive gate voltage), holes are transferred to the high-mobility layer and electrons remain trapped on the photogeneration layer, thus giving rise to a photoconductive gain. The gain being sublinear, a high dynamic range is achieved. Electron and hole recombination takes place upon applying a negative gate pulse, which erases the pixel. Thanks to the high mobility of holes, a competitive 30 fps response is demonstrated.

The second class is based on the *photosensitive voltage divider* concept (Figure 6.6). The voltage at the central node of the series connection of a light-sensitive element and a *blind* element changes its value according the illumination of the light-sensitive element, whose resistance changes upon illumination(57), (58), (59), (60). In Refs.(58), (59), (60) this node is connected to the gate of a TFT, which acts as a transimpedance amplifier converting voltage into current. Memory functionality has been embedded in Ref.(57) (where the blind element of the voltage divider is an organic non-volatile resistive switch), in Ref.(59) (thanks to the floating gate of the voltage-to-current converting TFT), and in Ref.(60) (where the light-sensitive element of the voltage divider is a phototransistor with suitable trapping states at the semiconductor–dielectric interface). Details are as follows.

Nau et al. introduced a pixel made by the series connection of an OPD and an organic non-volatile resistive switch (ORS).(57) The pixel can be seen as a voltage divider between the OPD and the ORS, the latter being a bistable two-terminal device whose resistance can permanently switch between a low and high resistive state upon proper voltage addressing. Upon illumination, the voltage drop across the OPD diminishes and a write operation is performed on the ORS, which permanently changes its resistive state. The ORS thus embeds memory functionality into the pixel, because the resistance

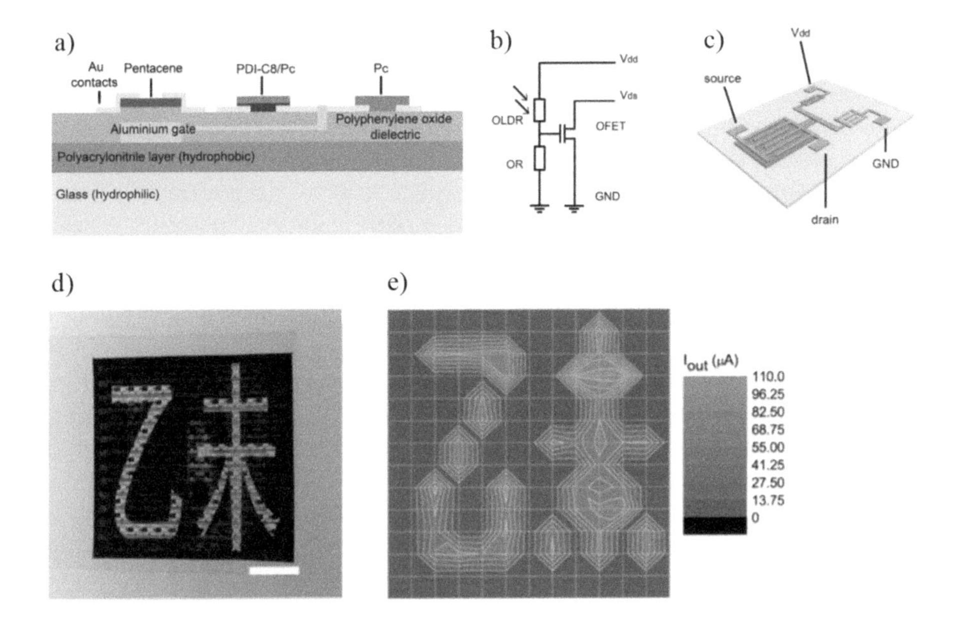

FIGURE 6.6
Unconventional imager based on the photovoltage divider concept. Panel a) shows the pixel cross section, panel b) the its electric diagram and panel c) a 3D pixel sketch. Panel d): chinese characters are employed as shadow mask to test imager capabilities. Panel e): the current map upon continual illumination shows a current difference of over six orders of magnitude between exposed and unexposed pixels. Reproduced with permission from [58]. Copyright Wiley-VCH (2015).

of the pixel in the dark depends on the state of the ORS. A 32×32 imager was fabricated on a flexible plastic sheet and 4 bit gray-scaling capabilities demonstrated. In addition, thanks to the rectifying current–voltage characteristic of the OPD, the latter also acts as an addressing element so that an active matrix configuration is obtained.

Wang et al. proposed an alternative architecture based on a three-component pixel based on a voltage divider (Figure 6.6)[58]. The latter is comprised of the series connection of an organic resistor and an organic light-dependent resistor; upon illumination, a voltage variation occurs at the node connecting the resistor and the light-dependent resistor, and this signal is fed to the gate of a TFT. A 12×13 array was demonstrated; thanks to the peeling technique, the array can be transferred to ultra-thin substrates and may be bended while retaining its functionality. It is worth noting that since the transistor is used to amplify the photovoltage signal of the voltage divider and convert it into a current, this pixel actually lacks an addressing element and the realized matrix has a passive matrix topology. Wang et al. later introduced an improved architecture where the TFT has a floating gate: by judiciously choosing the photoactive material for the light-dependent resistor, sensitivity to the NIR was obtained and a memory effect embedded into the TFT[59]. A 6×5 prototype was demonstrated.

Kim et al. developed a 10×10 imager on ultra-thin (1–3 μm) skin-compatible substrate[60]. The pixel topology is based on the voltage divider between a light-sensitive organic transistor and an IGZO-based transistor. The node of the voltage divider is connected to the gate of a third IGZO-based TFT that implements the voltage-to-current conversion and amplification at the pixel level. To achieve a very high device density (220 cm^{-2}), a deep ultraviolet irradiation (DUV) technique is employed as a high-resolution patterning tool. Upon illumination, photo-generated electrons are trapped at the organic semiconductor–dielectric interface. In order to increase the trapping time, the gate of the organic TFT is held at a positive voltage. Charge detrapping and erasing takes place upon negative gate biasing.

Finally, and even though it is not actually based on organic semiconductors, it is worth reporting the work of Cesarini et al., who demonstrated a 5×5 imager based on TiO_2 as the active material, with a pixel size of 650×400 μm and a horizontal (vertical) pitch of 525 μm (700 μm)[61]. TiO_2 can be solution processed and also printed (e.g., by screen printing) so that scalability can be foreseen for this realization. The working principle relies on the phenomenon of n-type photodoping induced by the UV exposure of titania: the amount of absorbed photons is translated into a resistivity change and permanently stored as a material electrical property, in contrast to conventional photodetection architecture based on electron–hole photogeneration. As a consequence, there is no actual need for an addressing element, and an imager can be conceived with a simple crossbar architecture. To avoid sneak paths, unaddressed columns and rows have to be kept at ground. This realization is very attractive thanks to its simplicity and very high fill factor. Sensitivity to UV combined with visible blindness is very attractive in fields such as chemical sensing, combustion monitoring, forensic applications, and missile warning.

6.6 Conclusions

The development of an organic imager by means of scalable printing techniques could open the way to large-area sensing, not only of visible but also NIR and, most notably, X-ray photons for biomedical radiographic applications. Passive matrices are relatively easy to develop, but the sneak path problem and the consequent limited number of pixels make them less attractive. As a matter of fact, only two realizations can be found in the literature[26], [27].

Active matrices are the ideal solution but require the integration of a switch, typically a transistor, and a photodetector, and this actually poses some issues. Firstly, the technology of organic transistors is still evolving and has only recently reached performances comparable, at least in terms of carrier mobility, to α-Si, for instance. TFT noise performance, which is very important too, is scarcely investigated in the literature, so that a comparison with the competing technology of solution-processable inorganic semiconductors (oxides, most notably) cannot be done. In addition, the switch in the pixel has to operate at low voltages, and this represents an additional potential issue (or may result in TFT performance degradation, for example, in terms of contact resistances). Furthermore, the integration of two different processes, one with an OPD and one with a TFT, onto the same substrate may require a trade-off and may result in non-optimal device performances once they are integrated into an array. This is even truer if one has to restrict the employed techniques to those that are scalable. The situation so far is that a fully printed organic imager has still to be developed. To circumvent these issues, many realizations have combined an array of OPDs with an inorganic switching matrix. To further simplify the production process, the photoactive organic medium is deposited as a continuous layer on top of the switching matrix, thus conferring optoelectronic functionalities to a chip optimized for electronic functionalities. This approach has yielded the largest arrays, up to the Mpixel complexity range(28), (44).

Unconventional architectures may represent, at the actual state of development of printed organic electronics, an interesting opportunity because the pixel can be realized by more homogenous elements with respect to the OPD and the TFT of a standard active pixel, thus decreasing integration issues. This is true both for pixels based on the photosensitive voltage divider concept, comprised of (photo)transistors and (photo)resistors, and for pixels based on phototransistors alone. Indeed, the only realization entirely based on OSCs and scalable techniques exploits an array of phototransistors(56). The relatively limited array size (4×4) underlines how difficult, from a processing point of view, the development of an organic printed imager is.

As a final note, we observe that an organic device analogous to inorganic CCD is missing in the literature. To date, only in Ref.(62) has the transfer photocharge from a reverse-biased capacitor to an adjacent one not yielded the demonstration of an array.

References

1. J.-L. Bredas, S. R. Marder, *The WSPC Reference on Organic Electronics: Organic Semiconductors*; World Scientific: Singapore, 2016; Vol. 1–2.
2. K. S. Park, J. Baek, Y. Park, L. Lee, J. Hyon, Y.-E. Koo Lee, N. K. Shrestha, Y. Kang, M. M. Sung, *Adv. Mater.* 2017, 29, 1603285.
3. K.-J. Baeg, M. Binda, D. Natali, M. Caironi, Y.-Y. Noh, *Adv. Mater.* 2013, *25*, 4267.
4. D. Natali, M. Caironi, *Photodetectors*; Woodhead Publishing: Sawston, UK, 2016; pp. 195–254.
5. M. Caironi, Y.-Y. Noh, *Large Area and Flexible Electronics*; Wiley-VCH: Weinheim, Germany, 2015.

6. J. Semple, *Semicond. Sci. Technol.* 2017, *32*, 123002.
7. V. C. Bender, T. B. Marchesan, J. M. Alonso, *IEEE Ind. Electron. Mag.* 2015, *9*, 6.
8. S. G. Higgins, B. V. O. Muir, J. Wade, J. Chen, B. Striedinger, H. Gold, B. Stadlober, M. Caironi, J. S. Kim, J. H. G. Steinke, A. J. Campbell, *Adv. Electron. Mater.* 2015, *1*.
9. X. Y. Chin, G. Pace, C. Soci, M. Caironi, *J. Mater. Chem. C* 2017, *5*, 754.
10. S. G. Bucella, A. Perinot, M. Caironi, *IEEE Trans. Electron Devices* 2017, *64*, 1960.
11. A. Perinot, P. Kshirsagar, M. A. Malvindi, P. P. Pompa, R. Fiammengo, M. Caironi, *Sci. Rep.* 2016, *6*.
12. X. Guo, Y. Xu, S. Ogier, T. N. Ng, M. Caironi, A. Perinot, L. Li, J. Zhao, W. Tang, R. A. Sporea, A. Nejim, J. Carrabina, P. Cain, F. Yan, *IEEE Trans. Electron Devices* 2017, *64*, 1906.
13. S. Mandal, G. Dell'Erba, A. Luzio, S. G. Bucella, A. Perinot, A. Calloni, G. Berti, G. Bussetti, L. Duò, A. Facchetti, Y. Y. Noh, M. Caironi, *Org. Electron. Phys. Mater. Appl.* 2015, *20*, 132.
14. S. G. Bucella, J. M. Salazar-Rios, V. Derenskyi, M. Fritsch, U. Scherf, M. A. Loi, M. Caironi, *Adv. Electron. Mater.* 2016, *2*.
15. G. Dell'Erba, A. Luzio, D. Natali, J. Kim, D. Khim, D.-Y. Y. Kim, Y.-Y. Y. Noh, M. Caironi, *Appl. Phys. Lett.* 2014, *104*, 153303.
16. S. H. Lee, Y. Xu, Y. Y. Noh, *Large Area and Flexible Electronics*; 2015; pp. 381–410.
17. S. M. Menke, N. A. Ran, G. C. Bazan, R. H. Friend, *Joule* 2017, *0*.
18. R. D. Jansen-van Vuuren, A. Armin, A. K. Pandey, P. L. Burn, P. Meredith, *Adv. Mater.* 2016, 28, 4766.
19. G. Pace, A. Grimoldi, M. Sampietro, D. Natali, M. Caironi, *Semicond. Sci. Technol.* 2015, *30*.
20. G. Pace, A. Grimoldi, D. Natali, M. Sampietro, J. E. J. E. Coughlin, G. C. G. C. Bazan, M. Caironi, *Adv. Mater.* 2014, *26*, 6773.
21. G. Azzellino, A. Grimoldi, M. Binda, M. Caironi, D. Natali, M. Sampietro, *Adv. Mater.* 2013, *25*, 6829.
22. G. Dell'Erba, A. Perinot, A. Grimoldi, D. Natali, M. Caironi, *Semicond. Sci. Technol.* 2015, *30*.
23. A. Grimoldi, L. Colella, L. La Monaca, G. Azzellino, M. Caironi, C. Bertarelli, D. Natali, M. Sampietro, *Org. Electron. Phys. Mater. Appl.* 2016, *36*.
24. A. Pierre, A. C. Arias, *Flex. Print. Electron.* 2016, *1*, 43001.
25. D. Braun, G. Yu, *MRS Proc.* 1999, *558*.
26. X. Xu, M. Davanco, X. Qi, S. R. Forrest, *Org. Electron. Phys. Mater. Appl.* 2008, *9*, 1122.
27. X. Xu, M. Mihnev, A. Taylor, S. R. Forrest, *Appl. Phys. Lett.* 2009, *94*.
28. K. Nishimura, Y. Sato, J. Hirase, R. Sakaida, M. Yanagida, T. Tamaki, M. Takase, H. Kanehara, M. Murakami, Y. Inoue, 2016, *2*, 110.
29. X. Tong, S. R. Forrest, *Org. Electron. Phys. Mater. Appl.* 2011, *12*, 1822.
30. C. K. Renshaw, X. Xu, S. R. Forrest, *Org. Electron. Phys. Mater. Appl.* 2010, *11*, 175.
31. K. Swathi, K. S. Narayan, *Appl. Phys. Lett.* 2016, *109*.
32. S. W. Jeong, J. W. Jeong, S. Chang, S. Y. Kang, K. I. Cho, B. K. Ju, *Appl. Phys. Lett.* 2010, *97*, 1.
33. T. Someya, Y. Kato, S. Iba, Y. Noguchi, T. Sekitani, H. Kawaguchi, T. Sakurai, *IEEE Trans. Electron Devices* 2005, *52*, 2502.
34. I. Nausieda, K. Ryu, I. Kymissis, A. I. Akinwande, V. Bulović, C. G. Sodini, *IEEE Trans. Electron Devices* 2008, *55*, 527.

35. I. Nausieda, K. Ryu, I. Kymissis, A. I. Akilnwande, V. Bulović, C. G. Sodini, *Digest of Technical Papers: IEEE International Solid-State Circuits Conference*; 2007; p. 72–73.
36. G. H. Gelinck, A. Kumar, D. Moet, J.-L. Van Der Steen, U. Shafique, P. E. Malinowski, K. Myny, B. P. Rand, M. Simon, W. Rütten, A. Douglas, J. Jorritsma, P. Heremans, R. Andriessen, *Org. Electron. Phys. Mater. Appl.* 2013, *14*, 2602.
37. P. E. Malinowski, P. Vicca, M. Willegems, S. Schols, D. Cheyns, S. Smout, M. Ameys, K. Myny, S. Vaidyanathan, E. Martin, A. Kumar, J.-L. Van Der Steen, G. Gelinck, P. Heremans, *IEEE Photonics Technol. Lett.* 2014, *26*, 2197.
38. R. A. Street, M. Mulato, R. Lau, J. Ho, J. Graham, Z. Popovic, J. Hor, *Appl. Phys. Lett.* 2001, *78*, 4193.
39. R. A. Street, J. Graham, Z. D. Popovic, A. Hor, S. Ready, J. Ho, *J. Non. Cryst. Solids* 2002, *299–302*, 1240.
40. T. Rauch, M. Böberl, S. F. Tedde, J. Fürst, M. V. Kovalenko, G. Hesser, U. Lemmer, W. Heiss, O. Hayden, *Nat. Photonics* 2009, *3*, 332.
41. T. N. Ng, W. S. Wong, M. L. Chabinyc, S. Sambandan, R. A. Street, *Appl. Phys. Lett.* 2008, *92*, 1.
42. H. Seo, S. Aihara, T. Watabe, H. Ohtake, T. Sakai, M. Kubota, N. Egami, T. Hiramatsu, T. Matsuda, M. Furuta, T. Hirao, *Jpn. J. Appl. Phys.* 2011, *50*.
43. D. Baierl, L. Pancheri, M. Schmidt, D. Stoppa, G.-F. Dalla Betta, G. Scarpa, P. Lugli, *Nat. Commun.* 2012, *3*, 1175.
44. S.-J. Lim, D.-S. Leem, K.-B. Park, K.-S. Kim, S. Sul, K. Na, G. H. Lee, C.-J. Heo, K.-H. Lee, X. Bulliard, R.-I. Satoh, et al., *Sci. Rep.* 2015, *5*, 7708.
45. T. Takahashi, Z. Yu, K. Chen, D. Kiriya, C. Wang, K. Takei, H. Shiraki, T. Chen, B. Ma, A. Javey, *Nano Lett.* 2013, *13*, 5425.
46. G. H. Gelinck, A. Kumar, D. Moet, J.-L. P. J. van der Steen, A. J. J. M. van Breemen, S. Shanmugam, A. Langen, J. Gilot, P. Groen, R. Andriessen, M. Simon, et al., *IEEE Trans. Electron Devices* 2016, *63*, 197.
47. J. C. Blakesley, R. Speller, *Med. Phys.* 2008, *35*, 225.
48. A. Iacchetti, M. Binda, D. Natali, M. Giussani, L. Beverina, C. Fiorini, R. Peloso, M. Sampietro, *Nucl. Sci. IEEE Trans.* 2012, *59*, 1862.
49. M. Binda, D. Natali, M. Sampietro, T. Agostinelli, L. Beverina, *Nucl. Instrum. Methods Phys. Res. A* 2010, *624*, 443.
50. Q. Chen, T. Hajagos, Q. Pei, In *Annual Reports on the Progress of Chemistry 2011, Vol. 107, Section C: Physical Chemistry*; Webb, GA, Ed.; Royal Society of Chemistry, London, 2011; pp. 298–318.
51. G. N. Ankah, P. Büchele, K. Poulsen, T. Rauch, S. F. Tedde, C. Gimmler, O. Schmidt, T. Kraus, *Org. Electron. Phys. Mater. Appl.* 2016, *33*, 201.
52. L. Basiricò, A. Ciavatti, T. Cramer, P. Cosseddu, A. Bonfiglio, B. Fraboni, *Nat. Commun.* 2016, *7*.
53. L. Zhang, T. Wu, Y. Guo, Y. Zhao, X. Sun, Y. Wen, G. Yu, Y. Liu, *Sci. Rep.* 2013, *3*, 1080.
54. H. Chen, N. Cheng, W. Ma, M. Li, S. Hu, L. Gu, S. Meng, X. Guo, *ACS Nano* 2016, *10*, 436.
55. X. Liu, E. K. Lee, D. Y. Kim, H. Yu, J. H. Oh, *ACS Appl. Mater. Interfaces* 2016, *8*, 7291.
56. A. Pierre, A. Gaikwad, A. C. Arias, *Nat. Photonics* 2017, *11*, 193.
57. S. Nau, C. Wolf, S. Sax, E. J. W. W. List-Kratochvil, *Adv. Mater.* 2015, *27*, 1048.
58. H. Wang, H. Liu, Q. Zhao, C. Cheng, W. Hu, Y. Liu, *Adv. Mater.* 2016, *28*, 624.

59. H. Wang, H. Liu, Q. Zhao, Z. Ni, Y. Zou, J. Yang, L. Wang, Y. Sun, Y. Guo, W. Hu, Y. Liu, *Adv. Mater.* 2017, *29*, 1.
60. J. Kim, J. Kim, S. Jo, J. Kang, J. W. Jo, M. Lee, J. Moon, L. Yang, M. G. Kim, Y. H. Kim, S. K. Park, *Adv. Mater.* 2016, *28*, 3078.
61. M. Cesarini, M. Binda, D. Natali, *Org. Electron. Phys. Mater. Appl.* 2017, *49*, 100.
62. C. P. Watson, D. M. Taylor, *Appl. Phys. Lett.* 2011, *99*.

7

Tactile Sensors for Electronic Skin

Fabrizio A. Viola and Pierro Cosseddu

CONTENTS

7.1 Somatosensory System 149
7.2 Electronic Skin 151
7.3 Capacitive Tactile Sensors 153
7.4 Resistive Tactile Sensors 155
  7.4.1 Piezoresistive Sensors 158
  7.4.2 Pyroresistive Tactile Sensors 160
7.5 Piezoelectric and Pyroelectric Tactile Sensors 162
7.6 Field-Effect Transistors as Tactile Sensors 165
7.7 Multifunctional Tactile Sensors 172
References 175

7.1 Somatosensory System

The human tactile system, unlike others sensory systems, is not located in a specific apparatus but is distributed throughout different parts of the human body, such as skin, muscles, skeleton, and so on. This system plays a fundamental role in modeling how humans interplay with their surroundings in terms of pain, temperature, and, finally, touch.

The skin (*cutaneous system*) is the largest organ in the human body and is a very important part of the tactile system; it performs the functions of protecting the body from bacteria, viruses, and potentially damage by external substances; thermoregulation (i.e., the ability to adapt our body to a great diversity of climates, including hot-humid, hot-arid, and warm), it helps to maintain our structural integrity and it achieves the function of sensation. These functions are enabled by a complex network of sensors (i.e., *somatosensory neurons* or *tactile receptors*) receiving specific inputs from the surrounding environment about object properties such as size, roughness, softness, temperature, and so on. The skin consists of different layers: the *subcutaneous tissue* (also called *hypodermis* or *subcutis*), the *dermis*, and the *epidermis.*

- *Subcutaneous tissue*: This connects the dermis with the other underlying tissue. It is used chiefly for fat storage, operating as filling and as a reserve of energy.
- *Dermis*: This consists of connective tissue and is connected to the epidermis through a basement membrane. Most of the tactile receptors are located in this layer of skin.
- *Epidermis*: This is the external layer of skin, so its fundamental function is to provide a barrier to the external environment. It is formed of five different layers (cornified layer, translucent layer, granular layer, spinous layer, and basal layer).

It is possible to distinguish two categories of tactile receptors, based on the *adaptation* of the electric response of a receptor neuron. According to [1], adaptation is "the decline of the electric responses of a receptor neuron over time in spite of the continued presence of an appropriated stimulus of constant strength."

- *Slow adaptation* (SA) or *tonic receptors* respond with a train of nerve impulses for all the duration of the external stimulus.
- *Rapid adaptation* (RA) or *phasic receptors* are sensitive only to dynamic stimuli.

In general, the most common division is based on the type of modality that they can detect: *mechanoreceptors* for responses to mechanical stimuli, *thermoreceptors* for responses to thermal stimuli, and *nociceptors* for responses to pain.

A mechanoreceptor is a receptor that can sense mechanical pressure or shear forces. These kinds of receptors can be divided into different categories based on their structure [2].

Meissner's corpuscles (RA I units): These have a small receptive field (3–4 mm). The receptive field of a receptor is the area (in general, a circular or ellisoidal area) in which a puntiform stimulus can excite it and has high sensitivity to dynamic forces in the range 10–50 Hz. They are located in different areas of the hairless skin, especially in lips and fingers, with a spatial density of 140 units/cm^2. They have a cylindrical structure covered by different layers of connective tissue and are involved in the detection of low-frequency vibrations, flutter, slip, and motion detection.

Pacinian corpuscles (RA II units): These units are mostly responsible for rapid adaptation in human skin. They are in fact sensitive to high-frequency forces (40–400 Hz) and they have a large receptive field (> 20 mm). Pacinian corpuscles are located throughout the body in various areas in the subcutaneous layer of the skin, and they are involved in the detection of surface roughness and small vibrations.

Merkel cells (SA I units): These are densely located in hairless skin (70 units/cm^2). They have small receptive fields and a high sensitivity to low-frequency dynamic skin deformation (<5 Hz) and to static forces. Merkel cells are involved in fine form detection and texture discrimination.

Ruffini endings (SA II units): Incorrectly classified in the past as thermoreceptors, these mechanoreceptors are activated by static stimuli applied perpendicularly and respond well when skin is stretched. Ruffini endings are located in the deep layers of the skin (subcutaneous tissue) with a spatial density of 10 units/cm^2. They can sense the stretching of the skin, especially around joints and fingernails, and they contribute to finger position, motion detection, and the perception of the shape of an object.

On the other hand, thermoreception is also a very complex sensory system. Thermoreceptors have, in general, a free nerve ending morphology (like Merkel cells) and they are the primary thermosensory units. They can be divided into two groups:

- *Cold receptors*: These thermoreceptive nerve endings have a diameter of 1.5–3.0 μm and innervate the layer between the dermis and the epidermis. They have a maximum sensitivity in the range of 25°C–30°C
- *Warm receptors*: These are more superficial and smaller than cold receptors (diameter of 1–2 μm), so in general they innervate the epidermis. Warm receptors transduce information through nerve pulses more slowly than cold ones. They have a maximum sensitivity in the range of 40°C–42°C.

At constant skin temperature in the normal range (35°C–37°C), all the cutaneous thermoreceptors generate a series of nerve impulses with a constant frequency. When the skin temperature falls, warm receptors exhibit inactivity, but in cold receptors the frequency of the nerve pulses increases. On the contrary, when the skin temperature rises, warm receptors increase the frequency of the pulses and cold receptors are inoperative [3]. Unlike mechanoreceptors, thermoreceptors have a small superficial density on fingertips (2–3 units/cm^2 for cold receptors and 2 units/cm^2 for warm receptors), and their density rises on lips (~20 units/cm^2).

7.2 Electronic Skin

As already mentioned, skin tactile receptors are excited by different external stimuli or our brain; in general, our nervous system is able to encode the information acquired by tactile receptors and give different sensations such as pressure, heat, cold, vibrations, and so on. The attempt to create an *artificial skin* or *electronic skin* (*e-skin*), a skin-like tissue that can simulate the receptive behavior of human skin, is motivated by the possibility of fabricating a multi-sensitive interface for different kinds of applications, such as robotics, prosthetics, and so on.

Referring to the characteristics of human skin, it is possible to summarize the requisites of artificial skin in terms of their mechanical requirements and thermal requirements.

Artificial mechanoreceptor characteristics:

- Spatial resolution, fingertip: > 140 units/cm2
- Force resolution: ~1 mN
- Force range: 0.001–10 N
- Bandwidth: 0–1000 Hz
- Response time: ~1 ms

Artificial thermoreceptor characteristics:

- Spatial resolution, lips: > 20 units/cm^2, other parts 5 units/cm^2
- Thermal resolution: ~0.02°C
- Temperature range: 0°C–55°C
- Temperature variation: 1°C/min

Besides the sensing properties described in previous paragraphs, artificial skin should have specific mechanical properties in order to mimic the mechanical behavior of the human skin layers. Some interesting works [4–5] have demonstrated that the *Young's modulus* (*E*) of the skin varies between 0.42 and 0.85 MPa for torsion tests and 4.6 and 20 MPa for tensile tests, and it is well known that the skin is stretchable up to 70% strain [6] and, consequently, can be adapted to the movements of human's body by bending and stretching.

For these reasons, in recent years the employment of flexible electronics, mainly based on the employment of organic semiconductor-based devices fabricated over plastic substrates, has gained huge attention and has been considered a viable approach to the fabrication of electronic skin systems over large areas.

Flexible and stretchable electronics [7] allows the fabrication of devices able to be bent and to be accommodated over irregular surfaces such as elbows and knees. Among the many approaches reported in the literature, Roger's group has pioneered this research field by exploiting materials with suitable mechanical properties that are able to mimic the mechanical properties of the skin, proposing an approach that led to the development of a variety of conformable *epidermal electronic systems* [8]. In addition, *organic electronics* may represent a valid candidate for the realization of tactile devices, thanks to its great potential for large-area fabrication techniques and the possibility of using flexible, stretchable, and ultra-thin materials. In this chapter we will introduce an overview of the most common approaches in recent years to the realization of an artificial electronic skin system by employing organic electronics.

7.3 Capacitive Tactile Sensors

A capacitive sensor takes advantage of the structure and working principles of capacitors. As shown in Figure 7.1, this type of sensor consists of two metallic, or conductive, plates with a dielectric material in between them. Capacitance can be expressed by Equation 7.1 in the common case of parallel-plate capacitors.

$$C = \varepsilon_0 \varepsilon_r \frac{A}{d} \tag{7.1}$$

where:

ε_0 and ε_r are the permittivity of the free space and the relative permittivity of the dielectric, respectively
d is the distance between the two conductive plates
A is the area of the plates

In the case of pressure sensors, a variation in d, due to an applied normal pressure on the structure, can change the value of the capacitance C if the reduction of the distance between the plates is large enough. Furthermore, changes in A are typically used to sense *shear forces.*

Capacitive sensors are generally used in different applications such as prosthetics and robotics thanks to their simple architecture and the ability to achieve high spatial resolutions [9–11]. For these kinds of sensors, the choice of the dielectric is fundamental to optimize their sensitivity. In particular,

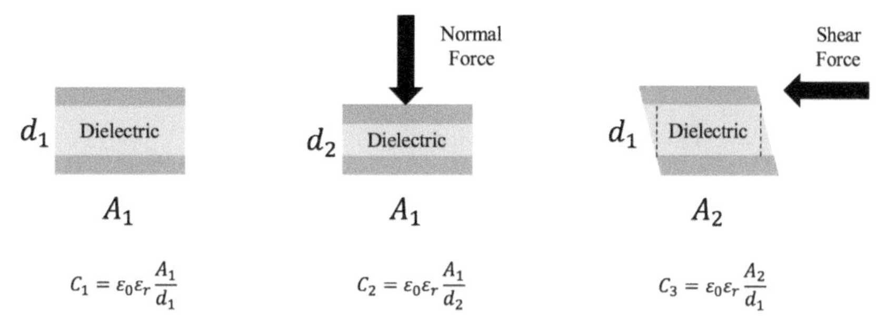

FIGURE 7.1
Capacitive device for transducing sensory stimuli in skin.

elastomers have high elasticity and deformability, with an elastic modulus typically two orders of magnitude lower than that of silicon, and high-yield strain [12]. For this reason, these polymers are widely used as an alternative to dielectric inorganic material.

An interesting approach has been proposed by Vandeparre et al. [13]. In this work, a conformable capacitive pressure sensor is fabricated by using a particular elastomeric foam dielectric and stretchable electrodes. The dielectric used is a flexible, microcellular polyurethane (PU) foam with ultrathin gold electrodes evaporated on it. The particular foam structure of the dielectric allows the sensors to be stretched, thus maintaining the electrical conduction properties of the metal film (Figure 7.2). The fabrication procedure is very simple: a two-component mixture of PU foam (Smooth-On FlexFoam-iT!®) is bar-coated on a glass carrier. After 2 h of cross-linking at room temperature, the top and bottom surfaces of the foam are coated by a patterned metal structure in order to obtain an array of 25 nm-thick gold electrodes, with an area of 1 cm^2 for each sensor.

Figure 7.2 shows the structure and the response of the proposed structure. The device is able to detect pressure variations from 10 to 300 kPa with good reproducibility and is able to work even after a wide strain (~75%). The main drawback of the proposed approach is that patterning through a shadow mask does not achieve the typical high resolutions required for mechanical sensors for tactile applications.

A different solution has been presented by Hu et al. [14], who developed a flexible capacitive sensor for the detection of strain and pressure. The most important advantage of this approach is the use of highly compliant materials for the fabrication of electrodes, made of a composition of ultra-thin silver nanowires (AgNws) over an elastomeric polyurethane substrate. In this case, the PU is not used as dielectric but as a carrier of the electrodes and the

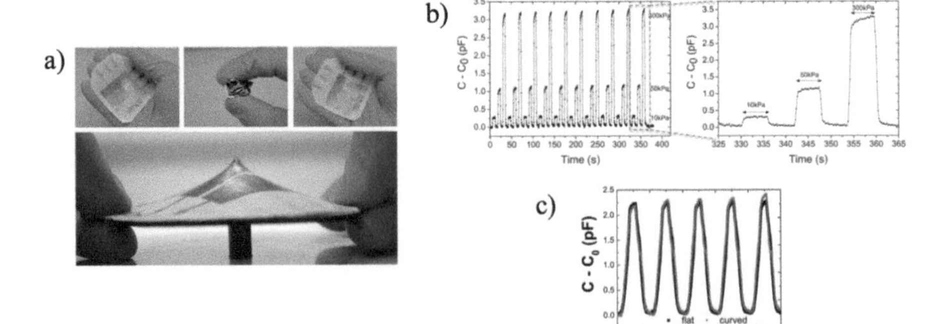

FIGURE 7.2
(a) Pictures of a matrix with 3 × 3 sensors; the structure can sustain crumpling as sharp indentation. (b) Change in capacitance for three different pressures. (c) Change in capacitance before (black line) and during (gray line) a 75% compressive strain [13].

sensitive material. One more very interesting feature of this device is that the electrodes are also completely transparent (Figure 7.3).

Silver nanowires were deposited by spray coating through a contact mask over a glass substrate and then annealed for 30 min at 190°C. PU liquid compound (Clear Flex®) was dropcasted over the AgNW on glass substrate. The resulting AgNW–PU composite sheet was peeled off and the sensors were fabricated by an elastomeric dielectric between two AgNW–PU composite electrodes with the conducting surface inward. As shown in Figure 7.3, if there is mechanical stimulation of the central sensor in a matrix of 10×10 pixels, the change of the capacitance of the addressed pixel is several times bigger than the neighboring sensors. Authors have demonstrated that this kind of approach can enable the detection of pressure variations in the range of 1–100 kPa, the typical pressure range of soft touches, hand grips, and finger presses with a spatial resolution of 1.5 mm^2.

The original and interesting work presented by Mannsfeld et al. [15] in 2010 demonstrated that it is possible to increase the sensitivity and the response speed of an elastomer-based capacitive sensor acting on its structure. In this work, three different types of polydimethylsiloxane (PDMS)-based capacitive structures were tested: pyramidal, linear, and planar (unstructured). The experimental results are given in Figure 7.4. The structured pyramidal and linear PDMS exhibit a higher sensitivity than the unstructured PDMS with the same thickness, especially for pressure values >0.5 kPa. The pressure sensitivity is the slope of the curves and can be written as

$$S = \delta\left(\frac{\Delta C}{C_0}\right)\Big/\delta p \tag{7.2}$$

where:

ΔC is the variation of capacitance under an applied pressure
C_0 is the capacitance in the resting state
δp is the applied pressure

All the types of structures show negligible hysteresis and can be cycled more than 1000 times. Authors have also demonstrated that these kinds of capacitors can detect loads as small as 3 Pa, so they can be used for sensor areas that require high sensitivity and resolution for very low pressure. On the contrary, the sensitivity in the medium–high pressure range is very low.

7.4 Resistive Tactile Sensors

The working principle of resistive sensors is based on the modulation of the resistance of a particular material due to pressure stimulation (*piezoresistive* sensors) or due to temperature stimulation (*thermistors* and *resistance*

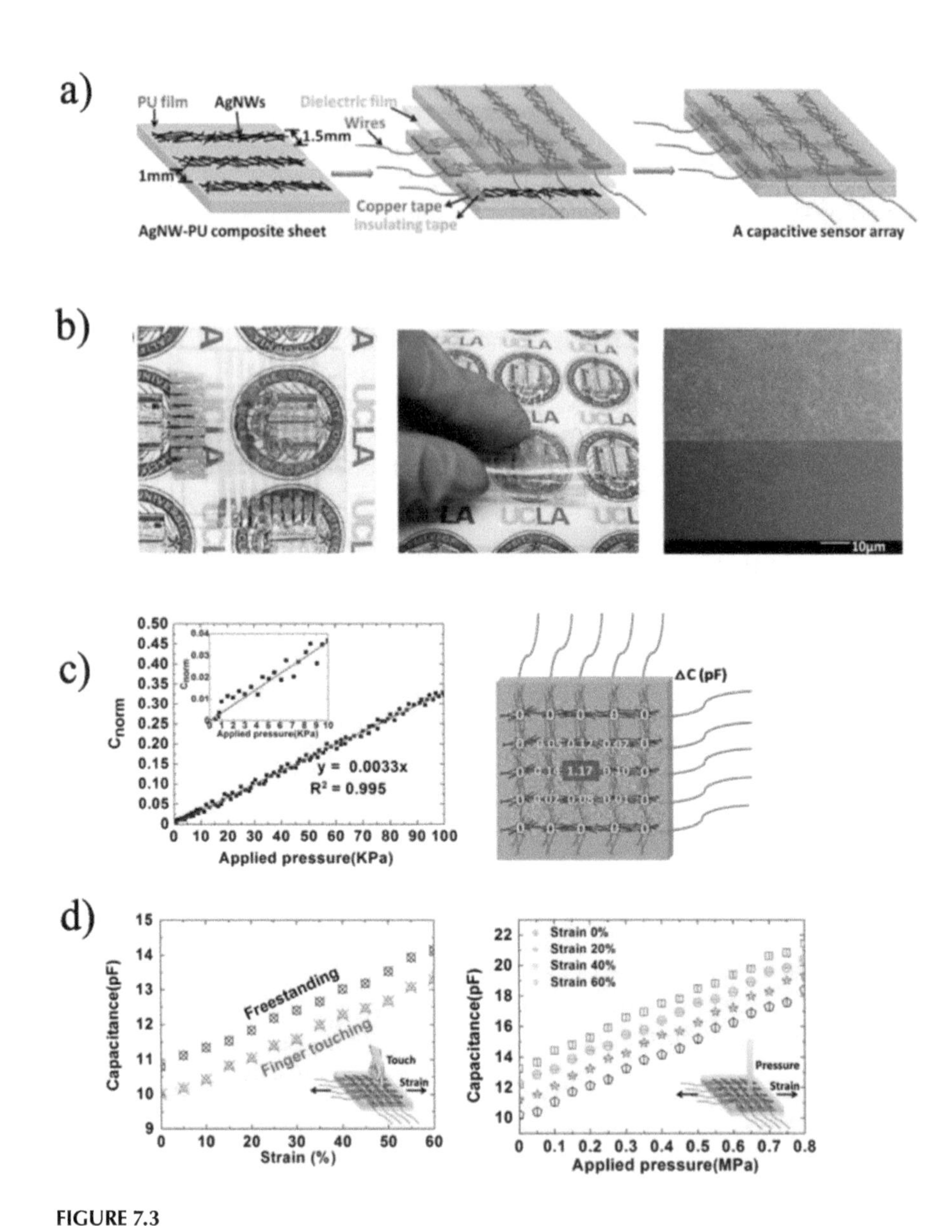

FIGURE 7.3
(a) Schematic illustration of the fabrication procedure of the capacitive array. (b) Image of a matrix with 10 × 10 transparent sensors and SEM images at the interface between AgNW and PU. (c) Change of capacitance due to an applied pressure and mapping of the measured capacitance changes where a pressure was applied on the central sensor. (d) Change in capacitance combined with figure touch versus uniaxial stretching strain, and change in capacitance versus independently applied transverse pressure and uniaxial elongation [14].

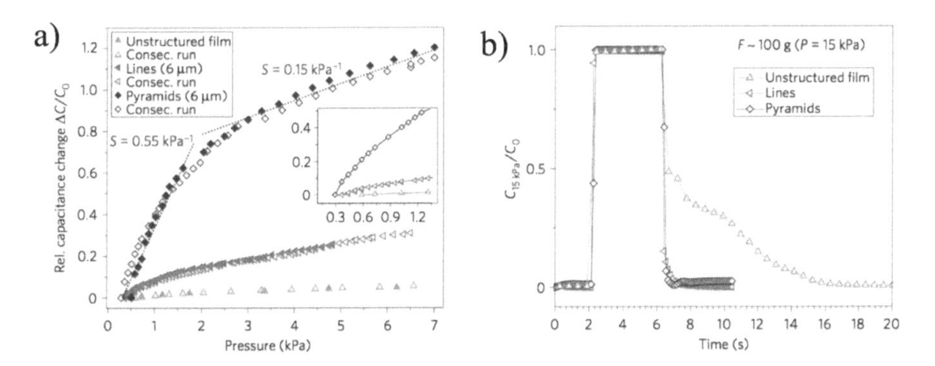

FIGURE 7.4
(a) Change in capacitance for different types of microstructurated PDMS films and their relative sensitivity. (b) Relaxation and resting state curves for pyramidal, linear, and unstructured films [15].

temperature detectors [RTDs]). Piezoresistive sensors are typically a pressure-sensitive element such as conductive polymers or elastomers. Resistive temperature sensors usually exploit the temperature dependence of classic metals or conductive polymers. According to Equation 7.3, resistance modulation can be easily detected by an electrical measuring system thanks to the voltage current characteristic of a simple resistive element.

$$V = RI \tag{7.3}$$

where:

R is resistance
V is the applied voltage across the resistance
I is the measured current

The change in resistance can be affected by different factors, such as

- Variations in contact resistance (R_C) between two conductive elements
- Variations in the geometry (length or width) of the sensing element (especially for strain sensors)
- Variations in the resistivity of a material due to a change in its chemical bonds
- Variations in the resistivity (ρ) of a conductive element due to temperature stress

Resistive devices have the advantage of requiring less electronic readout complexity and are less susceptible to noise than capacitive sensors. However, the power consumption is much higher and they have a lower frequency response [16].

7.4.1 Piezoresistive Sensors

Variations in contact resistance (R_C) depend on changes in the contact area between two conductive elements due to an externally applied pressure. This phenomenon is governed by Equation 7.4 [17]:

$$R_C \propto F^{-1/2} \tag{7.4}$$

In the literature, it has been demonstrated that sensors based on contact resistance are advantageous thanks to their high sensitivity, especially at low pressure, as well as to their large working range, low temperature sensitivity, and fast response [18]. Dabling et al. [19] reported an interesting study of the behavior of these kind of sensors; the authors demonstrated that their performances can suffer due to hysteresis, drift, and loss of sensitivity.

Gong et al. [20] proposed an wearable pressure sensor based on gold nanowires (AuNWs). The device can achieve high sensitivity (> 1.14 kPa^{-1}), a fast response time, and good stability. The sensing mechanism is related to the contact resistance of the AuNWs, which are sandwiched between two interdigitated arrays. In particular, when pressure is applied to the device, the number of AuNWs bridging the top and bottom electrodes increases and the contact resistance consequently decreases (Figure 7.5).

In the last decade, the combinations of different materials (e.g., composites based on conductive fillers dispersing into a non-conducting polymer matrix) are attracting more attention for many applications because of their unique electrical and mechanical properties. For example, Abyaneh et al. [21] have reported the piezoresistive behavior of a Zn–PDMS composite. The authors demonstrated that with this kind of composite it is possible to obtain high changes in resistance under uniaxial pressure due to variations of the tunneling effect between the Zn particles. Interesting results are shown in [22], where Dang et al. presented a study of the piezoresistive behavior of multiwall carbon nanotubes (MWCNTs) dispersed into silicone rubber. One of the main aspects of this survey is that the sensitivity of the composite can be tuned by changing the concentration of MWCNTs.

Very recently, Someya's group reported a transparent bending-insensitive pressure sensor based on composite nanofibers of CNTs and graphene

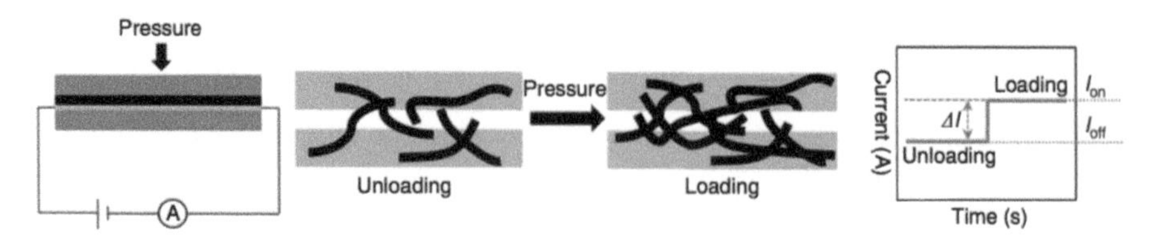

FIGURE 7.5
Schematic illustration of the AuNWs sensor based and current changes in responses to external pressure [20].

dispersed in fluorine rubber [23]. The fabrication process is based on the electrospinning process of a mix of fluorinate copolymer, CNTs, and graphene in order to reduce the rigidity and total thickness of the pressure sensor. The solution is deposited over different flexible plastic substrates. The bending-insensitive behavior is related to the nanofibrous structure that can change its alignment and perfectly follow the deformation, thus minimizing the strain in the single fibers. The authors deeply investigated the individual role of CNTs and graphene in the mixture and demonstrated that a simple mixture of CNTs and a fluorinated copolymer can be used as a pressure sensor, albeit showing a very low sensitivity. On the contrary, a mix of the copolymer and graphene cannot be used as a conductor or a pressure sensor. Only the composite of both CNTs and graphene can achieve a high enough sensitivity to be used as a pressure sensor. The working principle is based on the modulation of the conductivity of the sample due to the formation of conductive paths into the sensor's framework. The device, thanks to its low thickness (~1.4 μm) can be accommodated onto the surface of human skin and can detect and monitor pressure distribution, especially in the low range of 0.6–1.5 kPa (Figure 7.6).

Strain sensors have been used in many areas such as robotics, biomedical applications, bioengineering, wearable devices, and so on. The transduction mechanism is based on the variation of resistance of a material due to a modification in its structure caused by strain stress. In general, the sensitivity of a strain sensor is referred to as the *gauge factor* (GF), defined as

$$\mathrm{GF} = \frac{\Delta R}{R_0 \varepsilon} \tag{7.5}$$

where:

ΔR is the variation of the initial resistance R_0

ε is the strain

Recently, it has been demonstrated that, by using strain sensors based on composite materials, it is possible to increase the value of their GF. For instance, Xiao et al. [24] proposed a high-GF strain sensor based on ZnO nanowire/polystyrene on a PDMS film. This particular incorporation of ZnO as conductive material has led to a GF value of 116, which is comparable to those of traditional inorganic strain sensors. However, the high initial resistance ($R_0 \approx 10$ GΩ) necessitates the use of an high-precision instrument, thus greatly limiting the employment of this approach in practical applications. Recently, Gong et al. [25] presented a high-strain sensor composed of polydimethylsiloxane (PDMS) and polyaniline (PANI) for various applications such as tactile strain sensors. The authors explain that the variation of resistance during strain is affected not only by the deformation of the PANI film (the variation of geometry parameters l and w) but also by the formation and modification of micro-cracks in the sensor structure. The obtained GF

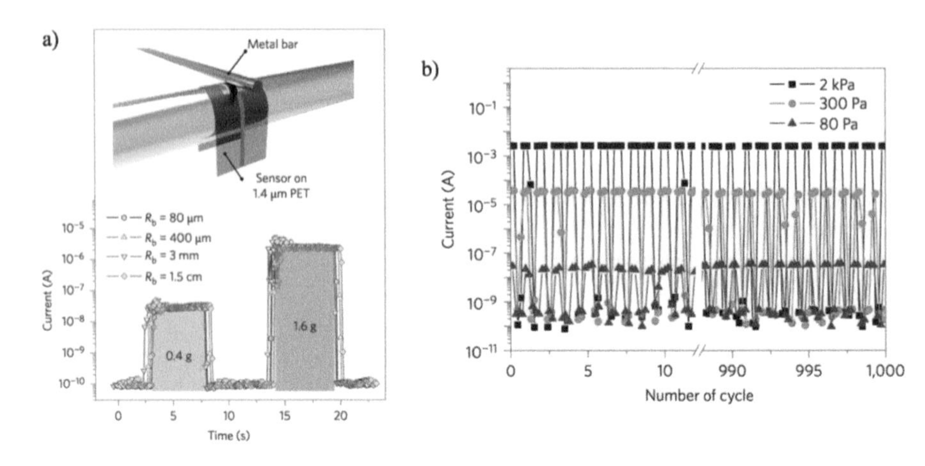

FIGURE 7.6
(a) Piezoresistive response of the device for different normal forces (0.4 g and 1.6 g). (b) On/off cyclic test of the sensor at different pressures [23].

(54 at 50% strain) is higher than that of other reported flexible sensors, and the proposed approach does not require a complex measurement system due to the high value of the measured current (≅ mA) at 1 V.

Piezoresistivity is also related to a change in chemical bonds in different materials [26]. Bae et al. [27] presented a piezoresistive strain sensor based on graphene fabricated on very thin, flexible plastic. The graphene was grown by chemical vapor deposition (CVD) and patterned by standard photolithography and reactive ion etching (RIE).

The device was able to sense tensile strain up to 7% and is completely transparent thanks to the optical properties of graphene. However, experimental results have shown a strong nonlinear behavior for strains higher than 1.8%. The authors explain this phenomenon by referring to the structure of graphene: the existence of two different piezoresistive regions is due to the formation of micro-cracks and defects for strains > 1.8%. On the contrary, in the linear region, the working mechanism is related to a modification of the carbon–carbon bond length during stretching.

7.4.2 Pyroresistive Tactile Sensors

In state-of-the-art temperature sensors, resistors whose electrical resistance greatly depends on temperature are of great interest. These sensors are usually divided into two categories: thermistors and RTDs. RTDs have a larger range of detectable temperatures, they are made of pure material, in general a metal, and they generally show a linear output characteristic. The work by Mattana et al. [28] represented an innovative step in the fabrication of RTD flexible sensors. In this work, the authors presented two temperature sensors fabricated with different techniques: gold resistors by standard photolithography and inkjet printing. In both cases, the sensors were able to detect temperature variations in the range of 10°C–80°C. Roger's group have developed a conformal resistive device for the thermal characterization of human skin [29]. The ultra-thin (~1.2 μm) and compliant structure of the presented sensor minimize the strain sensitivity of the device and maximize adhesion to the skin.

Thermistors have nonlinear characteristics and are usually made by polymers or ceramics with a particular piezoresistive behavior: negative (NTC) or positive (PTC) temperature coefficients. The relationship between the measured resistance and the temperature change can be expressed as

$$R(T) = R_0\left(1 + AT + BT^2 + CT^3\right) \tag{7.6}$$

where T represents the temperature in °C and R_0 is the value of the resistance at $T = 0$°C. For small temperature variation ΔT around T, this relation can be linearized as follows:

$$R(T + \Delta T) = R(T)(1 + \alpha_T \Delta T) \tag{7.7}$$

Where $\alpha_T = \frac{1}{R(T)}\frac{\Delta R}{\Delta T}$ is the sensitivity (temperature coefficient of resistance [TCR]) of the sensor in units of °C.

In addition to the good reliability and reproducibility of temperature sensors, a good linearity of the temperature-dependent resistance and high sensitivity is generally required.

Dankoco et al. [30] reported a thermistor developed to achieve surface temperature measurement of the human body. The device must be supplied under a bias voltage of 1 V and must present a high nominal resistance. The device is composed of silver deposited on a polyimide substrate (Kapton HN). The temperature sensor has a sensitivity of 2.19×10^{-3} °C^{-1} obtained in the typical range of tactile applications (20°C–60°C) and a nominal resistance of 2.032 kΩ at 38.5°C

Very recently, another example of skin-mountable temperature sensors was presented by Vuorinen et al. [31]. The devices, shown in Figure 7.7, were fabricated with inkjet-printed graphene/PEDOT: PSS ink on top of a skin-formable bandage-like substrate, which also provides good adhesion to skin. Using this approach, the authors managed to obtain a device that can monitor temperature changes directly on human skin with a TCR value higher than 0.06%/°C under optimal conditions (35°C–45 °C). Even if this device does not yet compete in terms of sensitivity with already existing temperature sensors, it could be used in its present form as a simple fever indicator on human skin. This is due to the simple fabrication process, which enables the low-cost fabrication of epidermal electronics with the added value of disposability. The scalable and simple manufacturing process combined with the stretchable, functional materials makes it possible to manufacture epidermal

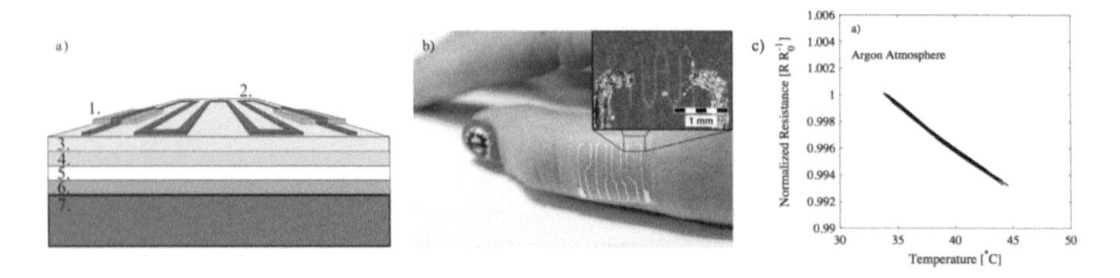

FIGURE 7.7
(a) Multilayer device structure: (1) silver conductors, (2) graphene/PEDOT:PSS temperature sensors, (3) PU substrate, (4) adhesive layer, (5) protective paper, (6) PET film, (7) cooling/heating element. (b) Photograph of the device on human skin. (c) Change in resistance in argon atmosphere [31].

temperature sensors, which are comfortable to use and have an excellent skin–device interface.

Another interesting approach has been reported by Bao's group [32]. In this work they presented a Ni microparticle-filled binary polymer composite used as temperature sensor. This material showed a much higher sensitivity than other types of flexible thermoresistors based on amorphous polymers. The resistivity of this composite material is affected by the temperature, but the strong PTC effect is limited to a range of 35°C–42°C, which makes the device particularly suitable for human body temperature.

7.5 Piezoelectric and Pyroelectric Tactile Sensors

The words *piezoelectricity* and *pyroelectricity* find their etymological origin in the ancient Greek words πιέζω (piezo) and πυροζ (pyros), which mean "to press" and "fire," respectively, and ελεκτρον (electron), which was a word associated with those materials known for their electrostatic properties. Piezoelectricity and pyroelectricity are the abilities to generate a voltage in response to an applied force or a temperature variation.

The piezoelectric effect can be described by a linear relationship between the *electric displacement* (*D*), the *piezoelectric coefficient* (*d*), and the applied mechanical stress (σ).

$$D_i = d_{\text{im}} \sigma_m \tag{7.8}$$

On the other hand, the pyroelectric effect is related to the external thermal stress (ΔT).

$$D_i = p_i \Delta T \tag{7.9}$$

where p is the *pyroelectric coefficient*, defined by Equation 7.10:

$$p_i = \frac{\partial D_i}{\partial T} \tag{7.10}$$

Even though the first reference to the pyroelectric effect can be found in the writings of Theophrastus in 314 BC, research did not become more precise until the nineteenth century, with Sir Brewster and the brothers Curie. But only after the studies of Voigt on the physical crystallography of materials, published in 1910, the relationship between the piezoelectric and pyroelectric effects in materials without a center of symmetry in their structure become clear.

Solids are characterized by three-dimensional arrangements of atoms generally locked into their positions. A *crystalline solid* is a material that has a repetitive structure in the three dimensions (x, y, z), has a long-range order, and has different physical properties in different directions (*anisotropy*). On the other hand, an *amorphous solid* does not present a long-range order in its structure and has the same physical properties in all directions (*isotropy*). Crystalline solids can be divided into 32 crystal classes or six crystal systems (*triclinic system*, *monoclinic system*, *orthorhombic system*, *tetragonal system*, *hexagonal system*, and *isometric system*). Only 21 crystallographic classes are non-centrosymmetric and show the piezoelectric effect, except for the isometric class 432 in which the piezoelectric charges along the <111> axes cancel each other out. Among the piezoelectric crystal classes, 10 also show the *pyroelectric* effect; these crystals contain a non-zero polarization when a temperature variation induces changes in the total dipole moment. Over a certain temperature T_C (called the *Curie temperature*), the pyroelectric effect vanishes and the crystals transform into the *paraelectric* state. As shown in Figure 7.8, only a few pyroelectric classes have *ferroelectric* properties—that is, the characteristic of some pyroelectric materials to have a *spontaneous* polarization that

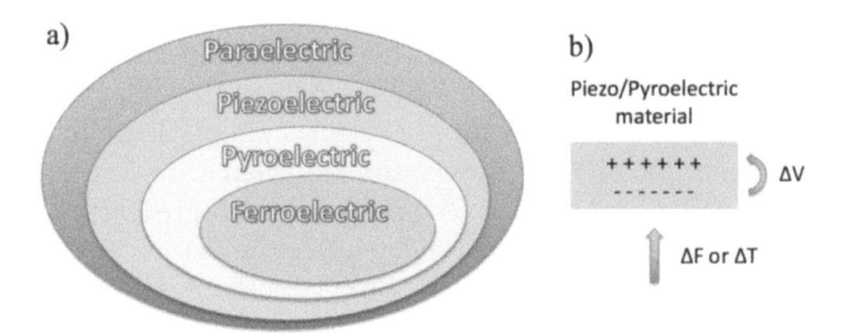

FIGURE 7.8
(a) Schematic representation of different classes of solid materials. (b) Piezo-/pyroelectric response under a specific external stress.

can be reversed by an external electric field. Piezoelectric and pyroelectric materials can be classified into three categories:

- *Crystals*: triglycine sulfate (TGS), lithium tantalate ($LiTaO_3$), etc.
- *Ceramics*: lead-zirconate titanate (PZT), etc.
- *Polymers*: polyvinyl fluoride (PVDF) and its copolymers (PVDF-TrFE), etc.

Compared with resistive devices, the advantages of using a piezo-/pyroelectric sensors are that these materials exhibit a good high-frequency response and high sensitivity [16]. However, these sensors can not work over their Curie temperature T_C and have unreliable static sensing properties.

Khan et al. [33] presented a pressure sensor for artificial skin applications based on a piezoelectric polymer, polyvinylidene fluoride-trifluoroethylene (PVDF-TrFE), over a thin polyamide (PI) substrate. The device consisted of 4×4 sensitive elements made by sandwiching the PVDF-TrFE between two conductive silver plates. The fabrication process started from the deposition of the bottom silver electrode by screen printing. Then, the sensitive elements were deposited over the bottom electrodes by screen printing or spin coating and sintered for 4 h at 130°C. After that, the silver-topped electrodes were patterned on the piezoelectric polymer. Piezo-/pyroelectric polymers and PVDF and PVDF-TrFE need polarization (or poling) through a high voltage across the polymer in order to introduce and increase their piezo-/pyroelectric behavior. The amplitude of the voltage depends on the thickness of the polymer to be poled; in the literature, it is reported that 100 V/µm at room temperature or 70 V/µm at 80°C are required [34]. The device was tested at different frequencies and different force values (0.5–4.0 N). The output voltages had the same quasi-linear behavior for the screen-printed and spin-coated sensors. Furthermore, the screen-printed sensors showed a higher sensitivity, and the authors explained that it could depend on the different thicknesses and uniformity obtained for spin-coated sensors.

In a different work, Yang et al. [35] presented the first application of a ceramic PZT pyroelectric nanogenerator (PNG) for detecting the temperature of a fingertip. The innovation of this approach is based on the use of an energy harvesting self-powered nanotechnology that provides the power instead of an external battery or supply system. The PZT micro-/nanowires were obtained from the bulk PZT mechanically, then a single micro-/nanowire was fixed on a glass substrate and poled under a high-voltage power supply. A PDMS thin layer was used to encapsulate the device. Figure 7.9 shows the variation of the temperature on the PNG and the measured output voltage. It is clear that the device is able to respond to dynamic changes in T with good reproducibility. The sensor was also tested with different temperatures and, as shown in Figure 7.9b, the response is linear in the range 300°K–325°K, which is the typical range for tactile applications. Power consumption is a fundamental issue for integrated e-skin systems; for this reason, the

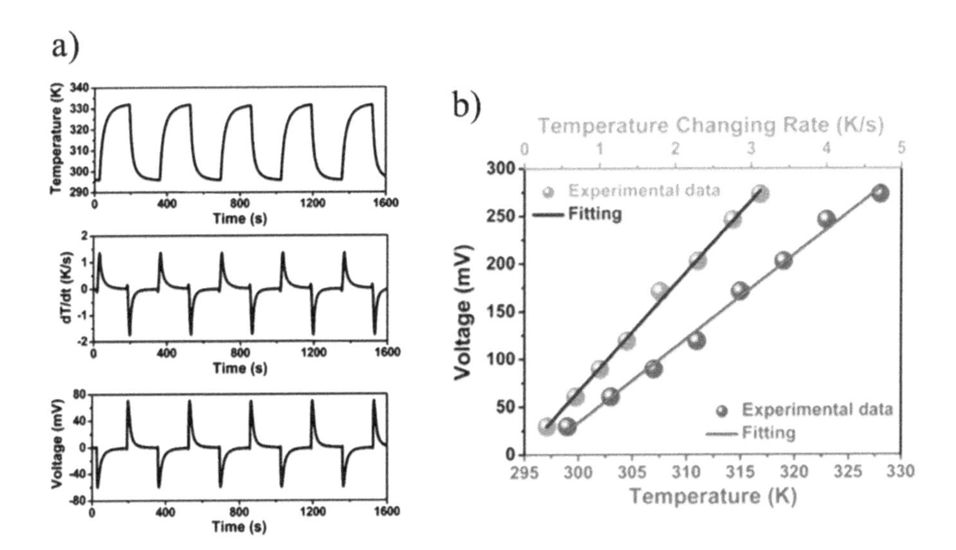

FIGURE 7.9
(a) Cyclic changes in temperature and the corrisponding differential curve, with the measured output voltage of the sensor. (b) Calibration curve of the PZT nanogenerator [35].

realization of self-powered systems is an important goal, especially for these kinds of tactile sensors. However, the response time of the device is very high (~0.9 sec) and the minimum detection limit is 0.4°K, which is higher than that requested for artificial skin applications.

7.6 Field-Effect Transistors as Tactile Sensors

A very large number of examples of tactile sensors reported in literature are based on field-effect transistors (FETs). The success of FET structures with respect to other electronic sensors is mainly related to the following aspects:

- FETs are multi-parameter devices, which allows the implementation of a wide range of transduction mechanisms.
- FET structures have a very low power consumption and, thanks to their field effect, have a good amplification of the response.
- With this kind of structure there is the possibility of implementing an array and matrix of sensors, coupled with the switching and addressing capabilities of these devices.

Referring to the output current of the transistor, I_{DS}:

$$I_{DS} = f\left(\mu, C_{ins}, W, L, V\right) \tag{7.11}$$

Note that several parameters play an important role in the final value of the current: the charge carrier mobility (μ) of the semiconductor, the insulating capacitance (C_{ins}) of the dielectric, the geometrical parameters width (W) and length (L) of the channel, and the operating voltages (V). All these parameters can in theory be employed for the transduction of external stimuli.

FETs as tactile sensors can be divided into two major groups. The first one includes approaches where the sensing element is the FET itself. The second group includes all the approaches where a FET structure is coupled with an external sensing element. In this case, the transistor acts as a transducing element.

One of the first works based on FETs mechanical sensors was proposed by Sekitani et al. [36]. They demonstrated that by applying a compressive strain in the active *pentacene* layer of an organic FET, an increase in the I_{DS} was observed. On the other hand, a decrease in the output current was induced during tensile strain. The authors assumed that this phenomenon could be attributed to a reversible change of the structural properties of the polycrystalline semiconductor film. Consider the equation of the hopping transport:

$$\mu \sim \mu_0 \exp\left(-\Delta E/k_B T\right) \tag{7.12}$$

where:

- μ is the semiconductor mobility during a strain
- μ_0 is the mobility without any tensile/compressive strain
- k_B is the Boltzmann's constant
- T is the temperature
- ΔE is the energy barrier for the hopping transport in the pentacene. It is evident that the strain has an effect on the semiconductor mobility, referring to the value of $\Delta E \propto -k_B T \ln\left(1+0.05\varepsilon\right)$, where ε is the strain expressed in percent. This behavior is confirmed by the value of the mobility during different strains, which is reported in Figure 7.10.

The correlation between the induced strain ε and bending radius R in a FET structure is described in [37] by the following formula:

$$\varepsilon = \frac{\left(d_1 + d_s\right)}{\left(2 * R\right)} \frac{\left(1+2\eta+\chi\eta\right)^2}{\left(1+\eta\right)\left(1+\chi\eta\right)} \tag{7.13}$$

where:

- d_1 and d_s are the thicknesses of the active layer and the substrate, respectively
- η is defined as d_1 / d_s
- R is the bending radius
- χ is the ratio between the Young's moduli of the active layer and the substrate (i.e., $\chi = Y_1 / Y_s$)

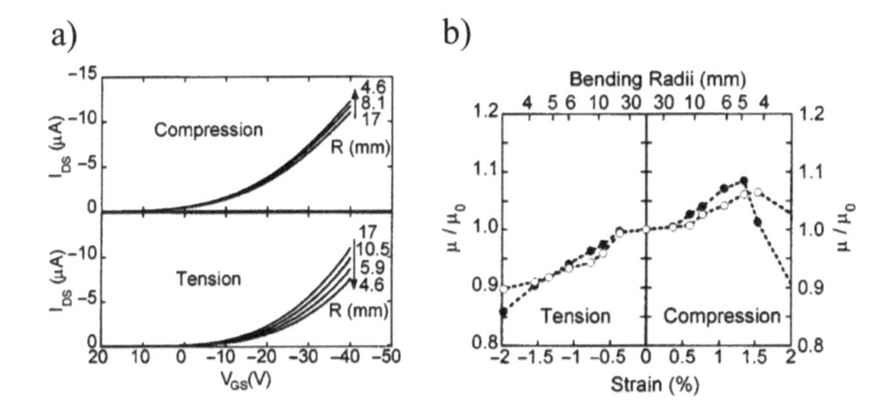

FIGURE 7.10
(a) Transfer curves of compressive and tensile strains. (b) Relative change in mobility of FETs plotted as functions of strain [36].

Scenev et al. [38] reported a deeper investigation into the effect of strain on the active layer in an organic FET. The authors performed an accurate morphological analysis of the active pentacene layer during strain stresses. Atomic force microscopy, shown in Figure 7.11, demonstrated that when a compressive strain is applied to the active layer of the FET structure, a decrease in the distance between grains is observed. Considering Equation 7.12, it is possible to assume that a compressive strain can reduce the energy barrier for hopping transport ΔE, and consequently an increase in mobility is achieved. On the other hand, the authors have demonstrated that strain stresses leave the crystal structure unaffected and the deformation of the active layer is completely reversible.

Particularly interesting is the work presented by Cosseddu et al. [39], in which it has been demonstrated that the morphological properties of the active layer have a strong correlation with the electrical behavior of

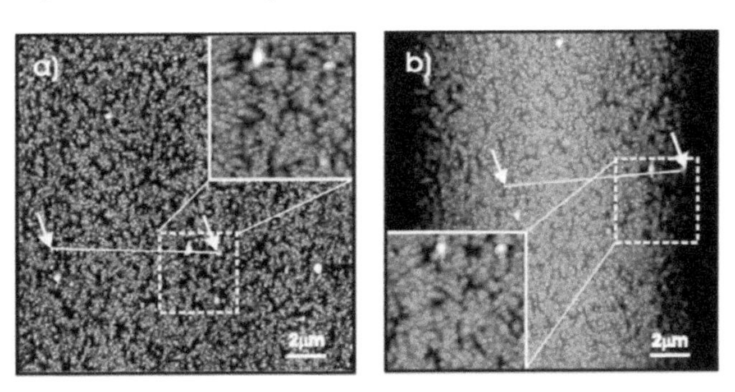

FIGURE 7.11
SFM micrographs of (a) a pristine pentacene layer, and (b) a 10% strained pentacene film in situ [38].

pentacene-based FET structures during compressive strain. In particular, modulating the pentacene morphology in terms of grain size, it is possible to tune the sensitivity to surface strain in a predictable and controllable way. However, the structures presented in [37–39] have a maximum tolerable strain of ~2%; it has been demonstrated that strains larger than 2% cause the formation of cracks within the gold source and drain the electrodes on top of the pentacene film due to the high thickness of the employed substrate (~175 μm).

In [40], the authors compare the sensitivity of transistors with different semiconductors (pentacene and P3HT). It has been demonstrated that the highly disordered structure of polymeric films such as P3HT lead to a decrease in the sensitivity of the device due to the transport properties of these polymers. On the other hand, pentacene-based devices show a higher sensitivity (12.4% kPa^{-1}), good reproducibility, and linear responses in the range 0.7–4.0 kPa. However, the recovery time is quite high (> 10 sec) and this can limit the dynamic response of the sensor.

The correlation between mobility and mechanical stress is not the only way to exploit a FET structure. For instance, Jung et al. [41] investigated and demonstrated the relationship between the variations of the mobility of a semiconductor due to variations in its temperature. The proposed device is a bottom-contact organic FET with a doped n-type silicon wafer that acts as substrate and gate. The gate dielectric is SiO_2 grown by thermal treatment and the active pentacene layer is deposited by thermal evaporation over a palladium source and drain contacts. The authors investigated the temperature dependence of the transport mechanism of pentacene both in subthreshold and saturation regimes in a very large range (273°K–453°K). Experimental results are shown in Figure 7.12. The saturation current shows negligible variations compared with those recorded in the subthreshold

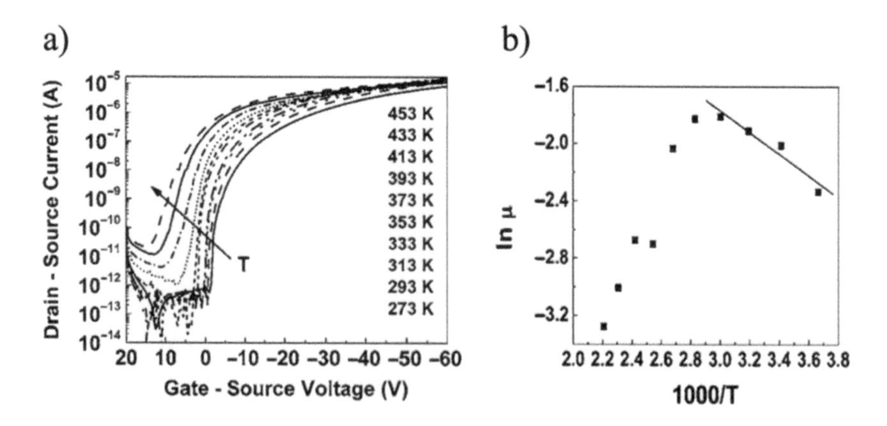

FIGURE 7.12
(a) Temperature-dependent transfer curve of a pentacene TFT. (b) Temperature-dependent mobility in the saturation region [41].

regime, where the current rises due to an increase in temperature from 273°K to 333°K. Interestingly, the device shows saturation in the response to temperatures higher than 333°K, which could have two possible causes: the desorption of the active pentacene layer at higher temperatures and/or an increase in the scattering of the pentacene carriers. The authors underlined the need for more studies. The temperature dependence of the output current is due to a variation of the effective mobility, which can be described in the following Arrhenius-like form:

$$\mu_{eff} \propto \exp\left(\frac{-E_a}{kT}\right) \tag{7.14}$$

where:

E_a is the activation energy
k is the Boltzmann's constant
T is the temperature

Trung et al. [42] proposed a FET-based temperature sensor with high thermal responsivity. The device exploits the thermosensitive properties of a nanocomposite layer of *reduced graphene oxide* (rGO) and a pyroelectric polymer (PVDF-TrFE). This layer is used as the active layer in the FET. Moreover, the authors demonstrated that by properly changing the thickness of the active rGO/PVDF-TrFE layer and by changing the concentration of rGO, it is possible to finely tune the sensitivity of the device. Also, the device shows a very high resolution (~0.1°C), which is comparable to that of human skin, and a very large range of detectable temperatures (30°C–80°C). Furthermore, the high flexibility and optical transparency of the employed materials make this device particularly suitable for artificial skin applications.

As already mentioned, FETs structures can act as the transducing part in a tactile sensor when coupled with an external sensing element.

In this case, the external stimulus, which can be mechanical (force, pressure, or strain) or thermal, is sensed not directly by the OFET semiconductor but by the indirect modulation of the OFET's electrical parameters. This kind of *indirect* sensing is a versatile approach and, thanks to the wide range of categories of sensing elements and fabrication techniques that can be employed, it is possible to achieve different sensitivities in different ranges, depending on the application's requirements.

One of the most common approaches is for sure the connection of a sensing element to the gate electrode in order to change the output current of the transistor, due to a variation in its gate capacitance. As already mentioned in Section 7.3, by microstructuring the framework of an elastomeric material as the dielectric layer of a capacitor, it is possible to increase the sensitivity in terms of $\Delta C / C_0$. Bao's group were the first to report a microstructurated PDMS layer as a gate dielectric in a FET structure. The first work, presented in 2010, is based on a single-crystal

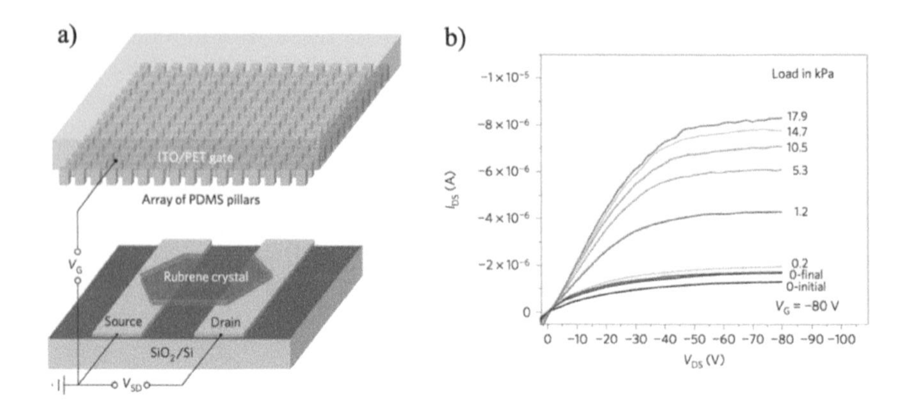

FIGURE 7.13
(a) Layout of pressure-sensing OFET and structured PDMS dielectric film. (b) IDS with different external pressures applied [15].

rubrene semiconductor (Figure 7.13) and a layer of PDMS pillars as the gate dielectric [15]. The sensor shows very high sensitivity, a large workable pressure range of 0.2–18.0 kPa, and negligible hysteresis. However, the use of highly n-doped silicon oxide wafer as the FET's substrate limits the flexibility and stretchability of the device, which are two fundamental requirements for e-skin application sensors. To overcome this problem, in 2013, the same group presented a fully flexible pressure sensor based on the FET structure with microstructure PDMS as the gate dielectric [43]. A flexible polymide (Kapton Hn) was used as the substrate, and PiI2T-Si (a high-mobility conjugate semiconductor polymer) was deposited over gold source and drain electrodes. The high sensitivity (8.2 kPa^{-1}) in the low pressure range 0–8 kPa and the fast response time (less than 10 ms) make this approach particularly interesting for artificial skin applications. Moreover, a linear response, but with a less sensitivity, was recorded in the range 8–50 kPa and the sensor had a detection limit of 0.02 kPa. However, the system requires a very high driving voltage, thus limiting the portability of the device.

Another kind of approach consists of modulating the effective V_{DS} applied to an OFET—for instance, with the employment of a conductive pressure sensor (as a conductive rubber) to be put in series with the source and drain electrode. Conductive rubber can decrease their electrical resistance when they are subjected to an external pressure. Someya et al. [44] were the first to adopt this particular solution. The OFET was realized on one side of a plastic substrate, while the rubber was laminated on the back side. A through-hole creates the connection between the source electrode and the pressure-sensitive rubber. The device is able to respond to pressure due to variations in the rubber's resistance, from 1 MΩ to 1 kΩ, which affects the output current I_{DS}.

Very recently, the same group presented a tactile sensor-based FET structure on an ultra-thin Parylene C substrate (~300 nm). This revolutionary approach exploits the biocompatibility of Parylene C, which makes the device imperceptible and means it can be transferred onto human skin without causing any dermal irritation [45]. The structure is a low-voltage FET in which the touch sensor is connected to the drain electrode. After a complete electrical characterization demonstrates the low-voltage behavior of the device, the transistor is tested as a tactile sensor. Its drain voltage depends on the conductivity of the touch sensor: in particular, when an object shorts the touch sensor's contact, the voltage applied to the drain electrode changes significantly. The device can easily detect and discriminate the touch of a conductive element or human finger.

A non-secondary problem with some of the previously reported structures [36–42] is that the external stimulus is directly applied to the semiconductor. This can lead to irreversible degradation of the active layer or of the source-and-drain contact. To overcome this problem, Lai et al. [46] proposed an interesting approach. In this work, a new FET structure called a *pressure-modulated organic field-effect transistor* (PMOFET) is presented. The peculiarity of PMOFETs is that the pressure-sensitive area is physically separated from the active area of the device. Basically, it is a floating gate transistor biased through a control capacitor called a *control gate*. The transduction mechanism is based on the modulation of a PDMS capacitor, which acts as the sensing element, connected to the floating gate due to a change in its thickness when it is exposed to mechanical stresses. According to Equation 7.15, this modulation can affect the floating gate voltage.

$$V_{FG} = \frac{C_{CF}}{C_{SUM}} V_{CG} + \frac{C_V}{C_{SUM}} V_C + \frac{Q_0}{C_{SUM}} \quad (7.15)$$

where:

V_{FG} is the floating gate voltage
C_{CF} is the control gate capacitance
C_V is the PDMS variable capacitance
Q_0 is the total charge in the floating gate (which depends on the fabrication process)
V_{CG} is the control gate potential
V_C is the potential applied to the top electrode of the PDMS capacitor
C_{SUM} is the sum of the all the capacitance in series with the floating gate

The structure and the response of the sensor are shown in Figure 7.14. A change in the floating gate voltage leads to a variation of the output current I_{DS}. Thanks to the field effect of the transistors, which acts as an amplifier, the percentage variation of the current is higher than that of the PDMS capacitor. Moreover, the device is able to sense forces with a resolution of 0.1 N in the range 0.1–5.0 N.

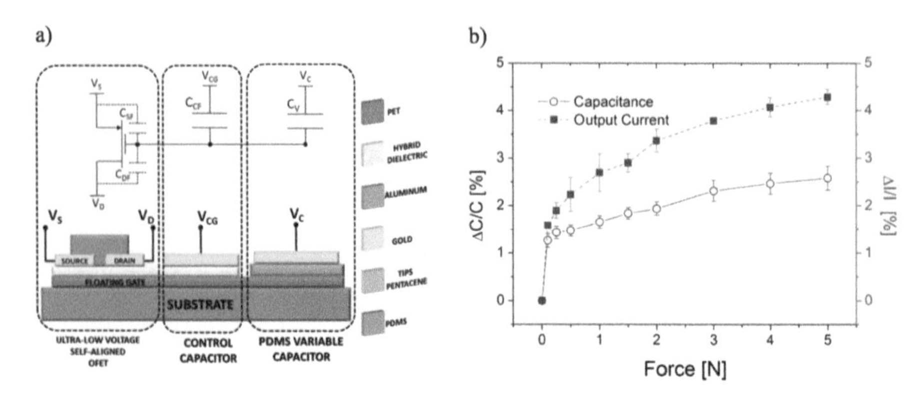

FIGURE 7.14
(a) Structure of the PMOFET and employed materials. (b) Variation in the output current of the PMOFET compared with the variation of the PDMS capacitance with the applied force [46].

7.7 Multifunctional Tactile Sensors

As already described in Section 7.2, artificial skin must have the ability to detect different parameters simultaneously in order to simulate the multi-functional nature of human skin. Normal, shear, and torsion force, temperature variations, stretching, and bending are, probably, the most important parameters that e-skin should be able to recognize.

One of the first approaches to fully flexible multifunctional sensors for artificial skin applications was the work presented by Graz et al. [47]. In this work, the authors presented an integration of flexible matrix cells for pressure- and temperature-sensing skin. Through a proper poling procedure, the authors demonstrated that is possible to maximize the piezoelectric and minimize the pyroelectric properties (and vice versa) of a single cell, thus tuning the sensitivity of each cell. This result is achieved by a combination of different ferroelectric materials; each multifunctional sensing element is based on pyroelectric ceramic nanoparticles (lead titanate) embedded in a ferroelectric polymer (PVDF-TrFE). Interfacing both sensory cells with organic transistors, it is possible to obtain an active matrix of sensors for flexible electronic skin.

Lee et al. [48] presented a highly sensitive and multifunctional resistive sensor for pressure and temperature monitoring for artificial skin applications. The sensing element is a hybrid PVDF thin film coupled with vertical zinc oxide (ZnO) nanorods over an inkjet-printed graphene-based electrode. Figure 7.15 shows a schematic illustration of the flexible multilayer device. ZnO nanorods were embedded in the PVDF polymer onto a rGO-treated flexible polyethylene terephthalate (PET) thin film. Graphene oxide aqueous ink was deposited over a flexible PET substrate via inkjet printing and reduced

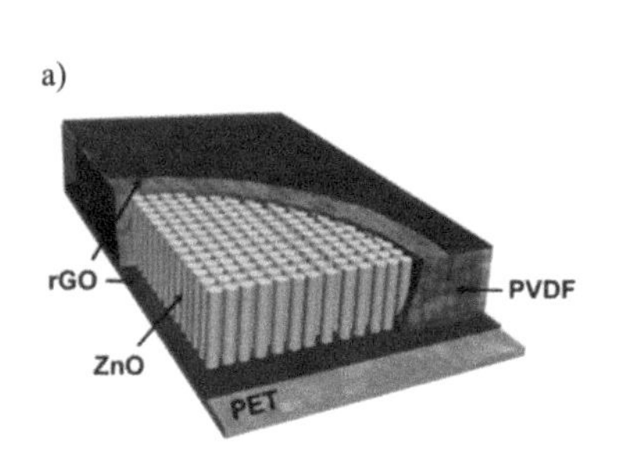

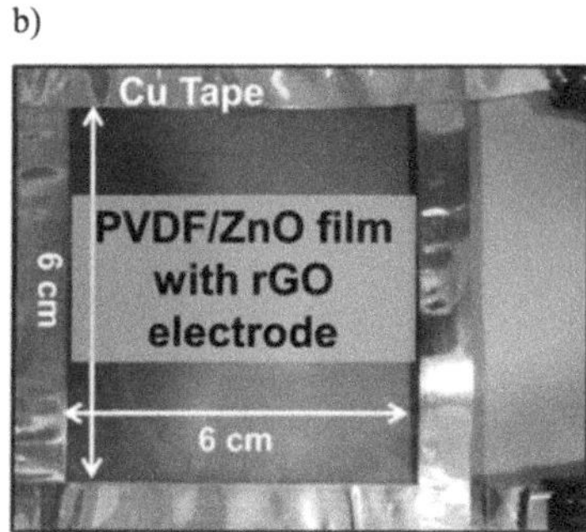

FIGURE 7.15
(a) Schematic representation of the sensor. (b) Photograph of the ZnO/PVDF composite device [48].

for use as an electrode. The ZnO nanorods were grown, followed by hydrothermal treatment. The 20 wt% PVDF solution was spin-coated onto the ZnO nanorods and then poled at a strong constant electric field of 300 kV/cm at 90°C to induce ferroelectricity in the PVDF. Monitoring the pressure, which affects the electrical resistance of the material, and the temperature, which can be inferred from the recovery time of the signal, has demonstrated the multimodal nature of the device. However, despite a pressure resolution of 10 Pa, which is several times lower than that of human skin, the detectable pressure range is 10–120 Pa, which does not cover the entire range requested for e-skin applications. On the contrary, the device is able to monitor temperatures between 20°C and 120°C, but there is no information about the thermal resolution.

Very recently, Zhao and Zhu reported an electronic skin with multifunctional sensors for temperature and pressure detection [49]. The system is based on a pair of platinum (Pt) ribbons connected into a Wheatstone bridge and covered by a PDMS elastomer. The transduction mechanism is based on the pyroresistive behavior of the platinum and the thermal conductivity of the PDMS. The resistance of the center ribbon (R_h), which acts as a pressure sensor, is 92 Ω and the resistance of the anular ribbon (R_c), which acts as a temperature sensor, is 933 Ω. R_h is heated to a higher temperature with respect to the ambient.

The ΔT between the center Pt ribbon and the ambient leads to a conductive heat transfer from the ribbon to the PDMS. When a pressure is applied on the elastomer, its thermal conductivity changes and this lead to a change in the ΔT between the center of the ribbon and the ambient. By calculating the variation of the temperature, it is possible to establish the externally applied pressure P. On the other hand, anular Pt ribbon is used as a simple pyroresistive sensor, and provides information on the ambient temperature. The system is able to simultaneously detect temperature and pressure variations with high sensitivity and resolution and negligible hysteresis.

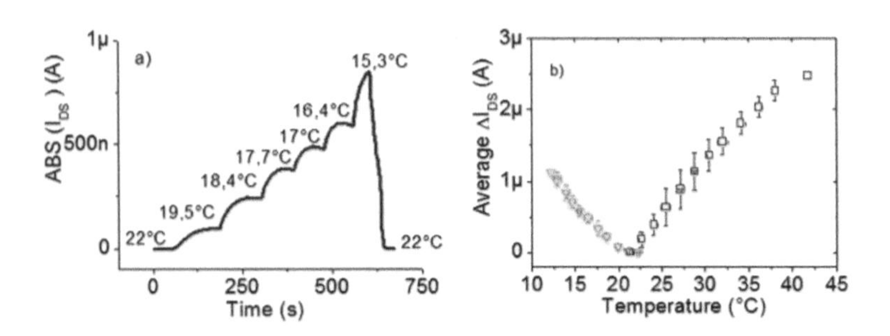

FIGURE 7.16
(a) Current variation induced by temperature. (b) Calibration curves obtained on one characterized OCMFET-based thermal sensor [50].

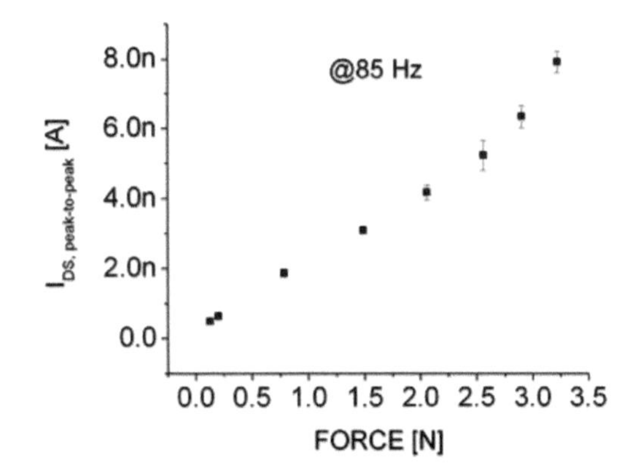

FIGURE 7.17
Calibration curves obtained on one characterized OCMFET-based force sensor [51].

A different approach has recently been presented by Cosseddu et al. and Spanu et al. [50–53], who demonstrated that by integrating an organic charge-modulated FET (OCMFET) architecture with a piezo-/pyroelectric polymer, it is possible to use the same device as both a temperature and pressure sensor. In these two cases, the active area of the device's floating gate is connected by a capacitor made out of PVDF. The sensor's working principle is very simple. If a temperature variation is induced on the device's sensing area, a charge separation will occur in the PVDF film, in turn leading to a perturbation of the charge in the floating gate. Using this approach, very similar to that of Lai et al. in [46], temperature variations will induce a modulation of the transistor threshold voltage in a reproducible way. The authors demonstrated that such an approach can be used for detecting temperatures from 10°C up to 50°C, as shown in Figure 7.16.

Interestingly enough, the same authors demonstrated that as PVDF is also a piezoelectric polymer, the same architecture can be employed to obtain high-performance pressure sensors. In this case, the working principle is very similar. Pressure applied to the PVDF capacitor will lead to a charge separation in the film, again leading to a variation in the transistor threshold voltage. Using this structure, the authors managed to detect very small pressures, in the range of tens of pascals, and obtained a linear and reproducible calibration curve for applied forces from 20 mN up to 3.5 N, as shown in Figure 7.17.

Moreover, such sensors can be employed to detect force variations up to 500 Hz.

Lai et al. [54] reported a different approach for the fabrication of OCMFETs as tactile sensors. Authors have demonstrated that OCMFET-based devices can be produced by means of large-area processes (inkjet printing and chemical vapor deposition) with significant performances and reliability. The fabricated devices were tested as temperature and pressure sensors. As temperature sensors, the devices showed good and reproducible responses and a sensitivity of 50 nA/°C in the range 18.5–50°C. As pressure sensors, the capability of the devices to operate in the pressure range $10^2 - 10^3$ Pa with reproducible performances was demonstrated.

More recently, Viola et al. [55] proposed an innovative multimodal tactile sensor based on a OCMFET coupled with a pyro/piezoelectric element (PVDF-TrFE). The overall thickness of the device is 1.2 um, being thus able to conform to any surface (including the human body), while keeping its electrical performances. Authors proposed a completed thermal and electromechanical charaterization of the sensor in the typical range requested for tactile applications.

References

1. C. L. Stanfield, *Principles of Human Physiology*, New York: Pearson, 2012.
2. E. R. Kandel et al., *Principles of Neural Science*, New York: McGraw-Hill, 2000.
3. M. Campero et al., "Unmyelinated afferents in human skin and their responsiveness to low temperature," *Neurosciences Letters*, 2010, vol. 470 (3), pp. 188–192.
4. P. G. Agache et al., "Mechanical properties and Young's modulus of the human skin in vivo," *Archives of Dermatological Research*, 1980, vol. 269 (3), pp. 221–232.
5. J. F. Manschot et al., "The measurement and modelling of the mechanical properties of human skin in vivo: I. The measurement," *Journal of Biomechanics*, 1986, vol. 19 (7), pp. 511–515.
6. C. Edwards et al., "Evaluation of biomechanical properties of human skin," *Clinics in Dermatology*, July–August 1995, vol. 13 (4), pp. 375–380.
7. T. Yamada et al., "A stretchable carbon nanotube strain sensor for human motion detection," *Nature Nanotechnology*, March 2011, vol. 6 (5), pp. 296–301.
8. D. H. Kim et al., "Epidermal electronics," *Science*, August 2011, vol. 333 (6044).
9. D. P. J. Cotton et al., "A multifunctional capacitive sensor for stretchable electronic skins," *IEEE Sensors Journal*, 2009, vol. 9 (12).

10. J. A. Dobrzynska and M. A. M. Gijs, "Polymer-based flexible capacitive sensor for three-axial force measurements," *Journal of Micromechanics and Microengineering*, 2012, vol. 23.
11. X. Guo et al., "Capacitive wearable tactile sensor based on smart textile substrate with carbon black/silicone rubber composite dielectric," *Measurements Science and Technology*, 2016, vol. 27.
12. S. P. Lacour et al., "Elastomeric capacitive sensors," *16th International Solid-State Sensors, Actuators and Microsystems Conference*, Beijing, China, June 2011.
13. H. Vandeparre et al., "Extremely robust and conformable capacitive pressure sensors based on flexible polyurethane foams and stretchable metallization," *Applied Physics Letters*, 2013, vol. 103 (204103).
14. W. Hu et al., "Elastomeric transparent capacitive sensors based on an interpenetrating composite of silver nanowires and polyurethane," *Applied Physics Letters*, 2013, vol. 102 (083303).
15. S. C. B. Mannsfeld et al., "Highly sensitive flexible pressure sensors with microstructured rubber dielectric layers", *Nature Materials*, 2010, vol. 9, pp. 859–864.
16. M. I. Tiwana et al., "A review of tactile sensing technologies with applications in biomedical engineering," *Sensors and Actuators A: Physical*, 2012, vol. 7, pp. 17–31.
17. R. S. Timsit, "Electrical contact resistance: Properties of stationary interfaces," *IEEE Transactions on Components and Packaging Technologies*, 1999, vol. 22 (1), pp. 85–98.
18. W. Y. Chang et al., "A large area flexible array sensors using screen printing technology," *Journal of Display Technology*, 2009, vol. 5 (6), pp. 178–183.
19. J. G. Dabling et al., "Static and cyclic performance evaluation of sensors for human interface pressure measurement," *34th Annual International Conference of the IEEE EMBS*, 2012, San Diego, CA.
20. S. Gong et al., "A wearable and highly sensitive pressure sensor with ultrathin gold nanowires," *Nature Communications*, 2014, vol. 5 (3132).
21. M. K. Abyaneh et al., "Giant piezoresistive response in zinc-polydimethylsiloxane composites under uniaxial pressure" *Journal of Physics D: Applied Physics*, 2008, vol. 41.
22. Z. M. Dang et al., "Supersensitive linear piezoresistive property in carbon nanotubes/silicone rubber nanocomposites," *Journal of Applied Physics*, 2008, 104 (024114).
23. S. Lee et al., "A transparent bending-insensitive pressure sensor," *Nature Nanotechnology*, 2016, vol. 11.
24. X. Xiao et al., "High-strain sensors based on ZnO nanowire/polystyrene hybridized flexible films," *Advanced Materials*, 2011, vol. 23, pp. 5440–5444.
25. X. X. Gong et al., "Flexible strain sensor with high performance based on PANI/PDMS films," *Organic Electronics*, 2017, vol. 42, pp. 51–56.
26. T. W. Tombler et al., "Reversible electromechanical characteristics of carbon nanotubes under local-probe manipulation" *Nature*, 2000, vol. 405, pp. 769–772.
27. S. H. Bae et al., "Graphene-based transparent strain sensor," *Carbon*, 2013, vol. 51, pp. 236–242.
28. G. Mattana et al., "Woven temperature and humidity sensors on flexible plastic substrates for E-textile applications," *IEEE Sensors Journal*, 2013, vol. 13 (10).
29. R. C. Webb et al., "Ultrathin conformal devices for precise and continuous thermal characterization of human skin," *Nature Materials*, 2013, vol. 12.

30. M. D. Dankoco et al., "Temperature sensor realized by inkjet printing process on flexible substrate," *Materials Science and Engineering* B, 2016, vol. 205, pp. 1–5.
31. T. Vuorinen et al., "Inkjet-printed Graphene/PEDOT: PSS temperature sensors on a skin-conformable polyurethane substrate," *Scientific Reports*, 2016, vol. 6 (35289).
32. J. Jeon et al., "Flexible wireless temperature sensors based on ni microparticle-filled binary polymer composites," *Advanced Materials*, 2013, vol. 25, pp. 850–855.
33. S. Khan et al., "Screen printed flexible pressure sensors skin," *25th Annual SEMI Advanced Semiconductor Manufacturing Conference* (ASMC), Saratoga Springs, NY, 2014, pp. 219–224.
34. R. S. Dahiya et al., "Deposition, processing and characterization of P(VDF-TrFE) thin films for sensing applications," *IEEE Sensors Conference*, Lecce, Italy, 2008, pp. 490–493.
35. Y. Yang et al., "Single micro/nanowire pyroelectric nanogenerators as self-powered temperature sensors," *ACS Nano*, 2012, vol. 6, pp. 8456–8461.
36. T. Sekitani et al., "Bending experiment on pentacene field-effect transistors on plastic films," *Applied Physics Letters*, 2005, vol. 86.
37. Z. Suo et al., "Mechanics of rollable and foldable film on-foils electronics," *Applied Physics Letters*, 1999, vol. 74.
38. V. Scenev et al., "Origin of mechanical strain sensitivity of pentacene thin-film transistors," *Organic Electronics*, 2013, vol. 14, pp. 1323–1329.
39. P. Cosseddu et al., "Continuous tuning of the mechanical sensitivity of Pentacene OTFTs on flexible substrates: From strain sensors to deformable transistors," *Organic Electronics*, 2013, vol. 14, pp. 206–211.
40. P. Cosseddu et al., "Strain sensitivity and transport properties in organic field-effect transistors," *IEEE Electron Device Letters*, 2012, vol. 33 (1).
41. S. Jung et al., "Temperature sensor using thermal transport properties in the subthreshold regime of an organic thin film transistor," *Applied Physics Letters*, 2007, vol. 90.
42. T. Q. Trung et al., "Flexible and transparent nanocomposite of reduced graphene oxide and P(VDF-TrFE) copolymer for high thermal responsivity in a field-effect transistor," *Advanced Functional Materials*, 2014, vol. 24, pp. 3438–3445.
43. G. Schwartz et al., "Flexible polymer transistors with high pressure sensitivity for application in electronic skin and health monitoring," *Nature Communications*, 2013, vol. 4.
44. T. Someya et al., "A large-area, flexible pressure sensor matrix with organic field-effect transistors for artificial skin applications," *PNAS*, 2004, vol. 101 (27).
45. R. A. Nawrocki et al., "300-nm imperceptible, ultraflexible, and biocompatible e-skin fit with tactile sensors and organic transistors," *Advanced Electronic Materials*, 2016, vol. 2.
46. S. Lai et al., "Ultralow voltage pressure sensors based on organic FETs and compressible capacitors," *IEEE Electron Device Letters*, 2013, vol. 34 (6).
47. I. Graz et al., "Flexible active-matrix cells with selectively poled bifunctional polymer-ceramic nanocomposite for pressure and temperature sensing skin," *Journal of Applied Physics*, 2009, vol. 106.
48. J. S. Lee et al., "Highly sensitive and multifunctional tactile sensor using free-standing ZnO/PVDF thin film with graphene electrodes for pressure and temperature monitoring," *Scientific Reports*, 2015, vol. 5.

49. S. Zhao et al., "Flexible bimodal sensor for simultaneous and independent perceiving of pressure and temperature stimuli," *Advanced Materials Technology*, 2017, vol. 2 (11).
50. P. Cosseddu et al., "A temperature transducer based on a low voltage organic thin-film transistor detecting pyroelectric effect," *IEEE Electron Device Letters*, 2014, 35 (12), 1296–1298.
51. F. Viola et al., "Flexible temperature sensors based on charge modulated organic thin film," *11th Conference on Ph.D Research in Microelectronics and Electronics, PRIME*, Glasgow, UK, 2015.
52. P. Cosseddu et al., "Tactile sensors with integrated piezoelectric polymer and low voltage organic thin-film transistors," *13th IEEE Sensors Conference*, 2014, Valencia 2–5 November 2014.
53. A. Spanu et al., "A high-sensitivity tactile sensor based on piezoelectric polymer PVDF coupled to an ultra-low voltage organic transistor," *Organic Electronics*, 2016, vol. 36, pp. 57–60.
54. S. Lai et al., "Floating Gate, Organic Field-Effect Transistor-Based Sensors towards Biomedical Applications Fabricated with Large-Area Processes over Flexible Substrates," Sensors, 2018, vol. 18 (688).
55. F. Viola et al., "Ultrathin, flexible and multimodal tactile sensors based on organic field-effect transistors," Scientific Reports, 2018, 8:8073.

8

Sensor Systems for Label-Free Detection of Biomolecular Interactions: Quartz Crystal Microbalance (QCM) and Surface Plasmon Resonance (SPR)

Şükran Şeker, M. Taner Vurat, Arın Doğan, A. Eser Elçin, and Y. Murat Elçin

CONTENTS

8.1 Introduction ... 181
8.2 Quartz Crystal Microbalance ... 183
8.2.1 Basic Principles of QCM ... 183
8.2.2 Measuring Principles ... 185
8.2.2.1 Active Oscillator Mode ... 185
8.2.2.2 Passive Mode ... 185
8.2.2.3 QCM-D™ ... 186
8.2.3 Electrochemical QCM ... 186
8.2.4 QCM-Flow Injection Analysis ... 188
8.2.5 Preparation of Crystal Surfaces ... 188
8.2.5.1 Polymer Coating ... 189
8.2.5.2 Langmuir–Blodgett Coating ... 191
8.2.5.3 Chemical Modifications ... 192
8.2.6 Biosensor Applications ... 193
8.2.6.1 Enzyme Biosensors ... 194
8.2.6.2 Immunosensors ... 194
8.2.6.3 DNA Biosensors ... 195
8.2.6.4 Cell Biosensors ... 196
8.3 Surface Plasmon Resonance ... 198
8.3.1 Working Principles of SPR Biosensors ... 200
8.3.2 Surface Plasmon Resonance–Mass Spectrometry (SPR-MS) ... 201
8.3.3 Surface Plasmon Resonance Imaging (SPRi) ... 201
8.3.4 Surface-Enhanced Raman Scattering (SERS) ... 202
8.3.5 Surface Plasmon Resonance–Quartz Crystal Microbalance (SPR-QCM) ... 202
8.3.6 Protocols ... 203
8.3.6.1 Priming the Sensor Chip ... 203

8.3.6.2 Sample Preparation..204
8.3.6.3 Buffer Preparation...204
8.3.6.4 Selection of a Suitable Chip 205
8.3.6.5 Immobilization of Sensor Chip....................................206
8.3.7 Data Analysis 207
Notes for QCM 208
Notes for SPR..209
References...209

Abbreviations

AW	acoustic wave
BAW	bulk acoustic wave
F	frequency
PZ	piezoelectric
QCM	quartz crystal microbalance
QCR	quartz crystal resonator
R	resistance
SAM	self-assembled monolayer
SEM	scanning electron microscopy
TSM	thickness shear mode
SPR	surface plasmon resonance
RIS	reflectometric interference spectrometer
SPR-MS	surface plasmon resonance–mass spectrometry
SPRi	surface plasmon resonance imaging
SERS	surface-enhanced Raman spectroscopy
SAW	surface acoustic wave
FTIR	Fourier transform infrared spectroscopy
EIS	electrochemical impedance spectroscopy
RU	resonance unit
HEPES	4-(2-hydroxyethyl)-1-piperazine ethanesulfonic acid
NHS	N-hydroxysuccinimide

EDC	1-ethyl-3-(3-dimethylaminopropyl)-carbodiimide
EDTA	ethylenediaminetetraacetic acid
SDS	sodium dodecyl sulfate
PDEA	2-(2-pyridinyldithio)-ethaneamine hydrochloride
SULFO-MBS	m-maleimidobenzoyl-N-hydroxysulfosuccinimide ester
SULFO-SMCC	sulfosuccinimidyl 4-(N-maleimidomethyl)cyclo-hexane-1-carboxylate

8.1 Introduction

The identification of biomolecular interactions between peptides, proteins, nucleotides, DNA, cells, carbohydrates, small molecules, and other biomolecules plays a central role in the development of diagnostic tests and drug discovery studies. Therefore, there is considerable interest and need for the development of sensing systems to characterize biomolecular interactions. Notable among these systems are biosensors that have the ability to detect biological interactions. These biosensors are divided into the following groups: electrochemical, piezoelectric, and optical [1]. A schematic representation of the classification of the biosensor systems is shown in Figure 8.1.

Commonly used assays in pharmaceutical screening or diagnostics usually involve labeling steps such as fluorescent, radioactive, chemiluminescent, or enzymatic labeling. This labeling indicates the interaction between the ligand and its receptor. It is a costly and a time-consuming process. Additionally, the label or secondary reagents used in some assays can interfere with the biomolecular interactions by blocking the binding site or causing steric hindrance. These challenges severely affect the detection method and may lead to false positive or false negative signals [2]. As per these unfavorable examples, the label-free technologies have gained importance in the recent years. There are several label-free biosensor techniques commonly used for chemical and biological sensing, including acoustic wave (AW) resonators, cantilevers, electrochemical methods, and surface plasmon resonance (SPR) (Table 8.1). These systems can measure the physical or chemical changes in conductivity, mass, dielectric permittivity, viscoelasticity, and so on without using any label. Also, SPR enables the quantitative analysis of the reaction kinetics of biomolecular interactions.

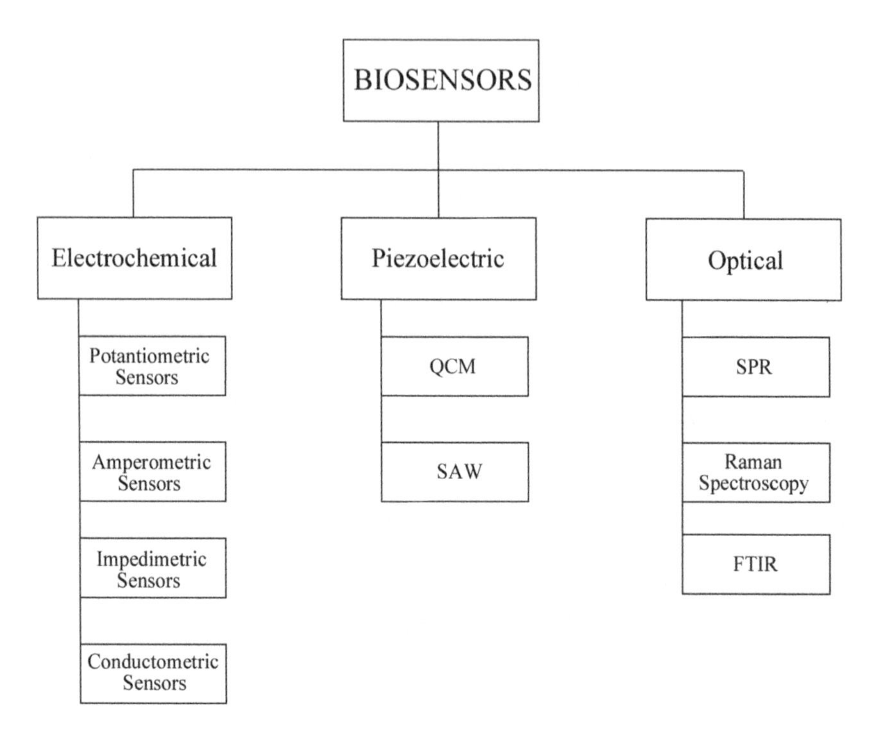

FIGURE 8.1
Classification of biosensors.

TABLE 8.1
Some Label-Free, Real-Time Biosensor Systems and Their Technical Properties

Method Name	Technology	Limit of Detection (ng/cm^2)	Instrument Name
Quartz crystal microbalance	Acoustic	0.5	E4 Auto (Q-Sense)
Surface plasmon resonance	Optical	0.01	Biacore 4000
Bio-layer interferometry	Optical	0.1	Octet RED384
Ellipsometry	Optical	0.1	LFIRE (Maven Biotechnologies)
Isothermal titration calorimetry	Calorimetric	n/a	iTC$_{200}$ (MicroCal)
Surface acoustic wave	Acoustic	0.05	sam5 (SAW instruments GmbH)
Arrayed waveguide grating	Optical	0.01	AnaLight 4D

Quartz crystal microbalance (QCM) and SPR are two very important label-free detection techniques that have been widely used in various applications, including biomedical research, chemical sensors, biosensors, food safety, and environmental monitoring. This chapter points to two label-free technologies capable of measuring biomolecular interactions with high selectivity and sensitivity without the need for labels or secondary reagents.

8.2 Quartz Crystal Microbalance

8.2.1 Basic Principles of QCM

QCM is a label-free, real-time mass-sensing detection technique for biological, chemical, and physical sensing applications. The QCM technique is based on the piezoelectric effect of quartz crystals. Piezoelectricity was first described by Paul Jacques Curie and Pierre Curie in 1880 during their studies on crystals (quartz, tourmaline, topaz, Rochelle salt, and cane sugar). The *piezoelectric effect*, also called the *direct piezoelectric effect*, is the ability to form an electric charge due to a change in the atomic structure of the material, forming a dipole moment (electric polarization) at the center of the material upon the application of mechanical stress to its surface [3]. One of the particular properties of the piezoelectric effect is reversibility. Conversely, the material mechanically deforms in response to an electrical field applied across the material, a phenomenon called the *inverse piezoelectric effect* [4].

Depending on the cut angle, there are several types of AW devices obtained from a mother crystal, such as bulk acoustic wave (BAW) or thickness shear mode (TSM), surface acoustic wave (SAW), flexural plate wave (FPW), Love wave (LW), and shear horizontal surface acoustic wave (SH-SAW) resonators. TSM mode oscillation causes a displacement parallel to the surface of the quartz [5]. Applying a voltage through the electrodes causes a shear deformation of the quartz wafer. The QCM system is a resonator operating in TSM vibrations, basically consisting of a thin piezoelectric AT-cut quartz crystal compressed between two metal (generally gold or silver) electrodes. The most commonly used electrode is gold, which is a noble metal resistant to oxidation and corrosion. AT-cut crystal oscillating in the TSM mode is obtained by cutting a $+35° \ 10'$ angle from the z-axis. The AT-cut quartz crystal is used for sensor applications as it has a temperature coefficient that is close to zero, $0°C \pm 50°C$, and has high sensitivity to any mass changes on the electrode surface [6].

When an electric field is applied across the electrode surfaces, the quartz crystal oscillates at its resonance frequency due to the piezoelectric property.

The resonance frequency of the crystal is related to changes in mass deposited on the surface through the Sauerbrey equation [7]:

$$\Delta F = -2F_0^2 \Delta M \Big/ \left[A \left(\mu_q \rho_q \right)^{1/2} \right] \tag{8.1}$$

where:

- ΔF is the frequency shift of the quartz crystal
- F_0 is the fundamental resonance frequency
- ΔM is the mass deposited on crystal surface
- A is the active electrode area
- μ_q is the shear modulus of the quartz (2.947 × 1011 gcm^{-1} s^2 for AT-cut quartz)
- ρ_q is the density of the quartz (2.648 gcm^{-3})

The Sauerbrey equation is valid in air atmosphere for evenly distributed thin rigid films such as metallic coatings and metal oxides, while it is invalid for thick, soft, or viscoelastic loadings such as polymers, cells, and biomolecular systems, since energy loss occurs due to the humidity generated during oscillation [8]. The first application of QCM was reported by King in 1964 for the monitoring of thin-film depositions in a vacuum or in gas-phase systems upon determining the linear relationship between the crystal's frequency and the mass deposited on the surface. To detect viscoelastic loading, the quartz crystal is submerged in liquid, which alters the resonance frequency due to damping during oscillation. In this case, the resonance frequency is related to the viscosity and the density of the liquid in contact with the quartz crystal surface. In addition, the dissipation factor and resistance are affected by energy losses of the shear wave that radiates from the non-rigid layer [9]. This phenomenon was first reported by Kanazawa and Gordon in 1985 [10]. The resonance frequency shift of the quartz crystal in contact with the liquid is given by

$$\Delta F = -F_0^{3/2} \left[\left(\rho_L \eta_L \right) / \left(\pi \rho_q \mu_q \right) \right]^{1/2} \tag{8.2}$$

where:

- ΔF is the measured frequency shift
- ρ_L is the density of the liquid
- η_L is the viscosity of the liquid

In principle, by using the relation of the oscillation frequency of the crystal and the mass deposition on its surface, the QCM sensors can be used for the characterization of adsorption/desorption during a reaction on the electrode surface of the crystal in the flow cell. The binding of a soluble analyte to a ligand, which is immobilized on the sensor surface, causes a decrease in the crystal's frequency. The change in frequency is proportional to the

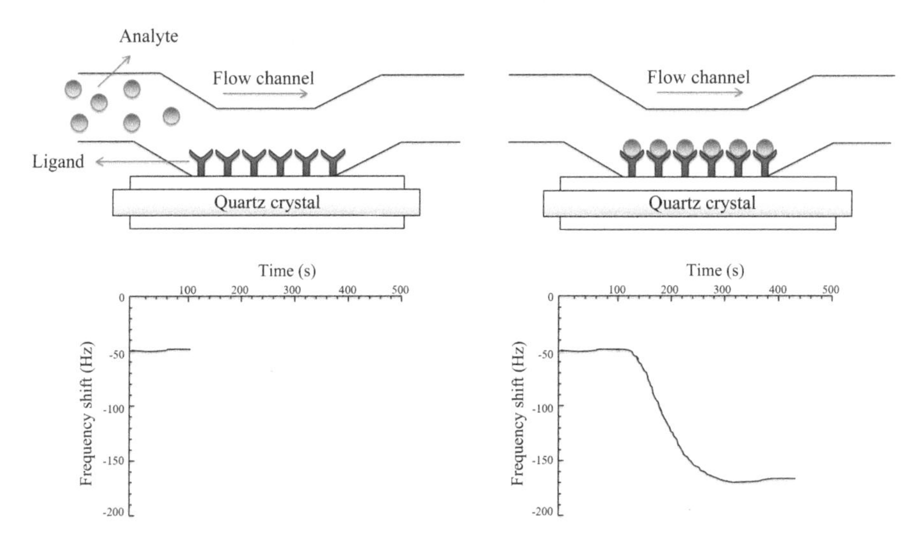

FIGURE 8.2
The relation between frequency shift and mass deposition: (a) Analyte deposition on a coated quartz crystal. (b) Binding a soluble analyte on the quartz crystal surface causes a decrease in the crystal's frequency.

concentration of the binding analyte. The frequency shift versus time curve is called a *sensorgram* (Figure 8.2).

8.2.2 Measuring Principles

8.2.2.1 Active Oscillator Mode

The active oscillator mode is a sensitive and cost-effective way to measure the shifts in crystal frequency caused by mass deposition or the viscoelasticity changes on the crystal surface. In this mode, the crystal is incorporated to a feedback loop of the oscillator circuit that is used for monitoring the shift of the resonance frequency of the crystal [11]. In addition to measuring the frequency, an amplitude-controlled oscillator circuit can be used to measure the amplitude. This helps to determine the viscoelastic properties of films on a quartz crystal and also provides information on the damping of the crystal [12].

8.2.2.2 Passive Mode

The impedance analysis of the shear oscillation is called the *passive mode* QCM. The impedance or network analyzers provide a more extensive analysis of the shear oscillation by measuring the electrical impedance or admittance of the quartz crystal near its resonance frequency. In passive mode, the oscillation is detected by the frequency of the applied AC voltage. However, impedance or network analyzers are bulky and expensive systems and are not appropriate for simultaneous multiple sensor characterization. Although the system is not short of advantages [13]:

1. There is no external circuitry that affects the electrical behavior of the system.
2. Parasitic effects can be eliminated by calibration in passive mode.
3. The conductance and the susceptance of the sensors are also measured.

8.2.2.3 QCM-D™

The other measuring principle is QCM with dissipation monitoring (QCM-D), which was commercialized by Q-Sense in 1996. This technique utilizes the rapid decoupling of sensor driving circuitry from the resonator. By monitoring the decay of the sensor oscillation, the resonance frequency and dissipation factor are acquired simultaneously [14]. Initially, the quartz crystal is excited with a frequency generator. Consequently, the source is switched off and the decay of the quartz oscillation is recorded. Simultaneously, the level of damping is measured with an amplitude-controlled oscillator circuit [11].

The dissipation factor (D), which is the inverse of the quality factor (Q), can be defined as the energy dissipation during the oscillation of resonator, as follows:

$$D = 1/Q = E_d/2\pi E_s \tag{8.3}$$

where:

Q is the quality factor that represents the energy dissipation regarding the stored energy in the oscillation per cycle [12]
E^s is the energy stored in the oscillating system
E^d is the energy dissipated during one oscillation

D is related to the viscoelastic property of the layer adsorbed on the quartz crystal surface and provides information concerning the energy dissipated during one oscillation after the adsorption of the analyte on the quartz surface. The ratio $\Delta D/\Delta F$ in turn provides information concerning the rigidity of binding at the interfaces [15]. A higher value of D displays a softer layer, whereas a lower D shows a relatively rigid and dense layer [16] (Figure 8.3).

8.2.3 Electrochemical QCM

The QCM system can be combined with an electrochemical technique, thus called *electrochemical QCM* (EQCM), requiring a frequency counter, a potentiostat, a dual-voltage power supply, and a voltmeter. The EQCM consists of a

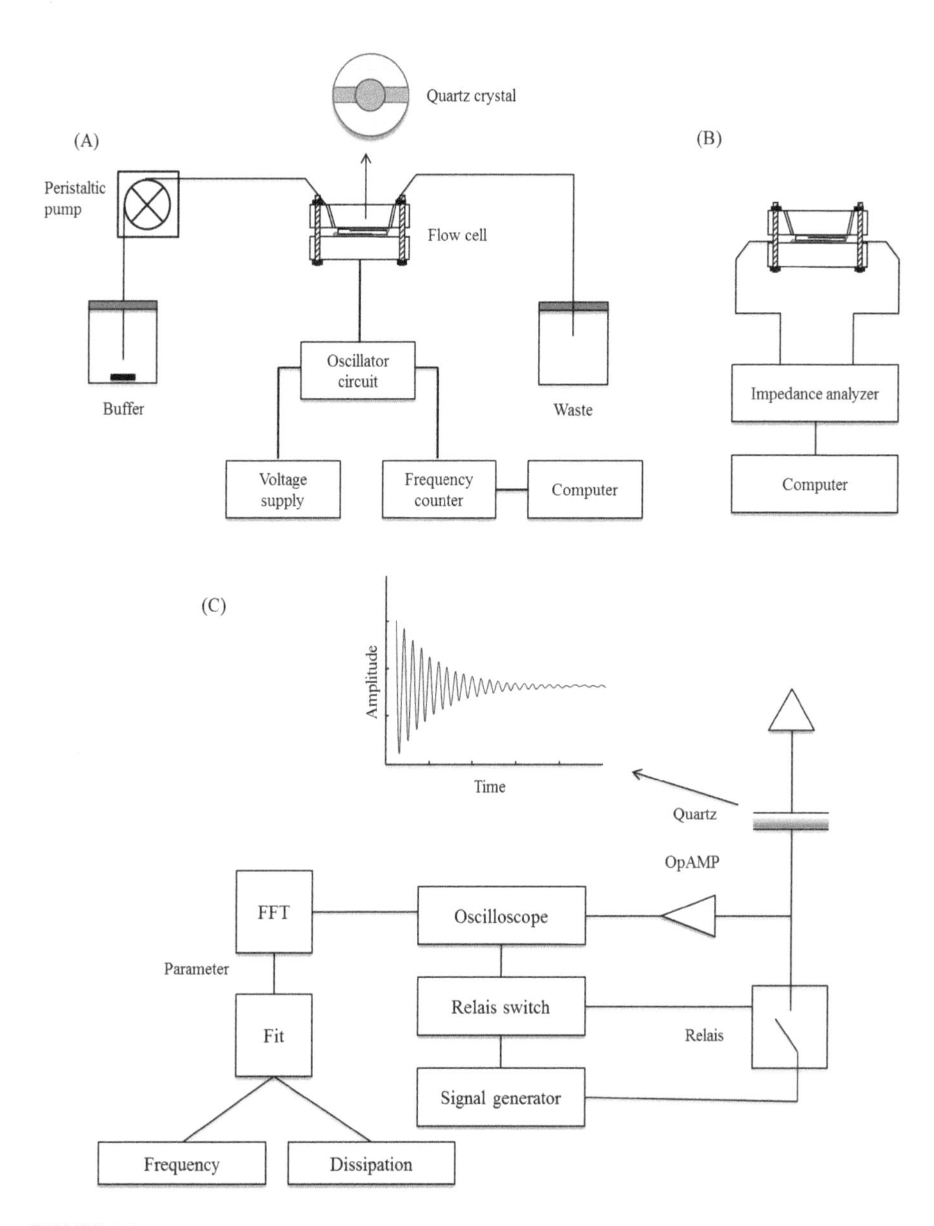

FIGURE 8.3
There are three basic measuring principles for the frequency response of quartz crystal: (a) active oscillator mode, (b) passive mode, and (c) QCM-D™ (Adapted and redrawn from [11]).

three-electrode electrochemical cell, including a reference electrode, a counter electrode, and a quartz crystal as a working electrode. During electrochemical reaction, mass deposition occurs on the crystal's surface, which is in contact with the electrolyte solution. The system simultaneously measures mass and current for voltametric studies. This technology has recently been used to study various surface reactions, such as the adsorption/desorption of adsorbate molecules, ion exchange processes in polymer films, and metal deposition onto the quartz crystal as a working electrode. Furthermore, EQCM can be performed to monitor the formation and to investigate the properties of polymer films formed by the electropolymerization technique. This system enables the measurement of change in potential, current, charge, and frequency at the working electrode.

8.2.4 QCM-Flow Injection Analysis

QCM sensors can also be integrated with a flow injection analysis (FIA) system that continuously monitors the binding of the analyte [17]. The QCM-FIA system can control the detection time and the concentration used in profiling kinetic studies with fully automated sensor applications. In the QCM-FIA system, a quartz crystal is placed between two O-rings into a flow cell. Since the mass-sensitive area is located in the center of the resonator, it should be noted that the O-rings are placed as far as possible from the quartz in order to provide minimal damping and to reduce the contact area of the O-ring. The flow cell has to be closed to prevent the reflections caused by longitudinal waves at the air–water interface. Further, the closed system reduces the formation of air bubbles [11]. The flow cell contains inlet and outlet channels, enabling the addition of the analyte solution, washing, or elution buffer at any desired time according to the assay protocol. For a homogeneous environment, the solution is mixed with a peristaltic pump. A computer-controlled peristaltic pump can be used to achieve discontinuous flow. The injection flow rate in the peristaltic pump is optimized to enhance the sensor sensitivity and affinity, resulting in effective sensor response [18]. Also, appropriate flow prevents excessive stress on the crystal and leakage of the liquid over the O-ring.

8.2.5 Preparation of Crystal Surfaces

Cleaning the surface prior to the coating procedure is a crucial process for removing both organic and inorganic contaminants. It avoids impurities on the surface with particulate-disturbing surface coating, modification, or sensing using chemical treatments such as piranha solution (a mixture of hydrogen peroxide and sulfuric acid), acetone, and so on.

In QCM studies, the piezoelectric sensor surfaces are coated or modified though various techniques. The coating of the crystal surface is necessary for creating active biomaterial surfaces with functional groups. A crucial step in QCM sensor studies is the selection of a suitable surface that allows the sensitive and selective adsorption of the molecules. The coating of the quartz crystal drastically enhances the selectivity and sensitivity of QCM responses. A chemically cleaned electrode surface can also be used without any surface coating.

There are three commonly used methods for the surface modification of biosensors: polymer coating, the formation of Langmuir–Blodgett coating, and chemical modification. Polymer coatings include glow discharge, dip coating, drop coating, spin coating, spray coating, electrospinning, and electrochemical polymerization.

8.2.5.1 Polymer Coating

Generally, in polymer coating methods, the film material is dissolved in an appropriate organic solvent. Following the deposition of the polymer on the quartz crystal, the solvent evaporates quickly from the surface and the film remains. The selectivity, reproducibility, response time, sensitivity, and reversibility of the QCM sensor relies on the chemical and physical properties of coated polymers.

In the glow discharge method, surface modification is carried out in a glow discharge under a monomer vapor phase, resulting in the direct deposition of a plasma-polymerized film (PPF) on the crystal surface. Briefly, the polymer used to modify the surface is placed in a glow discharge; it is bombarded by electrons, ions, and other species from the plasma [19]. The resulting films are extremely thin (<1 μm), flat and pin-hole free, and chemically and mechanically stable due to the cross-linked and highly branched structure (compared with conventional polymer films that contain linear chains of repeating units), allowing good and intensive adhesion of biological components onto the surface [20].

The dip coating process is an easier and low-budget way to prepare nanometer-thick films on quartz surfaces. It is based on the adsorption of the polymer on the surface from the polymer solution with suitable concentration. Briefly, the quartz crystal is dipped into a diluted polymer solution for a duration. Then, the excess solution is carefully removed from the crystal's surface with filter paper from the edge of the crystal. Finally, the crystal is kept under dry conditions (e.g., in air). The disadvantage of the process is that the thickness differs between the edges and the center of the coated layer due to the evaporation effect. The main dip coating parameters are the concentration, the dipping time, the withdrawal velocity, and the viscosity of the solution, which play key roles in preparing the film with the desired thickness and surface morphology for various applications [21].

Drop coating is a relatively simple technique performed with a small volume of polymer solution (several microliters or less) directly applied as a droplet on the center of the electrode either once or more times. The drop-coated crystal is then left to dry in ambient conditions or at a higher temperature. Later, the solvent is evaporated to form a film layer on the electrode surface. Although the process appears simple, some optimization is needed for obtaining a uniform film coating with controlled thickness and roughness.

The spin coating method is a widely used polymer coating technique for biosensor applications. In the spin coating method, the quartz crystal is placed onto the support of a rotating system (a spin coater). A polymer solution is prepared in a suitable solvent and dropped onto the center of the quartz crystal. The polymer is spread over the crystal while it is rotated at an appropriate speed and rotation time. During rotation, the solvent evaporates and leaves a polymer film over the crystal's surface. At the end of the rotating process, the thin and uniform film becomes homogeneous due to centrifugal forces. The film's thickness can be adjusted with the rotation speed, the spin duration, and the viscosity of the polymer solution [22].

In spray coating methods, the polymer film is formed by an airbrush system equipped with a nozzle that sprays the polymer solution on the center of the quartz crystal's surface. The spray pressure is controlled during the deposition of the polymer precursors [23].

The electrospinning method is a well-established experimental process in the production of diverse forms of fibrous structures [24]. Nanofiber-coated sensor systems have a higher surface-area-to-volume ratio and higher porosity compared with continuous film-based sensors, which has advantages such as high sensitivity and fast response times [25]. The system uses a high-voltage electrostatic field and a syringe pump to obtain an electrically charged polymer solution (Figure 8.4). The electrostatic field is used to draw a jet from the solution. The solution alone or with a functional material is transferred along a glass pipe to a needle. The highly charged polymer solution travels toward the quartz crystal located on the oppositely charged collector, and the solvent evaporates to form a nanofiber coating on the crystal surface.

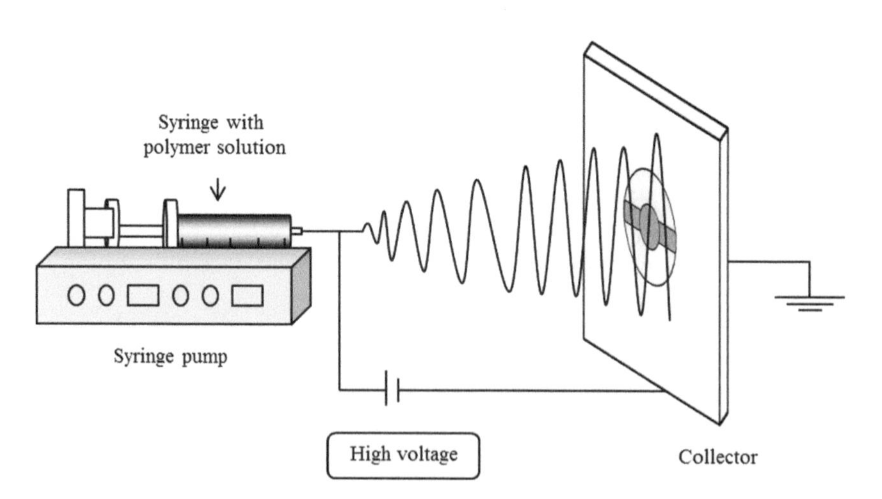

FIGURE 8.4
Schematic description of the preparation of electrospun-coated quartz crystal.

Electropolymerized films of conducting polymers have been widely used for EQCM sensor coatings [26, 27]. Conducting polymers such as polypyrrole, polyaniline, polyindole, and poly(N-methylpyrrole) have a π-electron backbone with extraordinary electronic properties including high electron affinity, low ionization potential, low-energy optical transitions, and electrical conductivity [28]. Conducting polymers can be synthesized using chemical (condensation or addition polymerization) or electrochemical (monomer oxidation) methods [29]. The most remarkable difference between the chemical and electrochemical synthesis of conducting polymers is that very thin films can be generated with well-controlled processes using electrochemistry compared with chemical polymerization [30]. Electrochemical polymerization uses a three-electrode configuration (reference, counter, and working electrodes) in a monomer solution, an electrolyte, and an appropriate solvent. In electrochemical polymerization, a current is applied to the solution, electrodeposition occurs on the working electrode (the electrode surface of the crystal), the monomer on the quartz is oxidized or reduced at the electrode of crystal to generate radical species that react with radical cations or other monomers, and this produces insoluble polymer chains on the electrode of the crystal. The thickness and topography of the coating can be manipulated by controlling the electrochemical polymerization conditions such as temperature, time, charge, solvent system, electrode system, and electrolyte.

8.2.5.2 Langmuir–Blodgett Coating

A Langmuir–Blodgett (LB) coating is formed by transferring monolayers from an air–liquid interface to the crystal's surface. Briefly, an amphiphilic compound is dissolved in an appropriate organic solvent (e.g., chloroform) and subsequently dispersed at the interface. As the solvent evaporates, the amphiphilic compounds are left spread over the liquid and an insoluble monolayer (a so-called Langmuir monolayer) of the compound at the air–liquid interface is formed. The resulting monolayer at the air–liquid interface is compressed by a moveable barrier to obtain a monolayer at the desired area per molecule. Finally, the LB coating is completed by immersing a substrate (hydrophilic or hydrophobic) in the amphiphilic compound-covered liquid (Figure 8.5) [31]. In a number of processes, water is chosen as the liquid due to its high surface tension. Multilayer structures are obtained by repeatedly dipping the same surface. The quality of monolayer is enhanced by optimizing the liquid conditions including temperature, pH, ionic content, immersion and dipping speeds, and substrate type. The film thickness and molecular orientation are better controlled than with casting methods. Although the process is relatively simple, rapid, and cost-effective, the two-dimensional structure of the LB monolayer may be disrupted by minor temperature changes or exposure to some solvent due to physisorption.

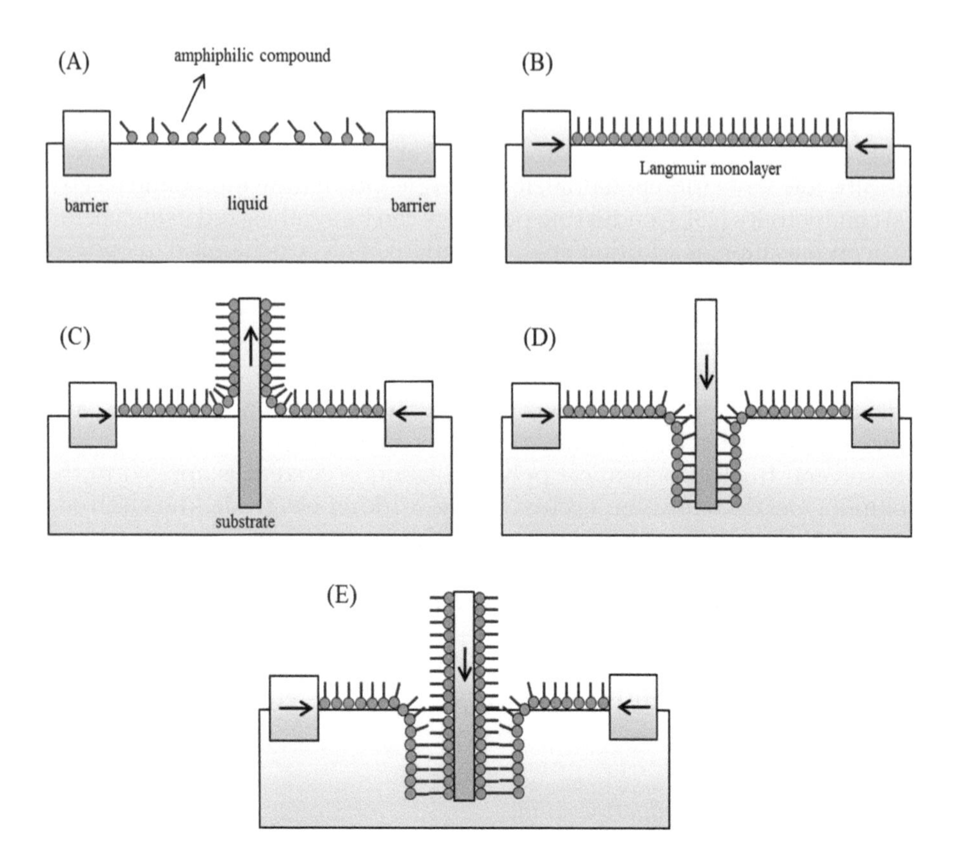

FIGURE 8.5
Representation of the LB coating method: (a) The amphiphilic compounds left spread over the liquid. (b) The LB monolayer at the air–liquid interface is compressed by a moveable barrier. (c) LB deposition on a hydrophilic substrate. (d) LB deposition on a hydrophobic substrate. (e) LB multilayer deposition on a hydrophilic substrate.

8.2.5.3 Chemical Modifications

Crystal surfaces are chemically modified by two major techniques. Self-assembled monolayers (SAMs) of alkanethiol are widely used for the chemical modification of a quartz crystal with a gold electrode to prepare biosensor surfaces with controlled thickness and well-ordered and stable layers. The alkanethiol SAM formation method is based on the strong chemisorption between the gold electrode and thiol head groups of alkanethiol molecules. Therefore, the SAM surfaces have high thermal and hydrolytic stability. The presumed adsorption chemistry is as follows:

$$R-\mathrm{SH}+\mathrm{Au}_n^{\circ} \rightarrow \mathrm{RS}^- - \mathrm{Au}^+ + \tfrac{1}{2}\mathrm{H}_2 + \mathrm{Au}_{n-1}^{\circ}$$

This equation demonstrates the oxidative addition of the S-H bond to gold atoms, which is succeeded by the reductive elimination of the hydrogen [32].

Alkanethiols have a carbon chain, a thiol head group that has high affinity for the gold surface, and a functional group at the tail that interacts with the biomolecules. The optical and electronic properties of the monolayer can be modulated by selecting alkanethiols with appropriate functional groups. The hydrophobic or hydrophilic groups on the SAM surface play a significant role in the immobilization of enzymes, proteins, or whole cells. The gold electrode of quartz crystals can be modified quite simply, which is one of the remarkable advantages of this method. The gold–sulfur interaction is performed by immersing a clean quartz crystal into a dilute solution (~1 mM) of the desired thiol at room temperature. The process optimization focuses on the roughness of the crystal surface, the solvent, the alkanethiol molecules, the temperature, and the concentration of the alkanethiol [33].

The other chemical modification method is based on self-assembly with organofunctional chloro- or alkoxysilane molecules on hydroxyl-terminated glass and metal surfaces; this is called *silanization*. Silanization protocols for quartz crystal with a gold electrode usually include rinsing steps with an acid solution such as piranha solution [34] or chromic acid treatment, which aims to produce high-density hydroxyl groups (-OH) on the surface before silanization [35]. The free hydroxyl groups react with the alkoxy groups on the silane molecules, thus forming a covalent Au-O-Si bond, which is thermodynamically and hydrolytically stable. In the silanization method, the surface is modified by dipping it into a solution containing a silane molecule. This method is very useful for biomolecule immobilization due to the simplicity of the process and the high stability of the modified surface. A number of very different silane derivatives for silanization can be found in the literature. The most commonly used silane molecules in biosensor applications are 3-aminopropyltriethoxysilane (APTES) and 3-aminopropyltrimethoxysilane (APTMS), which generate a self-organizing silane monolayer with terminal amine groups on its surface, using different cross-linkers (e.g., gluteraldehyde) for the further attachment of biomolecules.

8.2.6 Biosensor Applications

In many biosensor applications, biomolecules need to be immobilized on the quartz crystal surface in order to capture the target analyte, which obtains dynamic information relating to biomolecular interactions. The immobilization method of a biomolecule is an essential step in all sensing processes, retaining native conformation and binding activity. The immobilization method affects the quality of the stability, sensitivity, and reproducibility of the sensor response [6]. Various immobilization methods involving thiolation, silanization, and immobilization by cross-linking agents are used to bind the biomolecules on the crystal surface. At the end of a measurement, the sensor has to be regenerated to remove bound analyte from the crystal surface. For this purpose, a convenient elution buffer or high salt solution that does not damage the native structure of the biomolecules should be preferred [36].

8.2.6.1 Enzyme Biosensors

QCM enzyme biosensors are typically used for the evaluation of the catalytic properties of enzymes by measuring the mass deposition of the product molecule, which is formed through the enzymatic conversion of the substrate molecules. For that, the quartz crystal is coated with a layer that has the ability to detect the product of the enzymatic reaction. For example, the ammonium ion, which is an end product of the catalytic hydrolysis of urea by urease, is detected by a fullerene–cryptand-coated quartz crystal [37]. Quartz crystal coated with a nanofibrous layer of poly(lactic-co-glycolic acid) (PLGA) containing saturated fullerene C60 is used for the detection of the gluconic acid, the oxidation product of β-D-glucose by glucose oxidase [24]. In some studies using similar methods to measure enzyme activity, the substrate is immobilized on the quartz crystal to investigate the kinetic behavior of enzymatic hydrolysis. The frequency and dissipation shifts caused by substrate hydrolysis at various temperatures and enzyme concentrations are determined during an enzymatic reaction [38, 39]. QCM is also used in enzyme immobilization studies to determine the influence of surface chemistry on enzyme adsorption. The immobilization of *Candida antarctica* B lipase on a gold electrode modified with either methyl- (hydrophobic) or hydroxyl-terminated (hydrophilic) SAM surfaces is monitored using QCM [40].

8.2.6.2 Immunosensors

QCM is also used to quantify the interactions of antigens and antibodies on functionalized surfaces for the detection of viruses or bacteria via immunological reactions by monitoring changes in resonance frequency, resistance, and energy dissipation. In order to develop a successful piezoelectric immunosensor, the immobilization method must provide a more efficient binding and fast mass transport of the analyte and has to maintain the biological activity of the protein. The determination of an appropriate immobilization technique is a crucial point for each biological ligand for specific applications [41]. Protein A is a commonly preferred component for the immobilization of an antibody. The immobilization of protein A on the crystal surface is performed with different methods facilitated by direct covalent binding on the gold electrode via disulfide bridges of protein A. In this method, a solution of protein A (e.g., 1 mg/mL) is prepared in PBS and incubated on the clean gold electrode of the quartz crystal. The second method of protein A immobilization is based on simple physisorption on the polymer-coated surfaces, in which the gold electrode is coated with an polymer (e.g., polystyrene) and a solution of protein A is dropped and immobilized onto the coated crystal.

In the third method, before the immobilization of protein A on the quartz crystal, the gold electrode is incubated with an alkanethiol solution (e.g., 0.02 M cysteamine in water) to obtain a SAM on the crystal, after which the crystal is incubated with a glutaraldehyde solution (2.5% in water). The most efficient immobilization procedure utilizes SAMs with a covalent binding method for protein immobilization on the quartz crystal surface, following cross-linker incubation.

In antibody immobilization processes, in order to reduce nonspecific adsorption at the crystal surface, the unreacted and nonspecific sites on the crystal surface are blocked by some non-active proteins (gelatin, casein, or bovine serum albumin) to increase the sensitivity of the QCM immunosensor. After protein A immobilization, a recognizing antibody is added onto the protein A layer and immobilized on the crystal surface. After each incubation step, the quartz crystal is rinsed with buffer to remove unbinding reagents [42].

In some cases, the recognizing antibody is directly immobilized on the SAM-modified quartz crystal. Alkanethiol containing terminal carboxylic acid groups is self-assembled onto the gold surface of a piezoelectric crystal. The activation of the carboxyl group is performed by using carbodiimide chemistry. This immobilization method provides a covalent attachment of the specific antibody via its primary amine group to the carboxyl group on the crystal's surface [43]. Additionally, an immunosensor based on the covalent immobilization of proteins, antibodies, or antigens on the polymer layer can also be utilized. Polyethylenemine (PEI) cross-linked with glutaraldehyde (GA) is the most commonly used polymer layer for the immobilization of antibodies via aldehyde groups of GA on a quartz crystal coated with PEI [44].

It should be noted that in cases in which the recognizing material (protein or antigen) cannot tightly attach to the surface in the immobilization method, treatment with a low or high pH and a high salt concentration (often used in regeneration solution) can induce the desorption of the adsorbed analyte during the immunosensor's use [42]. Furthermore, the coatings obtained using the immobilization method must be thin and uniform to take a particularly high sensitivity [45].

8.2.6.3 DNA Biosensors

QCM is a rapidly advancing technology used to study the detection of infectious, hereditary diseases and genetically modified organisms (GMOs) and DNA–protein or DNA–drug interactions. In order to design QCM DNA biosensors, a single-stranded oligonucleotide that selectively hybridizes with the unlabeled complementary strand from a solution is immobilized by a wide range of techniques. The selection of an appropriate method for oligonucleotide immobilization considerably reduces the detection limit of a functionalized crystal surface, leading to higher sensitivity and the analytical performance of the QCM sensor.

In general, in order to reduce the detection limit of the sensor, quartz crystals with higher resonance frequencies can be used. Further, the steric hindrance of the hybridization reaction caused by the coating material on the quartz crystal can be avoided to improve immobilization yields by using multilayers composed of polymers/proteins and nucleic acid surfaces [11]. One of the most widely used immobilization methods is based on the interaction of biotin and avidin or streptavidin, which is conjugated with pyrrole, dextran, or thiol for the immobilization of streptavidin [46]. A biotin-labeled oligonucleotide is covalently attached to an avidin or streptavidin-coated electrode surface [47]. The other immobilization process is chemical adsorption of DNA, labeled with a thiol or a disulfide group on the gold electrode of a quartz crystal by a SAM (Au-S bonding formation) [48]. Further, besides the selection of an appropriate immobilization method, the determination of the optimum number of base pairs in the probe oligonucleotide, the probe concentration, and the amplification of the target DNA using a polymerase chain reaction (PCR) are crucial issues in improving the detection limit of the sensor. Additionally, the conditions of immobilization and hybridization reactions including time, solvent, and temperature should be optimized for each application.

The biggest challenge in DNA biosensor studies is the nonspecific adsorption of target oligonucleotides. To overcome this obstacle, a blocking solution (e.g., 1 mM mercaptohexanol, $HS[CH_2]_6OH$) is used to prevent nonspecific adsorption and enhance the hybridization reaction following the probe immobilization. Post-treatment of the surface with mercaptohexanol as a blocking agent may enhance the hybridization efficiency by 30% [49].

8.2.6.4 Cell Biosensors

To evaluate the cellular responses to chemicals or different surfaces, a QCM cell biosensor based on the immobilization of living cells is used. QCM cell biosensors require a cell culture incubator that provides a viable environment for cell cultures (37°C, 5% CO_2, and 95% humidity). All equipment exposed to cells and cell culture media such as the crystal chamber and quartz crystal must be sterilized with ethanol or UV exposure prior to use. The quartz crystal can be disinfected by being immersed in a penicillin (200 U/mL) and streptomycin (200 mg/mL) solution and stirred for 2 h [50]. After the quartz crystal is connected to the crystal chamber and placed inside the cell culture incubator, the cell culture medium or liquid sample is released directly onto the quartz crystal surface via a micropipette, a syringe, or a peristaltic pump system.

The majority of cellular applications using QCM depend on the successful adhesion and spreading of cells to the quartz crystal's surface (Table 8.2). These applications typically include the preparation of different crystal surfaces and the monitoring of cell adhesion to the surfaces. Firstly, the QCM surface is coated or modified with an appropriate method. The harvested

TABLE 8.2

QCM Cell Biosensor Studies

Cell Type	Surface	Assay Description	Reference
Bone marrow mesenchymal stem cells	Polystyrene-coated quartz crystal	Response to vinblastine	[52]
DH82 macrophage cell line	Gold	Response to Zymosan A, PBs, SWCNT	[53]
Human breast cancer cells (MCF-7)	Gold	Response to staurosporine	[50]
A431 cells	Glass-deposited crystal	Response to EGF	[54]
BAECs	ZnO nanostructure-modified crystal	Determination of cell adhesion	[55]
NIH 3T3 fibroblasts	Gold electrode modified with FN and PDL	Determination of cell adhesion	[56]
Human mammary epithelial cells (HMECs) and MCF-7 cells	Hydrophilic gold surface	Determination of cell adhesion	[57]
Rat liver epithelial cells (WB F344) and lung melanoma cells (B16F10)	Fibronectin and polystyrene	Determination of cell adhesion	[58]
L929 mouse fibroblasts	Gold electrode and PMSH modified gold electrode	Determination of cell adhesion	[59]

Note: PBs, polystyrene beads; SWCNT, single-walled carbon nanotube; EGF, epidermal growth factor; FN, fibronectin; PDL, poly-D-lysine; PMSH, poly(MPC–co–2-[methacryloyloxy] ethylthiol)

cells are seeded on the crystal surface and allowed to attach. Then, the cell adhesion and its kinetics are evaluated by measuring the change in frequency, resistance, or dissipation in real time.

In past decades, QCM has been used to detect the effect of an agent on cellular response, including migration, cell division, or cell death induced by chemicals present at the sensor surface. This cellular biosensor has the potential for the evaluation of biologically active agents or macromolecules that cause changes in cellular attachment in real time. In general, the protocol of cell–agent interaction experiments can be summarized in two parts: the cell deposition on the QCM crystal and the subsequent agent exposure procedure, in which the cells are seeded and allowed to spread on the functionalized QCM surface. In some studies, the quartz crystal is coated with extracellular matrix proteins that can promote cell attachment [51]. In the second part, the adherent cells on the crystal are treated with a bioactive agent that alters the cytoskeletal properties of living cells, leading to microtubule alterations and a change in the mass and viscoelasticity of the cell layer, which is indicative of the different steps of the cell–surface interactions.

8.3 Surface Plasmon Resonance

SPR systems work on the optical biosensors principle. It is currently the most commonly used quantitative spectroscopic method for the study of interactions between biomolecules [60], especially for the interactions of proteins and protein-like biomolecules. The system allows for label-free and real-time analysis, providing highly accurate results [61]. Among biosensor systems, SPR and only a few other systems give information on association and dissociation rate constants [62]. Currently, there is an increasing demand for the monitoring of living cell responses using SPR systems [63].

Introduced in 1968, SPR has been used in a wide range of applications since the early 1990s [64]. It is possible to use the SPR technique in many scientific fields such as biochemistry, molecular biology, medicinal chemistry, molecular pharmacology, and biophysics. Analyses that can be performed with SPR systems are as follows [65]:

- Thermodynamic analysis
- Binding stoichiometry
- Affinity analysis
- Kinetic analysis
- Interaction mechanisms
- Active concentration assays

SPR-based analytical instruments were first introduced by Biacore (Uppsala, Sweden) in 1990. Different models were designed for research, the food industry, GMP and GLP conditions, and so on. Nowadays, many commercial SPR systems use an optical technique to analyze changes in the refractive index and reflecting angle of the light at the sensor surface.

There are several advantages of SPR compared with alternative modes of analysis that involve optical sensor systems. SPR applications are becoming much more widespread than their competing methods [66]. For example, ~50% of the antibiotic residue analyses in the food industry are currently performed using SPR systems [67].

SPR systems consist of several main components: a light source, a metallic layer (gold, silver, etc.), a flow channel, a polarizer, a chopper, a prism, a glass slide, a light detector, a photodiode, and a lock-in amplifier (Figure 8.6).

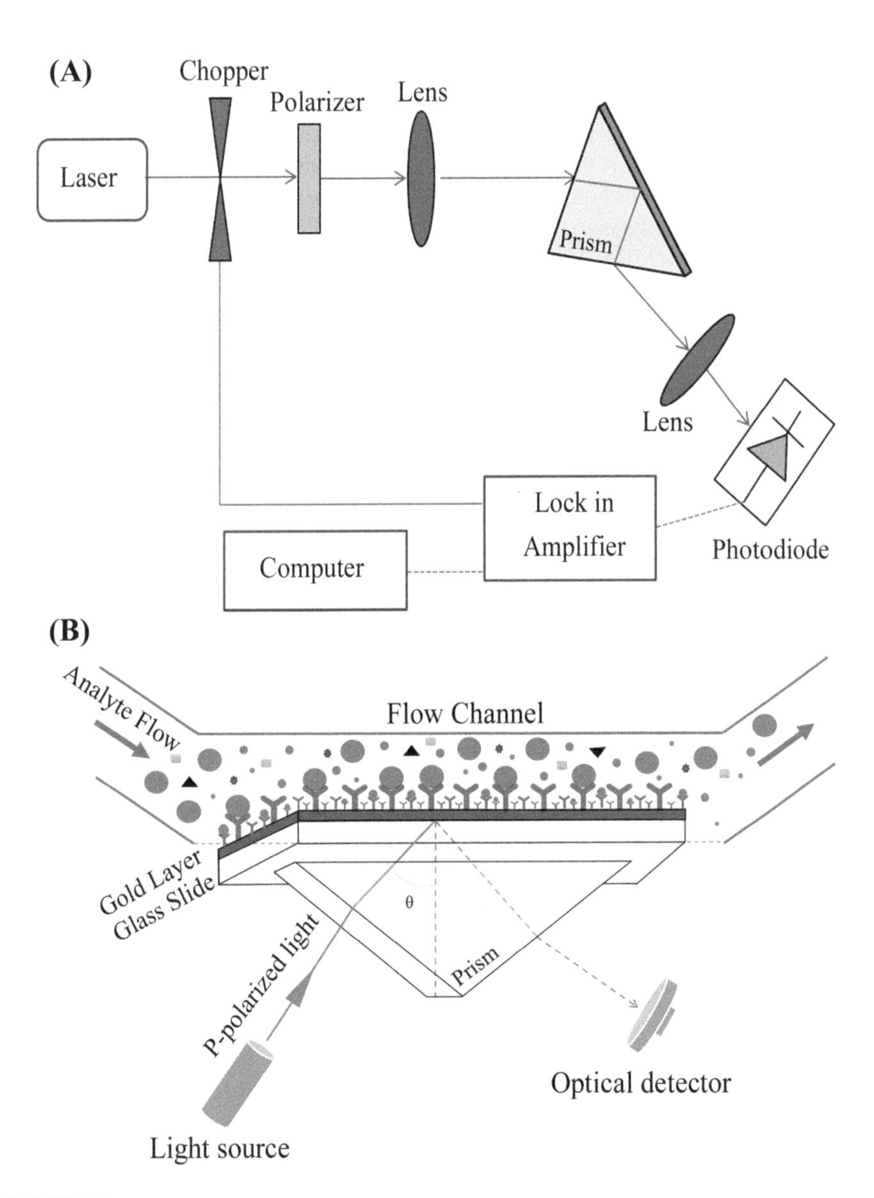

FIGURE 8.6
(a) Schematic representation of Kretschmann configuration SPR (adapted and redrawn from [68]). (b) Detailed layout of an SPR system coupled with a flow channel for the determination of an analyte in a sample. The SPR system consists of optical parts, light source, and detector. The sensor chip is coated with a gold film; the ligand binds to the gold surface; the buffer and analyte run through the flow channel. In this system, p-polarized light is transmitted at angle θ from the light source, which is reflected from the metal surface at a different angle, known as the *SPR angle*. Finally, the optical detector measures the change in the angle.

The incident beam is polarized in a parallel orientation to the plane of incidence, known as *p-polarization*. The chopper modulates the intensity of the incident light, and the lock-in amplifier is used to measure the light with reference to the chopping frequency; used together they improve the signal-to-noise ratio. The lens focuses a laser beam on the prism base surface, which is subsequently measured with an ultrasensitive photodiode. The Kretschmann configuration involves a thin metal film in direct contact with the prism base for the excitation of the surface plasmon (Figure 8.6) [68].

The most commonly used metallic layer in SPR systems is gold. Although silver as a noble metal creates a more effective system, gold serves as a better alternative [69]. Furthermore, the gold layer is modified with thiol groups to facilitate the covalent bonding of biomolecules, providing suitable conditions for biomolecular immobilization. Palladium, copper, and platinum are other alternatives for some systems [70]. Even though there is an ongoing debate regarding the effectiveness of gold and silver as the sensing layer, both metals have advantages over each other at different thicknesses and wavelengths.

The SPR technique utilizes monochromatic p-polarized light photons that are directed to the metallic surface. The photon beam transfers some of its energy to the electrons on the surface of the metal and reflects it at a different angle. When a photon hits the thin metallic surface at a certain angle, free electrons on the surface of the metallic layer absorb some of the energy. The reflecting photons have a different angle depending on different refractive indices [71]. The optical detector acquires changes in the light intensity and reflection angle, which is later converted into data [72]. Finally, concentrations of biomolecules are determined by using the association and dissociation levels obtained from sensorgrams.

8.3.1 Working Principles of SPR Biosensors

The SPR technique investigates the interactions between biomolecules. A ligand is immobilized on the metallic surface of the sensor chip; later, the sample is injected into the flow channel. When the ligand interacts with the analyte, the beam originating from the light source starts to reflect at a different angle, which is called the *SPR angle*. The optic detector acquires the changes in the reflection angles and provides data for sensorgrams. The analyte deposition on the sensor chip causes a change in the angle of the light incidence. The changes in the refractive index are recorded as sensorgrams in response units (RU) versus time [73]. In SPR experiments, 1 RU is equal to 1 pg analyte/mm^2. Also, 1 RU represents a 0.0001° change in the reflection angle [74, 75]. Figure 8.7 depicts a representation sensorgram that provides the following information: (1) interaction rate (association–dissociation), (2) analyte concentration, and (3) association-level data that constitutes the affinity constant.

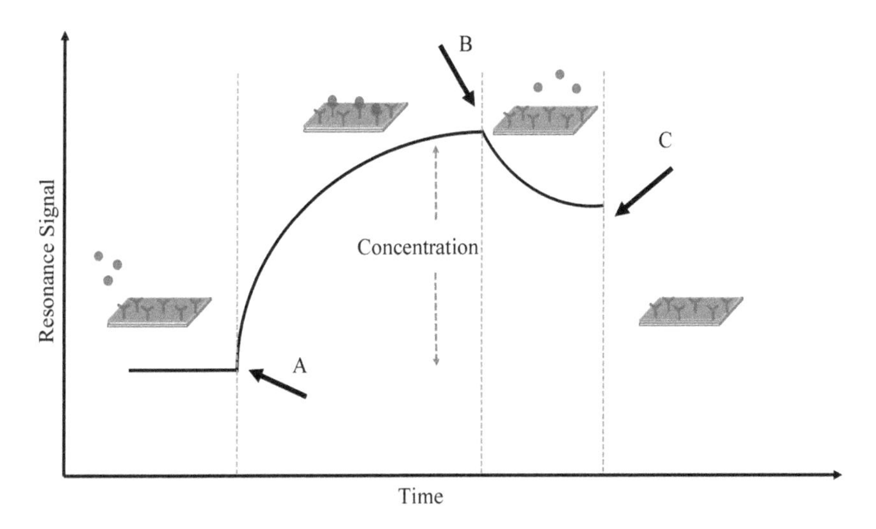

FIGURE 8.7
Schematic representation of the SPR sensorgram. The sensorgram is divided into four phases: baseline, association, dissociation, and regeneration, respectively. Point A represents the beginning of the sample injection and the interaction of the ligand and analyte, known as the *association point*. Point B represents the end of the injection and the continuation of the process with the buffer injection. At this point, the bound analytes are detached from the ligand by the flow. Point C represents the start of the regeneration.

Although SPR systems provide the option to work with a wide variety of biomolecules, such as nucleotides and peptides, it has some limitations as well (Table 8.3) [65, 76, 77]. SPR systems are integrated with other systems to overcome these limitations.

8.3.2 Surface Plasmon Resonance–Mass Spectrometry (SPR-MS)

SPR-MS is used for both complex qualitative and quantitative analyses [78]. Principally, the SPR system detects biomolecular interactions; on the other hand, MS detects the structural properties. SPR-MS is useful for the detection of the chemical composition of biomolecules. The most important advantage of these complex systems is high throughput analysis with reproducible results.

8.3.3 Surface Plasmon Resonance Imaging (SPRi)

SPRi is an SPR instrument integrated with a CCD video camera and is useful for visualizing the surface of the biochip. SPRi allows the measurement of real-time local changes at the biochip surface. SPR sensorgrams give detailed information about biomolecular interactions and molecular-binding kinetics, providing a useful approach for cancer cell research.

TABLE 8.3

Advantages and Limitations of SPR

Advantages	Limitations
• Highly sensitive.	• Immobilization of ligand without interfering with biomolecular interactions.
• Fast and less laborious technique.	• Only one sample can be analyzed at a time.
• Allows real-time detection of molecular interactions.	• Not applicable for concentration measurements.
• Enables simulation of the biological environment.	• Only works with small analytes.
• Acquires information on the analyte–ligand binding rate.	• Requires specific surface modifications.
• Automated SPR systems are suitable for reproducible analysis, which saves effort and time.	
• Low sample volume.	

8.3.4 Surface-Enhanced Raman Scattering (SERS)

SERS systems analyze the structure, shape, and size (thickness) of a surface. Controlling the surface properties is an important aspect of SERS analytical devices for surface research. SERS provides highly reproducible and robust substrates with nanosphere lithography. SERS systems as well as SPR can analyze biomolecular interactions and binding kinetics. In SERS, weak signals are strengthened by the addition of a metal nanostructure onto the SERS substrate and the fluorescence effect is reduced [79, 80].

SPR-based device systems are widely used for research purposes. Studies on the development and use of SPR-based systems are still in progress. Many research groups have focused on developing SPR systems and methods for sensitive and selective detection. A good example is the SPR-FTIR method, which examines the dynamic processes of living cells in real time [81].

8.3.5 Surface Plasmon Resonance–Quartz Crystal Microbalance (SPR-QCM)

QCM and SPR are effective label-free detection techniques that have uses in numerous fields (Table 8.4). Although both systems have different advantages

TABLE 8.4

Comparison between QCM and SPR

	Sensor Chip	Mass/Area	Sample Volume
QCM	Metal on quartz	1 ng/cm^2	50–200 μl
SPR	Metal on glass	0.1 ng/cm^2	10–100 μl

over each other, QCM and SPR systems are very useful for detecting association properties.

The analyses performed with combinations of these systems and other similar systems are much more accurate and reliable. The SPR, QCM, and EIS techniques are just a few of many examples of combined systems [82]. The combined application of SPR and QCM is an effective approach to evaluating surface characteristics. Combined SPR-QCM systems provide extensive information through the accurate measurement of both acoustic and optical properties. In SPR-QCM, SPR is used to monitor metal surfaces in real time, to detect molecular concentrations, and to measure adsorption kinetics. On the other hand, QCM is useful for determining the rigidity, viscosity, and deposition rate of layers.

Bund and coworkers have combined SPR and QCM systems for electropolymerization and to investigate the doping/dedoping behaviors of polymer films [83]. Combining SPR and QCM is a feasible technique for monitoring solution phase adsorption on surfaces, which allows the evaluation of chemical, electronic, viscoelastic, and optical properties [84].

8.3.6 Protocols

Although SPR experimental protocols may seem complex, they are highly applicable (Table 8.5). The main challenges of the system are metal surface selection, surface immobilization, and surface modifications. The main processes of SPR experiments are briefly summarized in the following sections.

8.3.6.1 Priming the Sensor Chip

The sensor chip is a reusable instrument. To assure validity, the chip must be cleaned with appropriate procedures before reuse. Cleaning is crucial, since coupling procedures involve many different buffers and chemicals. Buffers available for cleaning procedures are listed in the user manual for

TABLE 8.5

General Applications of SPR

Applications	Aims	Reference
Pharmaceutical analysis	Drug discovery Quality control Drug–target interactions Disease diagnostic applications	[85]
Food industry	Quantification of antibacterial and antimicrobial residues Detection of general allergens Detection of genetically modified organisms	[86]
Agricultural analysis	Quality analysis of water, soil, and fertilizers	[70]
Surface analysis	Evaluation of structure, thickness, and porosity	[65]

each respective system [87]. Distilled water or a buffer is injected following the attachment of the sensor chip. Air bubbles in the system can be removed by washing [88]. For the first couple of washes, the water must be at 25°C and later at 50°C. If the baseline does not stabilize after these steps, extra cleaning measures are due, which involve chemicals approved by the SPR system brand; for example, SDS, glycine, HCl, and sodium bicarbonate solutions.

8.3.6.2 Sample Preparation

Preparation of the sample is critically important for SPR studies as it directly affects the end results. First, baseline values for the sample and the running buffers must be equal [88]. Information regarding the solubility and association properties of the analyte in the buffer system is useful beforehand. In order to minimize dissimilarities and errors in measurement, all the buffers must contain the same concentration of chemicals, if any organic solvent is used, in preparing the solution. Also, higher sample injection volume leads to higher accuracy in readings.

8.3.6.3 Buffer Preparation

For any analysis with an SPR system, running and baseline buffers are necessary before and after the cleaning routines. There are a few standard buffers forming optimal environments for ligand–analyte interactions and consequently for SPR analysis [89].

8.3.6.3.1 Baseline Buffers

It is crucially important to establish a stable baseline in SPR systems. PBS and HEPES are the more commonly used buffers for baseline. These buffers should have a pH of 7.0. Tween® 20 may serve as a good surface activator. However, these surface activators interfere with the analyte–ligand interactions and cannot be used during the immobilization steps. A 0.2 μm sterile filter is useful for the preparation of the buffer, to eliminate any remaining solid particles.

8.3.6.3.2 Regeneration Buffers

Sensor chips in SPR systems are reusable, which prevents loss of time and money. However, in order to prevent any adverse effects on the subsequent analyses, cleaning must be highly effective. Different regeneration buffers should be preferred for each type of surface [90]. Regeneration buffers are categorized as follows:

- Divalent cation chelator
- High ionic strength
- Low pH
- High pH

The sensorgram data helps to determine buffer effectiveness, both during the regeneration step, and after the injection of the regeneration buffer. For example, if a buffer does not work in low pH as intended, a higher pH alternative can be used. Also, it is advisable to divide the injections into two parts: for example, two 1 min injections are preferred to a single injection that lasts 2 min. Trying mixtures of two buffers is useful if different buffers provide the same inadequate regeneration.

8.3.6.3.3 Running Buffers

Sample injection is performed with running buffers. HBS-EP, HBS-P, and HBS-N are usually used as the running buffer.

- HBS-EP: 0.01 M HEPES pH 7.4, 0.15 M NaCl, 3 mM EDTA, 0.005% v/v surfactant P20
- HBS-P: 0.01 M HEPES pH 7.4, 0.15 M NaCl, 0.005% v/v surfactant P20
- HBS-N: 0.01 M HEPES pH 7.4, 0.15 M NaCl

8.3.6.4 Selection of a Suitable Chip

There is a variety of chips used in the SPR system and each with a different mode of use. The chips are modified and pre-coupled by the proprietor. Chip types vary according to their area of use, life spans, and immobilization processes. The remainder of this section discusses the specifications of Biacore chips, which is the most commonly used SPR brand—specifically, Research Grade Chips, SA chips, NTA chips, and Pioneer Chips (Table 8.6).

- *Research Grade Chips (CM5)*: Pre-coupled with carboxymethylated dextran and suitable for general use.
- *SA Sensor Chips*: Pre-coupled with streptavidin

TABLE 8.6

General Applications of Biacore Sensor Chips

Sensor Chip Type	Applications
CM5 sensor chip	Suitable for general applications and research.
SA sensor chip	Suitable for the immobilization of biotinylated ligands.
NTA sensor chip	Suitable for the immobilization of His-tagged molecules.
B1 pioneer chip	Suitable for molecules with low immobilization capacities.
C1 pioneer chip	Suitable for binding very large molecules onto dextran matrices.
F1 pioneer chip	Suitable for molecules with low immobilization capacities.
L1 pioneer chip	Suitable for the direct binding of liposomes.

Notes: The sensor chips of the Biacore system and their applications are listed. Sensor chip surace features are discussed in Section 4 of the protocols.

- *NTA Sensor Chips*: Pre-coupled with nitrilotriacetic acid
- *Pioneer Chips (B1, C1, F1, L1)*: These sensor chips are often used in assays that bind large molecules. This group contains pre-coupled chips with various concentrations of dextran matrix, carboxymethyl, and lipophilic compounds.

8.3.6.5 Immobilization of Sensor Chip

Determining the appropriate immobilization method is crucial for SPR systems. Covalent immobilization and non-covalent capture are the two major immobilization methods.

8.3.6.5.1 Covalent Immobilization

Even though there are various covalent immobilization methods, amine, thiol, aldehyde, and maleimide couplings are the most commonly used groups for ligand immobilization. Optimally, immobilization should be completed in the shortest time possible and maintain over 95% purity. [91]. Determining the optimal amount for appropriate biomolecules is crucial in the immobilization process, as low amounts lead to better results [88]. High-purity biomolecules ensure high bonding capacity and analysis accuracy. The operator should check the RU value upon immobilization of the ligand. The immobilization process must be repeated if there is any remaining ligand on the chip.

8.3.6.5.1.1 Amine Coupling Amine coupling is the most widely used technique among covalent coupling methods. It is useful for immobilizing proteins and other biomolecules with protein-like structures [92]. Initially, dextran matrix is applied to the surface of the chip for activation with NHS–EDC, which takes 1–10 min [91]. Later, a protein-like ligand is injected into a buffer with low salt, which has a lower pH than the ligand's isoelectric point. A failed immobilization attempt can be improved by increasing the NHS–EDC activation concentration and the activation time.

8.3.6.5.1.2 Thiol Coupling Thiol coupling is a method based on exchange reactions between thiol and disulfide groups. The ligand should contain a free thiol group for immobilization. While amine coupling is useful for immobilizing protein-like molecules, thiol coupling facilitates the immobilization of molecules such as glycoproteins and polysaccharides [65]. As with amine coupling, activation must be performed prior to the immobilization. Following the activation of the sensor chip surface, a thiol coupling reagent such as PDEA provides reactive disulfide groups, which facilitate binding between the ligand and the chip's surface. Reactive disulfide bonds will interact with thiols upon injection of the biomolecules and form covalent bonds [73].

8.3.6.5.1.3 Aldehyde Coupling Aldehyde coupling uses hydrazone molecules, formed by the condensation of the hydrazide and aldehyde groups. Aldehyde coupling can be used to immobilize polysaccharides, glycoproteins, and glycoconjugates, and the orientation-specific immobilization of the functional groups that contain proteins. The NHS–EDC method is useful for surface modifications, as with amine and thiol groups. Reactive hydrazide groups interact with biomolecules, which contain aldehydes, and form covalent bonds, thus completing the coupling process.

8.3.6.5.1.4 Maleimide Coupling Maleimide coupling is a method similar to thiol coupling. The technique is based on thioether bonding between the ligand and the chip, facilitated by reactive maleimide. Sulfo-MBS and sulfo-SMCC cross-linkers are generally used for maleimide coupling.

8.3.6.5.2 Non-covalent Capture

Non-covalent (i.e., indirect) immobilization is an alternative to covalent immobilization methods and provides some advantages. In this method, it is possible to use less pure ligands, as the need to purify the ligand is unnecessary in some cases. However, in order to perform indirect immobilization, the surface should be suitable or modifiable for the specific ligand. Non-covalent immobilization agents must have high affinity and must be highly specific, and the immobilization levels of the ligand must be the same for each repeat experiment. This system forms a homogeneous surface on the sensor chip. The chip's surface is regenerated through the elution of the analyte that was previously bound to the ligand. The operator should confirm the integrity and activity of the ligand after each immobilization procedure.

8.3.7 Data Analysis

Primary data gathered from the SPR experiments are obtained as sensorgrams that can provide much valuable information regarding the interactions. At this point, even though it might be easy to evaluate the interactions between biomolecules, it is much harder to interpret the binding kinetics of these interactions [93]. Data gathered by the experiments are processed using software such as BIA Evaluation, Scrubber-2, CLAMP, and KD-Assistant. During data processing, sensorgrams of the reference solution and sample solution are separated from each other. The extraction procedure is repeated for each concentration. Even though automatic extraction is an option, manual extraction yields higher accuracy [87]. The next step involves the analysis and assessment of the results by using suitable kinetic models.

Notes for QCM

1. QCM measurements are extremely sensitive to vibrations, air flow, crystal position, temperature, and electrical contact, which cause long-term drift. Therefore, to minimize these effects, some measures are necessary:
   - The flow cell should be placed on a table that is not subject to vibrations.
   - Tape all the openings on the flow cell cap to prevent air flow.
   - While mounting the quartz crystal between two silicon O-rings, the flow cell and the quartz crystal should be completely dry. The position of the quartz crystal must be centered to the O-ring.
   - Temperature control is the most important aspect for reproducibility. The frequency of the crystal is very sensitive to the temperature, particularly if the crystal is in the electrolyte solution. Temperature plays a very significant role in long-term drift. Consequently, place the flow cell into a thermostatic bath or use a temperature control QCM system.
   - The wires that connect to the crystal holder should be in a resting position to reduce tension and possible movement.
   - If possible, the flow cell should be placed into a Faraday cage, to prevent external electrical interference that causes noise in measurements.
2. The crystals can be cleaned using 1:1:5 parts of NH_3:H_2O_2 (30%):H_2O. Since such cleaning solutions are severely caustic, they should only be used for short durations (1–5 min) to prevent damage to the electrodes. Following the cleaning procedure, the QCM crystal is rinsed with ethanol and deionized double-distilled water and dried in a gentle flow of nitrogen gas to obtain a clean and hydrophobic surface.
3. UV light at the specific wavelengths of 185 and 254 nm is used to remove the organic contaminants on the crystal's surface.
4. The frequency measurement begins after the change in frequency is less than 1 Hz/10 min.
5. Note that air bubbles in the flow cell may cause frequency jumps. If air bubbles occur, gently tap the flow cell with your fingertip until the bubbles pass through and exit the flow cell. Further, the flow rate should be stable in all experiments.

Notes for SPR

1. Samples should be degassed before use and the volume should be greater than 3 mL. In addition, the sample should be kept in closed vials to avoid sample loss.
2. Buffers must be degassed at room temperature upon preparation and before use.
3. SA and NTA sensor chips, which are processed via special pre-coupling methods, are expensive and unsuitable for reuse.
4. The NHS–EDC mixture ratio should be 1:1, with EDC (0.4 M) and NHS (0.1 M) in the coupling step. In addition, the NHS–EDC mixture should be prepared just before the immobilization process.
5. After the immobilization procedure, the operator should be careful not to touch the surface of the chip after placing it on the SPR device.
6. After the immobilization procedure, the sensor chip removed from the system should be placed into a 50 mL tube for reuse. Make sure to immerse the entire surface of the chip.
7. PDEA used in the thiol coupling process must be prepared with a fresh 0.1 M borate buffer at pH 8.5.
8. In non-covalent immobilization processes, ligand solution should not contain EDTA, and the running buffer should be 10 mM HEPES, pH 7.4, 150 mM NaCl, 50 μM EDTA, 0.005% surfactant P20.

References

1. Pohanka M., P. Skládal, and M. Kroèa. 2007. Biosensors for biological warfare agent detection. *Defence Science Journal* 57: 185–193.
2. Cooper M. A. 2003. Label-free screening of bio-molecular interactions. *Analytical and Bioanalytical Chemistry* 377: 834–842.
3. Curie J., and P. Curie. 1880. An oscillating quartz crystal mass detector. *Rendu* 91: 294–297.
4. Lippmann G. 1881. Principe de conservation de l'électricité. *Annales de physique et de* 5ª *chimie,* Serie 24: 145–178.
5. Bunde R. L., E. J. Jarvi, and J. J. Rosentreter. 1998. Piezoelectric quartz crystal biosensors. *Talanta* 46(6): 1223–1236.
6. Janshoff A., H. J. Galla, and C. Steinem. 2000. Piezoelectric mass-sensing devices as biosensors: An alternative to optical biosensors? *Angewandte Chemie International Edition in English* 39: 4004–4032.
7. Sauerbrey G. 1959. Verwendung von Schwingquarzen zur Wägung dünner Schichten und zur Mikrowägung. *Zeitschrift Physik* 155: 206–212.

8. Vashist S. K., and P. Vashist. 2011. Recent advances in quartz crystal microbalance-based sensors. *Journal of Sensors* 2011: 1–13.
9. Tuantranont A, A. Wisitsora-at, P. Sritongkham, and K. Jaruwongrungsee. 2011. A review of monolithic multichannel quartz crystal microbalance: A review. *Analytica Chimica Acta* 687(2): 114–128.
10. Kanazawa K. K., and J. G. Gordon. 1985. Frequency of a quartz microbalance in contact with liquid. *Analytical Chemistry* 57: 1770–1771.
11. Janshoff A., and C. Steinem. 2001. Quartz crystal microbalance for bioanalytical applications. *Sensors Update* 9: 313–354.
12. Lüthgens E., A. Herrig, K. Kastl, et al. 2003. Adhesion of liposomes: A quartz crystal microbalance study. *Measurement Science and Technology* 14: 1865–1875.
13. Arnau A. 2008. A review of interface electronic systems for AT-cut quartz crystal microbalance applications in liquids. *Sensors* 8: 370–411.
14. Lucklum R., and F. Eichelbaum. 2007. Interface circuits for QCM sensors. *Springer Series on Chemical Sensors and Biosensors* 5: 3–47.
15. Kulkarni A. P., L. A. Kellaway, and G. J. Kotwal. 2012. Application of quartz crystal microbalance with dissipation monitoring technology for studying interactions of poxviral proteins with their ligands. *Methods in Molecular Biology* 890: 289–303.
16. Liu G., and G. Zhang. 2013. Basic principles of QCM-D. In *QCM-D Studies on Polymer Behavior at Interfaces,* (Sharma S.K., Ed.) pp. 1–8. Springer Briefs in Molecular Science. New York: Springer.
17. Mishra G. K., A. Sharma, and S. Bhand. 2015. Ultrasensitive detection of streptomycin using flow injection analysis–electrochemical quartz crystal nanobalance (FIA-EQCN) biosensor. *Biosensors and Bioelectronics* 67: 532–539.
18. Ogi H., Y. Fukunishi, T. Omori, et al. 2008. Effects of flow rate on sensitivity and affinity in flow injection biosensor systems studied by 55-MHz wireless quartz crystal microbalance. *Analytical Chemistry* 80(14): 5494–5500.
19. Bilek M. M., and D. R. McKenzie. 2010. Plasma modified surfaces for covalent immobilization of functional biomolecules in the absence of chemical linkers: Towards better biosensors and a new generation of medical implants. *Biophysical Reviews* 2: 55–65.
20. Muguruma H. 2010. Plasma-polymerized films for biochip design. *Plasma Processes and Polymers* 7: 151–162.
21. Ramos-Jesus J., L. C. Pontes-de-Carvalho, S. M. B. Melo, et al. 2016. A gold nanoparticle piezoelectric immunosensor using a recombinant antigen for detecting Leishmania infantum antibodies in canine serum. *Biochemical Engineering Journal* 110: 43–50.
22. Kittle J. D., C. Wang, C. Qian, et al. 2012. Ultrathin chitin films for nanocomposites and biosensors. *Biomacromolecules* 13(3): 714–718.
23. Wong-ek K., O. Chailapakul, N. Nuntawong, et al. 2010. Cardiac troponin T detection using polymers coated quartz crystal microbalance as a cost-effective immunosensor. *Biomedizinische Technik/Biomedical Engineering* 55(5): 279–284.
24. Şeker Ş., Y. E. Arslan, and Y. M. Elçin. 2010. Electrospun nanofibrous PLGA/Fullerene-C60 coated quartz crystal microbalance for real-time gluconic acid monitoring. *IEEE Sensors Journal* 10: 1342–1348.

25. Jia Y., H. Yu, Y. Zhang, et al. 2016. Cellulose acetate nanofibers coated layer-by-layer with polyethylenimine and graphene oxide on a quartz crystal microbalance for use as a highly sensitive ammonia sensor. *Colloids and Surfaces B: Biointerfaces*, 148: 263–269.
26. Matsumoto K., B. D. B. Tiu, A. Kawamura, et al. 2016. QCM sensing of bisphenol A using molecularly imprinted hydrogel/conducting polymer matrix. *Polymer Journal* 48(4): 525–532.
27. Molino P. J., M. J. Higgins, P. C. Innis, et al. 2012. Fibronectin and bovine serum albumin adsorption and conformational dynamics on inherently conducting polymers: A QCM-D study. *Langmuir* 28(22): 8433–8445.
28. Gerard M., A. Chaubey, and B. D. Malhotra. 2002. Application of conducting polymers to biosensors. *Biosensors and Bioelectronics* 17(5): 345–359.
29. Ferreira D. C., A. E. da Hora Machado, F. de Souza Tiago, et al. 2012. Molecular modeling study on the possible polymers formed during the electropolymerization of 3-hydroxyphenylacetic acid. *Journal of Molecular Graphics and Modelling* 34: 18–27.
30. Guimard N. K., N. Gomez, and C. E. Schmidt. 2007. Conducting polymers in biomedical engineering. *Progress in Polymer Science* 32: 876–921.
31. Ying Z., Y. Jiang, X. Du, et al. 2007. A study of PVDF Langmuir–Blodgett thin film using quartz crystal microbalance. *Journal of Applied Polymer Science* 106(2): 1024–1027.
32. O'Dwyer C., G. Gay, B. Viaris de Lesegno, and J. Weiner. 2004. The nature of alkanethiol self-assembled monolayer adsorption on sputtered gold substrates. *Langmuir* 20(19): 8172–8182.
33. Chaki N. K., and K. Vijayamohanan. 2002. Self-assembled monolayers as a tunable platform for biosensor applications. *Biosensors and Bioelectronics* 17(1–2): 1–12.
34. Atashbar M. Z., B. Bejcek, A. Vijh, and S. Singamaneni. 2005. QCM biosensor with ultra thin polymer film. *Sensors and Actuators B: Chemical* 107: 945–951.
35. Suri C. R., and G. C. Mishra. 1996. Activating piezoelectric crystal surface by silanization for microgravimetric immunobiosensor application. *Biosensors and Bioelectronics* 11(12): 1199–1205.
36. Hauck S., S. Drost, E. Prohaska, et al. 2002. Analysis of protein interactions using a quartz crystal microbalance biosensor. In *Protein–Protein Interactions: A Molecular Cloning Manual,* Golemis E.A. and Adams P.D., Ed.) Cold Spring Harbor, NY: Cold Spring Harbor Laboratory Press, pp. 273–283.
37. Wei L. F., and J. S. Shih. 2001. Fullerene-cryptand coated piezoelectric crystal urea sensor based on urease. *Analytica Chimica Acta* 437: 77–85.
38. Hu G., J. A. Heitmann, and O. J. Rojas. 2009. In situ monitoring of cellulase activity by microgravimetry with a quartz crystal microbalance. *The Journal of Physical Chemistry B* 113: 14761–14768.
39. Bouchet-Spinelli A., L. Coche-Guérente, S. Armand, et al. 2013. Functional characterization of starch-degrading enzymes using quartz crystal microbalance with dissipation monitoring (QCM-D). *Sensors and Actuators B: Chemical* 176: 1038–1043.
40. Laszlo J. A., and K. O. 2007 Evans.. Influence of self-assembled monolayer surface chemistry on *Candida antarctica* lipase B adsorption and specific activity. *Journal of Molecular Catalysis B: Enzymatic* 48: 84–89.

41. Weltin A., S. Hammer, F. Noor, et al. 2017. Accessing 3D microtissue metabolism: Lactate and oxygen monitoring in hepatocyte spheroids. *Biosensors and Bioelectronics* 87: 941–948.
42. Michalzik M., J. Wendler, J. Rabe, et al. 2005. Development and application of a miniaturised quartz crystal microbalance (QCM) as immunosensor for bone morphogenetic protein-2. *Sensors and Actuators B: Chemical* 105: 508–515.
43. Vaughan R. D., C. K. O'Sullivan, G. G. Guilbault. 2001. Development of a quartz crystal microbalance (QCM) immunosensor for the detection of Listeria monocytogenes. *Enzyme and Microbial Technology* 29(10): 635–638.
44. Tsai W.-C., and I.-C. Lin. 2005. Development of a piezoelectric immunosensor for the detection of alpha-fetoprotein. *Sensors and Actuators B: Chemical* 106: 455–460.
45. Kurosawa S., J. W. Park, H. Aizawa, et al. 2006. Quartz crystal microbalance immunosensors for environmental monitoring. *Biosensors and Bioelectronics* 22(4): 473–481.
46. Mannelli I., M. Minunni, S. Tombelli, et al. 2005. Direct immobilisation of DNA probes for the development of affinity biosensors. *Bioelectrochemistry* 66: 129–138.
47. D'Agata R., P. Palladino, and G. Spoto. 2017. Streptavidin-coated gold nanoparticles: Critical role of oligonucleotides on stability and fractal aggregation. *Beilstein Journal of Nanotechnology* 8(1): 1–11.
48. García-Martinez G., E. A. Bustabad, H. Perrot, et al. 2011. Development of a mass sensitive quartz crystal microbalance (QCM)-based DNA biosensor using a 50 MHz electronic oscillator circuit. *Sensors* 11: 7656–7664.
49. Cho Y. K., S. Kim, Y. A. Kim, et al. 2004. Characterization of DNA immobilization and subsequent hybridization using in situ quartz crystal microbalance, fluorescence spectroscopy, and surface plasmon resonance. *Journal of Colloid and Interface Science* 278(1): 44–52.
50. Nowacki L., J. Follet, M. Vayssade, et al. 2015. Real-time QCM-D monitoring of cancer cell death early events in a dynamic context. *Biosensors and Bioelectronics* 64: 469–476.
51. Guan Z. Y., C. W. Huang, M. C. Huang, et al. 2017. Controlling multi-function of biomaterials interfaces based on multiple and competing adsorption of functional proteins. *Colloids and Surfaces B: Biointerfaces* 149: 130–137.
52. Şeker Ş., A. E. Elçin, and Y. M. Elçin. 2016. Real-time monitoring of mesenchymal stem cell responses to biomaterial surfaces and to a model drug by using quartz crystal microbalance. *Artificial Cells, Nanomedicine, and Biotechnology* 44(7): 1722–1732.
53. Wang G., A. H. Dewilde, J. Zhang, et al. 2011. A living cell quartz crystal microbalance biosensor for continuous monitoring of cytotoxic responses of macrophages to single-walled carbon nanotubes. *Particle and Fibre Toxicology* 8(4): 1–17.
54. Yang R., J. Y. Chen, N. Xi, et al. 2012. Characterization of mechanical behavior of an epithelial monolayer in response to epidermal growth factor stimulation. *Experimental Cell Research* 318(5): 521–526.
55. Reyes P. I., Z. Duan, Y. Lu, et al. 2013. ZnO nanostructure-modified QCM for dynamic monitoring of cell adhesion and proliferation. *Biosensors and Bioelectronics* 41: 84–89.

56. Da-Silva A. C., S. S. Soares, and G. N. Ferreira. 2013. Acoustic detection of cell adhesion to a coated quartz crystal microbalance: Implications for studying the biocompatibility of polymers. *Biotechnology Journal* 8(6): 690–698.
57. Zhou T., K. A. Marx, A. H. Dewilde, et al. 2012. Dynamic cell adhesion and viscoelastic signatures distinguish normal from malignant human mammary cells using quartz crystal microbalance. *Analytical Biochemistry* 421(1): 164–171.
58. Fohlerová Z., J. Turánek, and P. Skládal. 2012. The cell adhesion and cytotoxicity effects of the derivate of vitamin E compared for two cell lines using a piezoelectric biosensor. *Sensors and Actuators B: Chemical* 174: 153–157.
59. Watarai E., R. Matsuno, T. Konno, et al. 2012. QCM-D analysis of material–cell interactions targeting a single cell during initial cell attachment. *Sensors and Actuators B: Chemical* 171: 1297–1302.
60. Cho W., L. Bittova, and R. V. Stahelin. 2001. Membrane binding assays for peripheral proteins. *Analytical Biochemistry* 296: 153–161.
61. Yadav S. P., S. Bergqvist, M. L. Doyle, et al. 2012. Snapshot of rapidly evolving label-free technologies used for characterizing molecular interactions. *Journal of Biomolecular Techniques* 23: 94–100.
62. Hahnefeld C., S. Drewianka, and F. W. Herberg. 2004. Determination of kinetic data using surface plasmon resonance biosensors. *Methods in Molecular Medicine* 94: 299–220.
63. Vala M., R. Robelek, M. Bocková, J. Wegener, and J. Homola. 2013. Real-time label-free monitoring of the cellular response to osmotic stress using conventional and long-range surface plasmons. *Biosensors and Bioelectronics* 40: 417–421.
64. Schuck P. 1997. Use of surface plasmon resonance to probe the equilibrium and dynamic aspects of interactions between biological macromolecules. *Annual Review of Biophysics and Biomolecular Structure* 26: 541–566.
65. Singh P. 2014. *Surface Plasmon Resonance: Nanotechnology Science and Technology.* New York: Novinka Press, pp. 20–33.
66. Huet A.-C., T. Fodey, S. A. Haughey, et al. 2010. Advances in biosensor-based analysis for antimicrobial residues in foods. *Trends in Analytical Chemistry* 29: 1281–1294.
67. Mungroo N. A., and S. Neethirajan. 2014. Biosensors for the detection of antibiotics in poultry industry: A review. *Biosensors* 4(4): 472–493.
68. Sathiyamoorthy K., J. Joseph, C. J. Hon, and M. V. Matham. 2011. Photoacoustic based surface plasmon resonance spectroscopy: An investigation. In *International Conference on Applications of Optics and Photonics* (pp. 80010K–80010K). International Society for Optics and Photonics, pp. 359–382.
69. Liu H., B. Wang, E. Leong, et al. 2010. Enhanced surface plasmon resonance on a smooth silver film with a seed growth layer. *ACS Nano* 4(6): 3139–3146.
70. Hammond J. L., N. Bhalla, S. D. Rafiee, and P. Estrela. 2014. Localized surface plasmon resonance as a biosensing platform for developing countries. *Biosensors* 4(2): 172–188.
71. Wijaya E., C. Lenaerts, S. Maricot, et al. 2011. Surface plasmon resonance-based biosensors: From the development of different SPR structures to novel surface functionalization strategies. *Current Opinion in Solid State and Materials Science* 15: 208–224.
72. Besenicar M., P. Macek, J. H. Lakey, and G. Anderluh. 2006. Surface plasmon resonance in protein–membrane interactions. *Chemistry and Physics of Lipids* 141(1–2): 169–178.

73. Vachali P. P., B. Li, A. Bartschi, and P. S. Bernstein. 2015. Surface plasmon resonance (SPR)-based biosensor technology for the quantitative characterization of protein–carotenoid interactions. *Archives of Biochemistry and Biophysics* 572: 66–72.
74. Patching S. G. 2014. Surface plasmon resonance spectroscopy for characterisation of membrane protein–ligand interactions and its potential for drug discovery. *Biochimica et Biophysica Acta* 1838: 43–55.
75. Skoog D. A., F. J. Holler, and S. R. Crouch. 2007. *Principles of Instrumental Analysis*, 6th ed. Belmont, CA: Thomson Brooks/Cole.
76. Ling J., H. Liao, R. Clark, et al. 2008. Structural constraints for the binding of short peptides to claudin-4 revealed by surface plasmon resonance. *The Journal of Biological Chemistry* 283(45): 30585–30595.
77. Myszka D. G., and R. L. Rich. 2003. SPR's high impact on drug discovery: Resolution, throughput and versatility. *Drug Discovery World*, Spring: 49–55.
78. Krone J. R., R. W. Nelson, D. Dogruel, et al. 1997. BIA/MS: Interfacing biomolecular interaction analysis with mass spectrometry. *Analytical Biochemistry* 244(1): 124–132.
79. Sharma B., R. R. Frontiera, A.-I. Henry, et al. 2012. SERS: Materials, applications, and the future. *Materials Today* 15: 16–25.
80. Iliescu T., M. Baia, and D. Maniu. 2008. Raman and surface enhanced Raman spectroscopy on molecules of pharmaceutical and biological interest. *Romanian Reports in Physics* 60: 829–855.
81. Ziblat R., V. Lirtsman, D. Davidov, and B. Aroeti. 2006. Infrared surface plasmon resonance: A novel tool for real time sensing of variations in living cells. *Biophysical Journal* 90: 2592–2599.
82. Damos F. S., R. K. Mendes, and L. T. Kubota. 2004. Applications of QCM, EIS and SPR in the investigation of surfaces and interfaces for the development of (bio)sensors. *Química Nova* 27(6): 970–979.
83. Bund A., A. Baba, S. Berg, et al. 2003. Combining surface plasmon resonance and quartz crystal microbalance for the in situ investigation of the electropolymerization and doping/dedoping of poly(pyrrole). *The Journal of Physical Chemistry* B 107(28): 6743–6747.
84. Mohan T., K. Niegelhell, C. Nagaraj, et al. 2017. Interaction of tissue engineering substrates with serum proteins and its influence on human primary endothelial cells. *Biomacromolecules* 18: 413–421
85. Olaru A., C. Bala, N. Jaffrezic-Renault, H. Y. Aboul-Enein. 2015. Surface plasmon resonance (SPR) biosensors in pharmaceutical analysis. *Critical Reviews in Analytical Chemistry* 45(2): 97–105.
86. Reder-Christ K., and G. Bendas. 2011. Biosensor applications in the field of antibiotic research: A review of recent developments. *Sensors* 11(10): 9450–9466.
87. Nguyen B., F. A. Tanious, and W. D. Wilson. 2007. Biosensor–surface plasmon resonance: Quantitative analysis of small molecule–nucleic acid interactions. *Methods* 42(2): 150–161.
88. Tanious F. A., B. Nguyen, and W. D. Wilson. 2008. Biosensor–surface plasmon resonance methods for quantitative analysis of biomolecular interactions. *Methods in Cell Biology* 84: 53–57.
89. Schasfoort R. B. M., and A. J. Tudos. 2008. *Introduction to Surface Plasmon Resonance*. In *Handbook of Surface Plasmon Resonance*, Schasfoort R.B.M., and A.J. Tudos (Ed.). London: Royal Society of Chemistry, pp. 1–26.

90. Herberg F. W., and B. Zimmermann. 1999. Analysis of protein kinase interactions using biomolecular interaction analysis. In *Protein Phosphorylation: A Practical Approach*, 2nd ed. Oxford: Oxford University Press, pp. 335–371.
91. Guiducci C. 2011. Surface plasmon resonance based systems (handout). Ecole Polytechnique Fédérale de Lausanne, Switzerland, pp. 1–28.
92. Daghestani H. N., and B. W. Day. 2010. Theory and applications of surface plasmon resonance, resonant mirror, resonant waveguide grating, and dual polarization interferometry biosensors. *Sensors* 10(11): 9630–9646.
93. Myszka D. G., and T. A. Morton. 1998. CLAMP: A biosensor kinetic data analysis program. *Trends in Biochemical Sciences* 23(4): 149–150.

9

Low-Power Energy Harvesting Solutions for Smart Self-Powered Sensors

Albert Álvarez-Carulla, Jordi Colomer-Farrarons, and Pere Ll. Miribel

CONTENTS

9.1 Introduction 217
9.2 Energy Harvesting 218
  9.2.1 Types of Energy Harvesting Sources and Power Ranges 218
  9.2.2 MPPT Algorithms 223
  9.2.3 Fields of Application: From Bridge Monitoring to Medical Implants 226
  9.2.4 Powering Solutions for Human Wearable and Implantable Devices 233
9.3 Multi-Source Self-Powered Device Conception 237
9.4 Summary and Conclusions 240
9.5 Acknowledgments 240
References 241

9.1 Introduction

Interest in the recovery of available energy from the environment is constantly increasing. Some of the key drivers include present and future concerns regarding the greenhouse effect in relation to climate change and the tremendous benefits conferred by reducing power consumption in devices we use on a daily basis, such as mobile phones, cars, airplanes, handheld devices, sensors, and so on.

The possibility of converting the energy available in the human environment and the capacity to transform it into useful energy has resulted in the creation of infrastructure capable of recovering tens of megawatts in the form of electrical energy—for example, wind turbines or wave energy systems. This is so-called macro-scale energy harvesting and is perhaps best known to the general public. While most people pinpoint macro-harvesting, there is an important piece of energy hidden on the micro-scale. However, there is a lack of clarity in the meaning of recovery on the micro-scale. In this chapter and in the state-of-the-art literature, *micro-scale energy harvesting* refers to the recovery of energy in the range of nanowatts to milliwatts [1].

Evidently, the most appropriate means to implement micro-scale levels of energy recovery depends largely on the field of application. Prime examples are portable consumer products; the possibility of keeping a mobile phone, laptop, or PDA charged without the need for a battery, with all the attendant environmental advantages, gives us a clear perspective of the great potential of micro-energy harvesting. If electronic equipment could generally avoid the need for a battery, being charged instead by the user's own movements, by the difference in temperature between it and the environment, the existing ambient light, or the electromagnetic waves it produces in the environment, and if we extend this to millions of mobile telephones, we can extrapolate the enormous advantages of being able to develop efficient electronic systems without batteries.

In the same vein, it is important to highlight the benefits of intelligent sensors in all those applications able to adapt to small energy recovery levels. These smart sensors have the facilities to self-charge with harvested energy, measure environmental phenomena, process information, and transmit data from their location to a remote station. The fields of application are broad: from the monitoring of the structural condition of bridges, highways, intelligent buildings, aircraft, and so on [2], through distributed sensor networks applied to pervasive medical healthcare [3], to the surveillance of large environmental facilities, and so on.

In general, the conception and development of intelligent and autonomous sensors without battery requirements bring many advantages from the point of view of cost, maintenance, and so on. Accordingly, this chapter will review the principal sources of energy present in the environment that are capable of being recovered and converted to electrical energy, with attention to those orientated to the micro- or nanoscale. Moreover, the topic of maximum power point tracking (MPPT) algorithms will be introduced, which can be applied to any type of energy harvesting sources (EHSs). Finally, the reader will find a review of several application examples, from structural health monitoring (SHM) to implantable devices.

9.2 Energy Harvesting

9.2.1 Types of Energy Harvesting Sources and Power Ranges

There exist various sources for the recovery of energy present in the environment, and those sources as well as the type of conducting element determine the admissible levels of recoverable energy, as do the fields of application. Among the typical sources and energy transducers available in the environment, there are, for example, vibrations [4], heat [5], light [6], and radio waves [7]. The available energy per unit area, or volume, for each one of these sources depends heavily on the size, operating conditions, and technologies available [1]. Some values are introduced in Table 9.1 [8–15] for vibrational, ambient light, radio frequency (RF), and thermoelectrical sources. Vibrational energy is a typical approach to harvesting energy. Three main solutions are developed based on the type of transducer or micro-generator: piezoelectric (~200 μW/cm^3), electrostatic (~50–100 μW/cm^3), and electromagnetic (<1 μW/cm^3). There are wide-ranging scenarios where these types of harvesting are, or could be, applied basically to monitor structural integrity. We can think of different fields of applications in terms of the following: (1) stiff structures that themselves produce movement (e.g., ships, containers, mobile devices, housings of fans, escalators and elevators in public places, appliances, refrigerators, bridges, automobiles, building structures, trains), (2) elastic structures that exhibit elastic deformation of their walls (e.g., rotor blades, windmill blades, aircraft wings, pumps, motors, HVAC ducts, rotorcraft), and (3) soft structures with very low elastic modulus and high deformation ratios (e.g., different textiles, leather, rubber membranes, piping with internal fluid flow).

TABLE 9.1

Derived From

Energy Source		Performance
Ambient Light	Indoor 10μW/cm^2 (low illumination) Typical office 100μW/cm^2 Outdoor 10mW/cm^2 Full bright sun 10 mW/cm^2	Solar Cells (6830 lx 10W/m^2) Indoor solar cells (10lx to 1400 lx)
Vibrational	4 μW/cm^3 (human motion Hz range) 800 μW/cm^3 (machines-kHz ranges) These numbers depend heavily on size, excitations, technologies, etc. Typically: Piezoelectric ~200 μW/cm^3 Electrostatic ~50–100 μW/cm^3 Electromagnetic <1μW/cm^3	Microgenerators[11,12] 350μW[13] 22μW[14] 400μW[15]
RF	GSM 4μW/cm^3 WiFi 1μW/cm^3 These number depen heavily on frequency of operation and distance between base station and receiver.	< 1μW/cm^2 transmitter[12] ~1mW for proximate stations (inductive coils) @900MHz, 1.1m@24.98dBm (0.315W), ~20μW[16] @4MHz, 25mm (subcutaneous powering) >5mW[17]
Temperature	Human 25μW/cm^2-60μW/cm^2 Industry 10mW/cm^2	Thermoelectric generators 60μW/cm^2 Thermolife ΔT=5°C[18]

For industrial applications, it is possible to take advantage of *fixed-frequency* vibrations because AC-driven motors and pumps produce vibration harmonics from their drive frequency (e.g., 60 Hz in the United States and 50 Hz in Europe). For vehicles, the result is more *random* vibrations than in the other categories. Although nearly all vehicular applications provide significant vibration amplitudes and sufficient levels for energy harvesting, this energy is available more through random occurrences such as bumps/rough surfaces/dynamic frequencies than through one particular frequency of interest.

Different industrial solutions exist in the market, and we will introduce only a few examples. Advanced Linear Devices® has developed different energy harvesting modules; EnOcean GmbH defines self-powered sensor networks and monitor energy harvesting from natural motion; MicroStrain®, the Volture system from Mide Technology®, and the Perpetuum PMG® series harvest the vibrations in their installed environments to generate electricity for wireless sensors and other applications.

Thermal generators are based on the Seebeck effect and convert the temperature difference or heat into an electrical energy. An interesting example is the low-power thermoelectric generator (TEG) by Thermo Life® (Thermo Life Energy Corp, CA) [15], which has a surface area of 0.5 cm^2, is 1.6 mm thick, and can supply energy to a biosensor when in contact with the skin.

Interesting research has been presented by IMEC [16]. IMEC has demonstrated the integration of a wireless autonomous sensor system in clothes. The system is fully autonomous for its entire life and requires no servicing from the user, such us replacing or recharging the battery. The shirt has integrated electronics and can be washed in a regular washing machine. The device is powered from a rechargeable battery. The battery is constantly recharged, mainly through thermoelectric conversion of the wearer's body heat. The TEG is divided into 14 modules to guarantee user comfort. It occupies less than 1.5% of the shirt area and typically generates a power of 0.8–1.0 mW at about 1 V with regular sedentary office activity. However, if the user walks indoors, the power increases up to 2.7 mW at 22°C due to forced convection. The TEG is neither cold nor intrusive for the user. In colder environments where other clothes need to be worn on top of the shirt, the power generation is typically not affected. In summary, this is one of the major fields with a huge rate growth expected in the following years.

Ambient light is other typical example. In [6], the utilization of indoor cells for extremely low conditions of illumination and low voltages of operation is shown for a cell of 55 × 20 × 1.1 mm with a power of 5 μW at 10 lx to 200 μW at 1450 lx for a voltage drop of 2 V. RF harvesting is also applied but depends heavily on the frequency of operation and distance between the base station and the receiver, where the energy conversion takes place. Typical values for standard bands are 4 μW/cm^2 (GSM) and 1 μW/cm^2 (Wi-Fi). In the case of inductive coils, an interesting case is reported in [14], where a CMOS solution is implemented in a 0.18 μm technology, operating with an input carrier of 900 MHz and with an input voltage of 250 mV in the integrated antenna in the tag, which will theoretically generate an average power of 25 μW with an output voltage of 1.8 V to supply the tag electronics. Attention is especially focused on-body harvesting. The path of technological evolution is defining a new scenario in which it will be possible to monitor patients anywhere and at any time (Figure 9.1) [2], which is of increasing interest. The traditional approach, where patients are monitored during hospital or surgery visits, would be replaced by continuous and remote monitoring, which could have a great impact on patients' quality of life. This is the concept of *pervasive monitoring* [17, 18]. Different scenarios can be envisaged in the continuous search to meet technological challenges through the miniaturization, intelligence, and autonomy of biomedical devices [19], looking for new implantable devices, or capsules, such as ultrasound pill cameras. Here, one can see the importance of the autonomy of the devices and the importance of body harvesting, which is addressed in more detail in Section 9.2.3, along with new approaches. We will introduce some examples. The conversion, for instance, of natural bodily motion based on mechanical (vibrational) energy to electrical energy is one of the main research topics. In this context, the power that can be harvested when a human being walks or runs has been studied at different locations, with an average of 0.5 mW/cm^3 for hip, chest, elbow, upper arm, and head, and a maximum of 10 mW/cm^3 for the ankle and knee. The mechanical energy can be converted to electrical energy based, for instance, on electromagnetic, electrostatic, or piezoelectric principles. But with these approaches just a few microwatts are available [19]. The other classical source

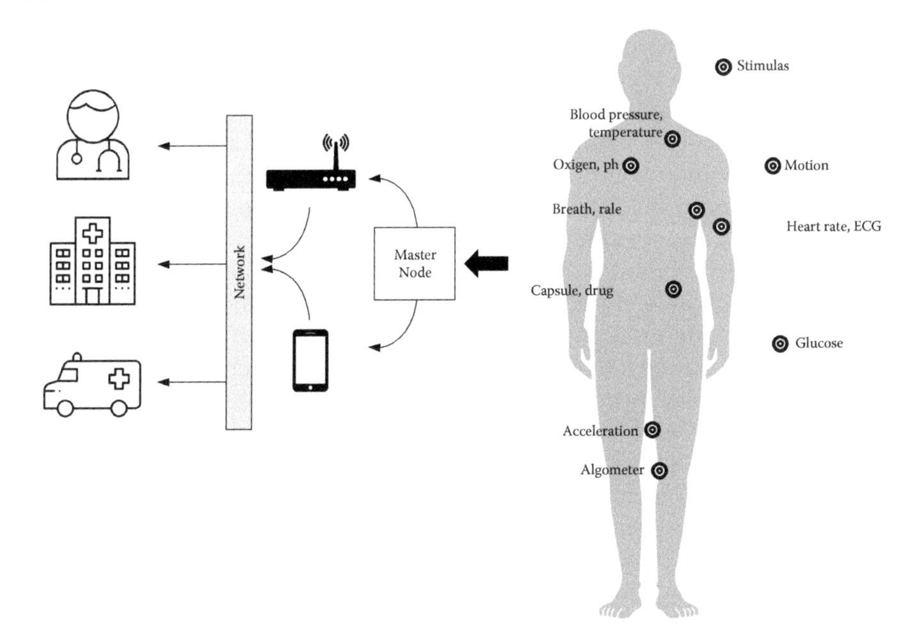

FIGURE 9.1
Typical full wireless body sensor networks (WBSNs).

of harvesting is body heat. Taking into account that the total amount of power that is wasted in the form of heat is around 100 W, the energy that can be recovered, typically, using the Carnot engine, is in the range of 2.4–4.8 W, but just a few milliwatts are actually available [18].

Furthermore, in the function of the field and the application, if the quantity of energy available for recovery is small, it is useful to be able to combine different sources at the same time. This way they are combined, and at the same time the system does not depend entirely on one source. A platform that combines different sources [20] in the environment to charge a micro-system has been presented in a form that could, at the conceptual level, be autonomous. The combination of four sources is envisaged as follows [1]: *vibrations*, based on the use of a piezoelectric generator as the conducting element; *light*, based on the use of interior solar cells; *RF signals*, through electromagnetic coupling between inductors, and finally, *thermoelectricity*. This system is implemented and described in more detail in Section 9.3 of this chapter. The ability to recover energy on the micro- or nanoscale requires the development of efficient conduction elements, and this is a topic of great interest [21] in which the great importance of energy modelization in the design of these new self-powered nanodevices and the conception of nanonetworks is envisaged [22]. Recent references exist in the field of microelectromechanical systems (MEMS), as much on the aspect of the sensor as of the energy conductor. In [23], a full review of new approaches is given by the authors. Special interest is focused on the field of energy harvesting for self-powered sensors and their expected impact against the greenhouse effect [24]. Recent developments have been published regarding new approaches for micro- and nano-harvesting purposes in the field of transducers based on the piezoelectric effect [25]. In [26], the authors present a report on piezoelectric thin films, looking for ultra-low-voltage applications and low frequencies of operation (50–500 Hz), with power densities as high as 10^{-4} W/cm^2 at 1 g acceleration. Other EHSs and transducers have been investigated. Some solutions based on the electrostatic effect have been reported recently. In [27], a sandwiched parallel-plate combination of two capacitors is implemented, with a final size of 1 cm^2, capable of operating at low accelerations and low frequencies, up to 50 Hz, with a maximum recovered power of 20 nW for a maximum acceleration of 10 ms^{-2}. This is a technological approach, but in [28] a circuitry approach to improve the recovered energy is presented that lowers conduction losses and increases the voltage across the plates. In the field of light, there are new technological approaches [29] and circuitry solutions to improve the MPPT technique, which maximizes the power that is harvested. This technique represents an significant improvement in recovered power when it is applied to systems recovering energy at the micro-scale. In such cases, it is necessary to develop MPPT techniques with lower power consumption, as in [30], where from an available power of 140 μW it is possible to recover 46.65 μW (an efficiency of 33.15%), and with power consumption in average lighting conditions (784–2152 lx) of 180 μW and efficiencies up to

39.51%. New technological advances in nanomaterials present new opportunities in the future development of the storage element, with special interest in grapheme because it presents high electrical conductivity and mechanical stability, ideal for new storage elements such as supercapacitors [31]. At the nano-scale there is interesting work being done regarding nanoantennas, which are dipole antennas 60 nm in width, operating in the THz range, where small metal–oxide–metal (MOM) tunneling diodes 50 nm in range are envisaged for the AC rectification to recover energy from the incoming signal [32]. Special interest, at the micro- and nanoscale, is focused on TEGs for energy harvesting purposes. In [33], Si/SiGe heterostructure nanoscale materials, grown on Si substrates, are derived to improve thermoelectric performance at room temperature, improving the Seebeck coefficient compared with a bulk p-type Ge generator, with comparable doping density. In general, at the level of the implementation of the electronics associated with these systems, they must present a trade-off between their level of intelligence, the level of energy they can recover, and their own consumption. For instance, in the case of TEG, a power management unit (PMU) is necessary, which generates a DC voltage level suitable for the rest of the electronics and loads present in a self-powered sensor node. In [24], a PMU is presented that is capable of setting up a low-input voltage from the TEG, as low as 150 mW to an output voltage, for ultra-low voltage circuitries of 0.85–0.90 V. In the case of vibrations coming from piezoelectric or electrostatic transducers, the PMU presents a sensible circuitry that is defined by the integrated rectifier, as in the case of RF harvesting [32]. There is an interesting possibility of combining full-wave active rectifiers [34], capable of working with low-voltage amplitudes and with high efficiencies up to 90%, with a charge-pump module to work with scavenged low voltages and generate regulated DC levels [35]. The charge-pump, as a DC–DC converter, is designed to track the maximum power point transfer, in the range of 10 μW to 200 μW, for specific load conditions, operating from 0.5 to 2.5 V with efficiency close to 50%. A more recent work by the same authors presents a different PMU with higher efficiencies, up to 93% with an MPPT also implemented, and a greater input voltage range of operation [36], from 0.44 to 4.15 V. In this case, the active rectifier also acts as a voltage doubler. In the field of RF harvesting, another key point is the design of the AC–DC stage [37]. A very interesting setup is presented in [38], where a full battery-free dynamic on/off time (DOOT) control with a Buck converter is implemented, with efficiencies up to 95% and static consumption as low as 217 nW.

9.2.2 MPPT Algorithms

All EHSs have a unique V-I characteristic curve that varies with environmental conditions—as light in photovoltaic cells (PVCs), temperature in TEGs, the amplitude and frequency of oscillation in piezoelectric generators (PEGs) in cantilever configuration, wind speed in wind turbine applications, and so on—and the intrinsic properties of the EHS itself—as the number of PVCs, the Seebeck coefficient in TEGs, the internal output resistance of PEGs, and so on. Different and unique V-I characteristics also define different and unique operating points where the EHS provides its maximum power capability. Figure 9.2 presents generic V-I and P-V curves produced by TEG generators. The extraction of the maximum power depends on environmental conditions and the intrinsic properties of each individual EHS.

Nowadays, PVCs and wind turbines are flagship EHSs for renewable energy systems (RES) in the macro-scale. These macro-scale EHSs are clean sources of energy that can provide several benefits as unlimited renewable sources, preserve and protect the environment, or improve public health. However, they have two major drawbacks: their low-efficiency conversion to electrical power and their high cost [39, 40]. In order to mitigate the high cost of the generated power, great effort has been made to improve the efficiency conversion of these EHSs. To achieve that, several techniques have been applied. C. G. Popovici et al. [41] show how efficiency conversion can be improved by controlling the temperature of the PVCs using air-cooled heat sinks. Another way is to use a solar tracking method that ensures that the time when the PVC is facing the sun is maximized, as described by V. Sumathi et al. [42]. One analogue solution applied to wind turbines consists in the use of yaw mechanisms to position the rotor to face the wind in order to maximize the power generated, as described by J. G. Njiri et al. [43]. Meanwhile, R. Tiwari et al. [44] describe how the outputted power can be controlled by adjusting the pitch or attack angle of the blades in order to maximize power during low wind speed periods and, as a secondary aim,

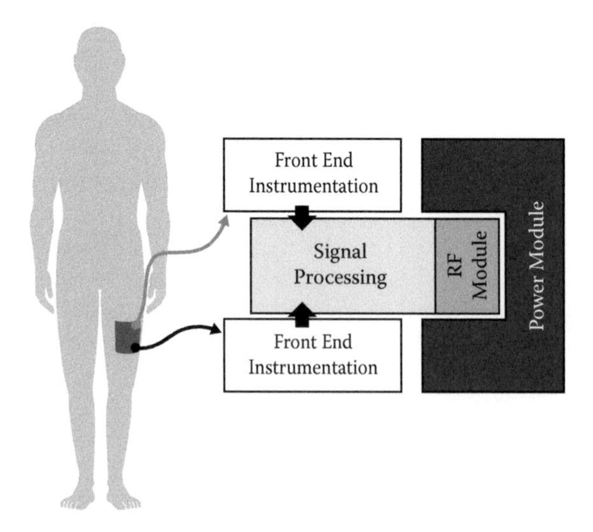

FIGURE 9.2
Typical V-I and P-V curves for TEGs: (a) different gradient temperature ($\Delta T_1 < \Delta T_2 < \Delta T_3$), and (b) different internal output resistance ($R_{S1} < R_{S2} < R_{S3}$).

protect the generator during high wind speed periods. These varieties of techniques are based on mechanical solutions to low-efficiency problems and are mostly available in macro-scale energy scenarios due to the amount of energy needed to perform the required mechanical actions. Another kind of solution to low-efficiency conversion during electrical conversion, common to all EHSs, is to use techniques that vary the equivalent electrical load impedance to set the operating point of the EHS where the power outputted is maximum. These techniques can be applied independently or in conjunction with the techniques cited previously, thus making them an option to be applied in both micro- and macro-scale scenarios. All these techniques are MPPT algorithms. MPPT algorithms are conceived to extract power from one or more power sources with maximum efficiency—that is, extracting as close to the maximum power as they can supply in the given environmental conditions. It is important to remark that this efficiency is related to the EHS and not to the module responsible for performing the MPPT algorithm itself or the global system.

The current rise of different off-grid phenomena, such as the Internet of Things (IoT), wireless sensor networks (WSNs), or the Trillion Sensor (TSensor) movement, strengthen the need to harvest energy from the environment to supply each one of the nodes in the network locally. In this way, the use and design of MPPT algorithms becomes a priority in order to operate with low amounts of energy during long periods and avoid the use of batteries. Table 9.2 [48–53] provides some of the most common MPPT algorithms applied in state-of-the-art off-grid applications. These different MPPT techniques are not exclusive to a given type of EHS, and the same technique can be used indistinctly with different sources, such as vibrational, thermal, RF, or light sources. Some examples follow.

Several industrial uses of these MPPT algorithms are present in state-of-the-art off-grid applications. D. Champier et al. [54] show how the energy wasted by biomass stoves in the form of heat can be harvested to supply small devices such as cellular phones, radios, or some LEDs. The authors present the use of a TEG placed in contact with the stove. A DC–DC converter is implemented with MPPT functionality driven by a microcontroller. The perturb and observe (P&O) technique is the MPPT algorithm adopted in this case. This technique measures the voltage and current outputted by the TEG, applying a perturbation on the duty cycle of the DC–DC converter. The system tracks the gradient of the generated power. The duty cycle perturbation of the DC–DC converter is derived in order to move the operating point toward the MPP. With this technique, the power is extracted from the TEG with an efficiency of 99%, and the system achieves an overall efficiency of 90% for extracted power levels, which ranges from below watt power values to tens of watts.

A recent implementation of the P&O technique has been presented by Z. Zeng et al. [55] for an RF energy harvesting system. In this case, Z. Zeng et al. implemented a reconfigurable Dickson rectifier [56] to convert WLAN 2.4 GHz RF energy to DC voltage. The proposed architecture rectifier enables reconfigurable

TABLE 9.2

Main Widely Used MPPT Algorithms in Off-Grid Applications

Technique	Based principle	Benefits	Drawbacks
Perturbation and Observation (P&O) [48]	Comparison of prior power outputted by the EHS with the current one after operating point perturbation.	• Ease of implementation.	• Oscillation around MPP. • Efficiency greatly affected by circuit noise. • Slow speed MPP setting.
Incremental Conductance algorithm (InC) [49]	Based on $\partial P / \partial V = 0$, aims to equalize the current conductance with incremental conductance.	• Higher tracking accuracy compared to methods that are more complex.	• More complex and computationally demanding compared to other techniques.
Incremental Impedance algorithm (InI) [49]	Same principle than InC but based on $\partial P / \partial I = 0$ and equalizing the current impedance with incremental impedance.	• Higher tracking accuracy compared to methods that are more complex.	• More complex and computationally demanding compared to other techniques.
Fractional Open Circuit Voltage (FOCV) [50]	Measurement of the open circuit voltage of the EHS to calculate and set the proper output voltage for MPP achievement.	• Simple and easy to implement.	• Unable to track short-term changes of the MPP without constantly interrupting the EHS operation for open circuit measurement reducing the power extracted.
Fractional Short Circuit Current (FSCC) [51]	Measurement of the short circuit current of the EHS to calculate and set the proper output current for MPP achievement.	• Simple and easy to implement.	• Unable to track short-term changes of the MPP without constantly interrupting the EHS operation for open circuit measurement reducing the power extracted.
Impedance Matching (ImpM) [52], [53]	The input impedance of a DC/DC converter is tuned selecting the switching frequency, the duty cycle and the value of the converter's inductance to equalize it to the EHS internal output impedance.	• Simple and easy to implement. • Does not oscillate around the MPP.	• Efficiency related to accuracy of setting the value of the inductance to the desired value. • Fixed switching frequency and inductance's value during design stage. Not possible to optimize the converter performance once it is implemented. • Discontinuous-conduction mode. At higher power levels, high current rating semiconductors are needed.

(Continued)

TABLE 9.2 (CONTINUED)

Main Widely Used MPPT Algorithms in Off-Grid Applications

Technique	Based principle	Benefits	Drawbacks
Extremum Seeking Control (ESC) [54]	Feedback control loop is used to apply a perturbation to the duty cycle of a converter, observe the power outputted by the generator and derive from it the next duty cycle perturbation to move operating point towards MPP.	• Does not oscillate around MPP. • Does not require a system model.	• More complex and computationally demanding compared to other techniques. • The existence of various parameters for the control loop increases the chances of malfunctions if no proper adjustment is performed.
Paraskevas' method [55]	The input impedance of a DC/DC converter is tuned by the MPPT module itself varying the duty cycle.	• Efficiency issues present at ImpM method due to accuracy setting the value of the inductance and switching frequency are overcome.	• Efficiency affected by the mismatch between the targeted input impedance and the variability of the internal output impedance of the generator due to environmental conditions.
Look Up Table (LUT) method [56]	Comparison of the generator's voltage and current with stored values of them into LUT at MPP for different environmental conditions.	• Fast speed MPP setting.	• Bulk storage memory proportional to the number of stored operational conditions. • Look up table specific for a given EHS.

cascaded stages to implement conversion ratios of 2, 4, 6, 8, 12, and 24 of voltage gain. Thus, when the available input power levels are low, the rectifier can be reconfigured to achieve large conversion ratios, and when large input power levels are present, the rectifier offers small conversion ratios in order to achieve better efficiency. This reconfiguration is carried out by an MPPT module that uses the P&O technique to set the operating point of the system close to the MPP. Once the rectifier is configured for a given conversion ratio, a microcontroller reconfigures the rectifier to two different conversion ratios, to the immediately lower and higher ratios. Then, the observer circuit compares the power outputted by these two configurations and indicates to the microcontroller which configuration extracts the most power. Finally, the microcontroller sets the pertinent rectifier configuration and repeats the operation, moving the operating point around the MPP. With this solution, the authors achieved a peak conversion efficiency of 38.4% for an available power range of –22–0 dBm.

Another example of MPPT application is presented by J. Hyun et al. [57]. In this study, the type of energy source is solar and the extracted energy during daytime is stored into a battery and then used to supply an LED lamp during nighttime. To implement it, the authors used a commercial off-the-self (COTS) component, the BQ25504 integrated circuit (IC) made by Texas Instruments (TI). This IC uses the fractional open-circuit voltage (FOCV) technique for MPPT algorithm implementation. The IC disconnects the EHS to measure the open-circuit voltage at the current environmental conditions. Then, after connecting the EHS to the system again, it modulates the equivalent input impedance of a boost converter to get the operating voltage at the output of the EHS to 80% of the open-circuit voltage measured previously. By implementing this MPPT module, the authors avoided efficiency drops in the system due to the voltage increase of the battery during charging. In the results, they show how the implemented system extracts a constant power level of 112 mW, while the same system without the MPPT module extracts an initial power level of 106 mW, which drops quickly until arriving at 16 mW.

In the field of aerospace, A. Álvarez-Carulla et al. [58] present a novel strain monitor circuit to be applied in SHM applications. This solution, the application of which is shown in Figure 9.3, is based on a piezoelectric material

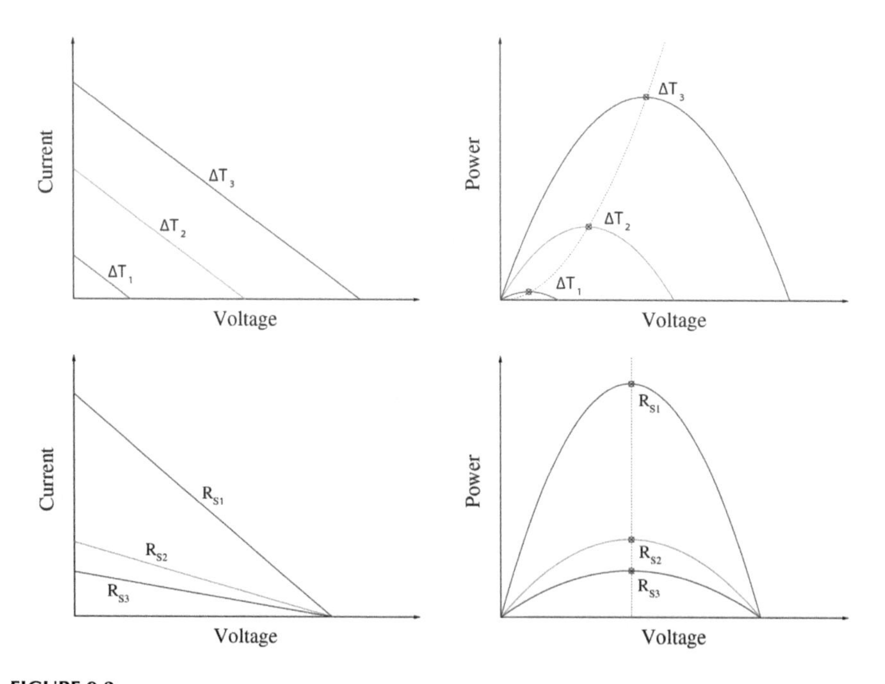

FIGURE 9.3
Diagram application of a piezoelectric-based wireless sensor node as a smart patch for aircraft wing strain monitoring. Figure extracted from [58].

that senses the strain suffered by the stiffened wing of an aircraft while it harvests energy from the mechanical vibration of the mechanical part. To perform this action, the system uses the FOCV technique to implement an MPPT algorithm. Due to the switching character of this technique, the authors exploit the measurement of the open-circuit voltage of the PEG to get a direct indicator of the suffered strain at the same time. With this mode of operation, the authors extracted a strain measurement while extracting energy from the PEG with an efficiency close to 100% over a range of environmental conditions.

9.2.3 Fields of Application: From Bridge Monitoring to Medical Implants

We have broadly introduced the types of sources that can recover energy from the environment, and among them we have introduced the monitoring of the mechanical conditions of infrastructures (e.g., bridges [59]) in the application of sensors in intelligent buildings [60], surveillance systems [38], and medical implants [18, 19]. The solutions implemented extend from the use of indoor solar cells [61], which utilize ambient light to control the environment in buildings, where the low consumption of electronics would permit the system to operate with only 100 lx; through solutions in which the electromagnetic energy present in the environment from the electrical network, such as radiation from radio stations, is utilized for mobile telephones [62]; to more revolutionary solutions such as an automatic sensor based on a miniature airflow energy harvester generator that consists of a wing attached to a cantilever, capable of generating up to 90 μW in conditions of airflow speed as low as 2 ms^{-1} [63]. An interesting work on very small self-powered sensors is reported in [64]. This is an approach measured in millimeters cubed rather than a bulky solution greater than 1 cm^3.

But particular interest is focused on self-powered devices destined for medical applications. In the last 5 years, interest in the development of so-called body sensor networks (BSNs) has increased [65]. As depicted in Figure 9.4, the miniaturization of the main electronics involved in systems such as instrumentation, signal processing modules, communications devices, and sensors, has opened up this field, where not only single wearable sensors but also implantable e-health solutions are being conceived and new trends are being introduced, such as pervasive monitoring [17]. This approach should represent lower costs for health systems thanks to the reduced number of visits and hospital stays patients will require. Different scenarios can be envisaged in the continuous search to meet technological challenges through miniaturization, including micro- and nanotechnologies, intelligence, and autonomy [18]. As has been pointed out, special interest has been taken in developments that address chronic illness [66], where traditional approaches are not capable of ruling out sudden-death events, particularly in the case of cardiovascular illness [67]. Opportunities to detect symptoms early in patients who are at risk, to monitor patients who are following a

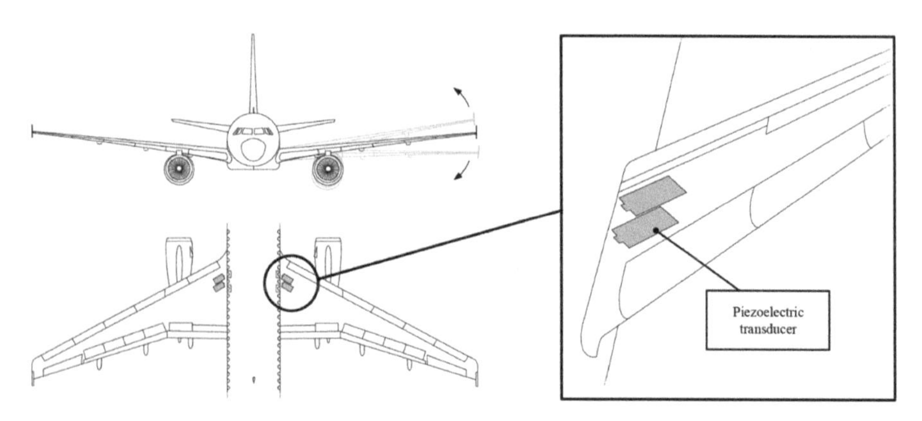

FIGURE 9.4
Miniaturization of the main electronics.

course of treatment, or to track how a disease is progressing offer the possibility to prevent the worst scenarios. In particular, it is of special interest to track the medical parameters of patients who are not following traditional clinical observation in hospital. The trick is to monitor patients throughout their daily routine. Another scenario focuses on elderly patients, who constitute a population at risk; as people live longer, the need for medical resources increases. The possibility of monitoring this substantial proportion of the population remotely from their homes will reduce medical costs [68]. However, BSN monitoring is also opening up new options of interest to hospitals. People who undergo surgery may need to be tracked via intensive monitoring immediately before and after the operation, not just while they are confined to bed. Furthermore, the possibility of monitoring the patient's specific intervention zone after surgery is increasingly becoming a topic of interest [69]. However, the development of such solutions is quite complex. Different papers have been published recently that focus on applications in different fields. These include, for instance, approaches that present multi-sensing platforms for patients who could be monitored in their daily lives at home [70]. A variety of sensors are placed in a monitored home laboratory where different activities are tracked to provide information regarding the motion of the patient; but this is not limited to an option based on wearable sensors. The authors combine bodily devices with ambient sensors programmed for specific algorithms that have the ability to identify only one activity at a time. Other approaches have focused on single units that consist of different multi-sensors [71]. In such approaches, a single unit placed on the chest of the patient is able to monitor body motion, activity intensity, and other parameters such as heart rate, provide electrocardiogram (ECG) data, and so on.

BSN solutions are not just conceived as discrete external resources. Specific and highly complex solutions are envisaged in the search for devices that are truly implantable in the body and not just positioned on the body. Power consumption is therefore one of the main issues to consider in a BSN [72]. Such systems have been powered, but the approach in external devices is different from that in implantable devices [73], as will now be discussed.

Ideally, the change toward removing the use of batteries should be a priority. Concerns about the role of batteries are also presented in [74], where computing platforms based on electronic textiles are considered as one field for future development. Batteries have a limited lifetime that may be as long as 10 years in some applications, but the ideal approach would be to remove this element and ensure the reliability and operability of the systems without the need to use an element that must be replaced. This is particularly salient when considering implantable devices. The possibility of ensuring a long-term working life for such devices, without the use of batteries, can improve the quality of life of patients without the need for any surgery. Nowadays, the concept of new e-health systems already imposes very low-power consumption restrictions on the electronic instrumentation, processing devices, and communication systems in order to extend the operating life of batteries. If the batteries are removed, then the system must be powered by different kinds of energy sources present in the immediate environment—that is, the body itself—or use specific solutions that will depend on the placement of the sensors. This situation produces new challenges for engineers. Systems that rely on just one power source could be a problem. If that energy source is not always available, the power could fail. So, different approaches should be considered. In one case, if a battery is still used, that power source could be combined with other possible scavenging sources in such a way that, in terms of the operating protocol of the system, the battery life is extended. This could be achieved by recharging the battery when enough energy is recovered, or the battery could be in an open-load configuration in which the required operating energy is supplied by the scavenger module. If the battery is removed altogether, the system must rely on a combination of different scavenger energy sources.

The types of sensors, their location, and the amount of data that must be processed and transmitted define the power budget that must be considered in order to define the power module.

The approach depicted in Figure 9.1 shows the monitoring objectives of such a system and the concept of a closed loop. In such a system, monitoring the patient is not the only objective; the system also has to be able to generate electrical stimulation. This is a new approach. In some sense, the monitoring sensor nodes must take measurements periodically, while sensors that also have a treatment functionality can operate either periodically or be event driven, which thereby defines different power consumption scenarios. The typical paradigm of a sensor node with functionality is an artificial pancreas [75]. Another key example is related to micro-nerve stimulation and recording, as in previous work [76], for which bioamplifier circuits and stimulator output stage circuits are being developed [77] for regenerative microchannel interfaces.

Many examples are currently being developed in the field of implantable devices. Some of them are presented in what follows, paying special attention to the power involved, the power supply, and the operating ranges. In the field of pacemakers, [78] presents a programmable implantable microstimulator with wireless telemetry for endocardial stimulation in order to detect and correct cardiac arrhythmias. That option lists the global power consumption as 48 μW, but it relies on a rechargeable battery based on RF coupling. An implantable device that is placed deep in the body is presented in Figure 9.5. Another interesting case is presented in [79], where a transcutaneous implantable device is developed based on an IR link just a few centimeters from the emitter. The device is completely autonomous, battery free, and powered through an AC signal operating at the industrial, scientific, and medical (ISM) radio band. It incorporates low-frequency operation at 13.56 MHz, with a global power consumption of 270 μW and an ASIC size of 1.4×1.4 mm^2. An interesting approach is presented in [80], where, for the particular case of glucose monitoring, a non-invasive solution is presented

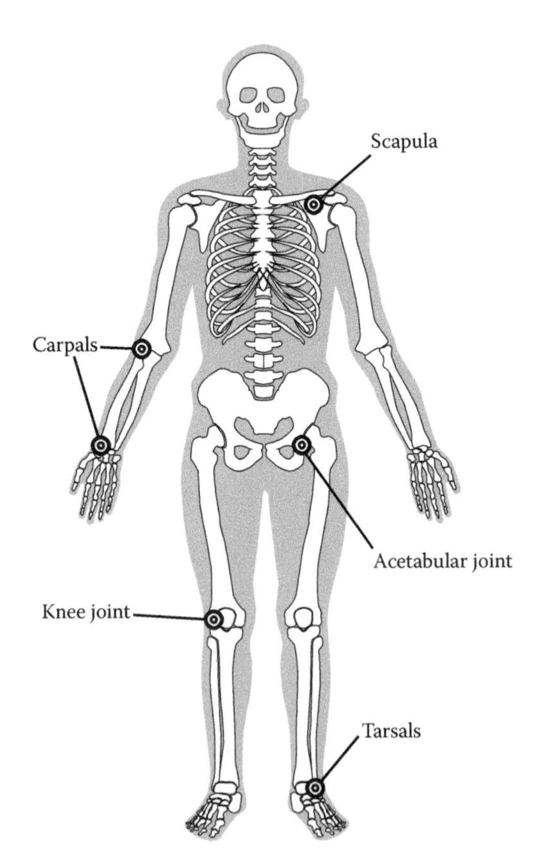

FIGURE 9.5
Implantables for biomechanical applications.

based on an active contact lens, whose size is limited to 0.5×0.5 mm^2. In that case, the implant is powered by a rectified 2.4 GHz RF power signal source at a distance of 15 cm with a global power consumption of 3 μW. Some specific examples of neuroscience applications are also introduced. In [81], a CMOS implementation is presented that operates as a stimulator of the dorsal root ganglion, which has been experimentally probed with rats. A battery-free solution is envisaged with an induced 1 MHz RF signal at a distance of 18 mm, at the standard medical implanted communication system (MICS) frequency of 402 MHz. Another example of an inductively powered neural system on a chip (SoC) is presented in [82]. In a 0.5 μm technology, a 4.9×3.3 mm^2 SoC is designed for a 32-channel wireless integrated neural recording system, with a power dissipation of 5.85 mW.

There are also other implantable devices that are more oriented toward biomechanical applications—in particular, the resources needed for monitoring prosthetic implants that are used in human beings to replace joints and bones (Figure 9.5). Such implants are designed to have a duration of 20 years, but degradation can lead to the necessity to replace them, with all the associated inconveniences and risks to the patients. The possibility of integrating electronics to monitor fatigue in the implants has clear advantages in protecting against premature mechanical failures. The amount of wear implants suffer depends on many variables, but in particular on the level of activity and the weight of the patient. The typical locations of such implants are knee joints, carpal and tarsal, scapula, and hip socket joints. Arthroplasty applications with monitoring electronics must ensure genuine autonomous long-term operation, in terms of capturing measurement data that is to be transmitted and powering the system. Interesting applications have been reported where self-powered systems are based on piezoelectric transducers that use mechanical deformation to sense the strain but are also harvesting power for the system [83]. In such cases, human motion is used to power the electronics and the battery is removed. In [84], a knee implant based on piezoceramics is presented that is able to deliver a maximum average electrical power of 12 mW at the tibial base plate. In [85], based on another approach, the estimated power is 1.8 mW. The power levels are low, and so the power consumption of the integrated electronics must be very low [86].

The design of an implantable device must contemplate different modules for their implementation. The energy levels that are involved in the development of an integrated solution will vary in terms of the final application and the level of intelligence of the device. A system that must work continuously is not the same as a system that works in bursts. Four modules are usually present in an implantable device: (1) the sensor(s) that fix(es) the type of measurement and the complexity of the front-end instrumentation, which is the signal conditioning; (2) the data processor; (c) the wireless module; and (d) the PMU. If the system has to close the loop, the stimulation electronics—based on DC–DC converters and drivers—are also involved in the power budget.

The design of the front-end electronics [87] depends on the type of medical signals that must be considered. It can be stated that these signals may vary from a few microvolts to several millivolts, with a frequency band that varies from a few Hz hertz to several kilohertz, as depicted in Figure 9.6. Typically, electroencephalogram (EEG) signals range from 1 to 50 μV and have low frequencies from 1 to 100 Hz. ECG signals present the highest voltage level: in the range of millivolts. Electromyogram (EMG) signals and action potentials present the highest frequencies: up to 10 kHz. The power consumption levels (in terms of energy per operation) depend on the electronics involved. If a microprocessor is needed, then a typical power consumption of 300 μW for commercial solutions, such as the Texas Instruments® MSP430 microprocessor, is far from the desired power budget. Favorable examples of specific microprocessors have been presented in [88] and [89]. In [88], the Phoenix Processor is presented in a 180 nm technology with a power consumption of 226 nW, but with an emphasis on the standby consumption, which is as low as 35.4 pW. It has an area of 915×915 μm^2 and operates at 0.5 V. It evolved from [89], where the subthreshold operating region is explored.

ADC conversion in terms of sample conversion rates and sub-microwatt transmitters (as a function of the distance between the implanted device and the exterior on the one hand and the data transmission rate on the other) are two of the key elements of these designs. Some pertinent examples show how research aims to find ADCs with lower power consumption. In [90], ultra-low-power front-end signal conditioning is implemented for an implantable 16-channel neurosensor array, with a maximum power

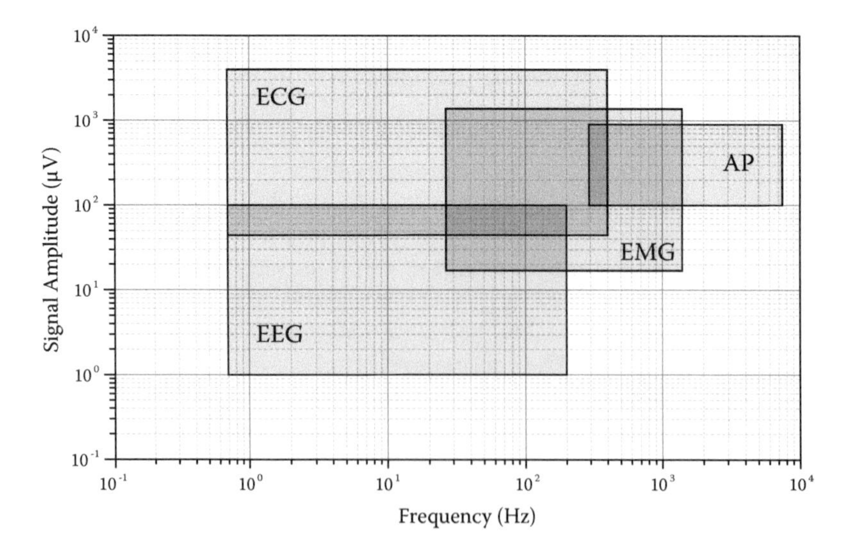

FIGURE 9.6
Some types of medical signals.

consumption of 50 μW per channel. The ADC is based on a commercial solution derived by Analog Devices®: a 12 bit AD7495 micro-SO8 standard packaged chip. The complete system has a total consumption of 12 mW. Specific examples of ADCs for implantable applications have also been derived, as in [91], where a 400 nW SAR ADC converter with 8 bits of ENOB at 80 kS/s is presented in a 180 nm technology. The implantable blood pressure sensing micro-system developed in [92] achieves 10 bit resolution with an integrated cyclic ADC converter with 11 bit resolution and a power consumption of 12 μW at 2 V. The pressure sensor has a full-power dissipation of 300 μW. In [93], an integrated ADC for a blood glucose monitoring implant presents even better performance at just 10.2 nJ per sample, with 10 ENOB, operating at very low frequencies. Another example, in this case in the field of neural signal acquisition, is presented in [94], where the ADC has a power consumption of 240 nW based on a boxcar sampling ADC [95] operating at 20 kS/s.

Regarding the wireless module, there are different approaches depending on the type of signal captured and the amount of data, related to the types of complex signal that define a wide range of data transmission constraints from a few samples by second, as in the case of heart ratings (typically 25 SPs), to several megabytes per second, such as medical imaging [96] or recent developments in endoscopic capsules [97, 98].

The review in [99] analyzes in detail the parameterization of the different implantable devices in terms of the three key variables: the size of the implantable, the power involved, and functionality and performance. Two main approaches to communications are considered depending on the placement of the implant in the body: inductive coupling [100], when there is a short distance between the implant and the exterior of the body, and far-field electromagnetic communication [101]. The approach for a subcutaneous implant based on inductive links, where coil misalignment and the effects of the geometry have a great impact on performance [102], is not the same as solutions operating at high frequencies following the MICS protocol [103, 104]. For inductive link circuits, the operating frequencies are in the range of a few megahertz, typically in the 13.56 MHz band following the ISM protocol, with characteristic data modulation methods for transmission, such as BPSK, PWM-ASK, ASK, or OOK, and a data transfer rate that is not very fast: from 100 kbps to 1.6 Mbps [105].

In [106], a high-speed OQPSK is presented with a bit rate of 4.16 Mbps, based on an inductive link, with high power levels and multiple coils, operating at 1 MHz for the power link and at the ISM 13.56 MHz frequency for the data. In [107], an implant OOK transmitter is presented that operates at 2.4 GHz in 180 nm technology and which is able to transmit at 136 Mbps with a power consumption of 3 mW. As is pointed out in [86], for the particular case of an orthopedic implant, the power consumption for the full electronics must be low. The total power consumption of the full electronics is not stated, but worthy of special attention are an DC–DC converter that recovers energy from PZTs and an ADC that presents a quiescent power consumption of 150 nW and an operating consumption of 12.5 μW @ 1.8 V for a sample frequency of 4 kHz and an ENOB > 7 bits.

9.2.4 Powering Solutions for Human Wearable and Implantable Devices

Thus far we have introduced the concept of energy harvesting and of the basic sources that permit the recovery of energy present in the environment (Figure 9.7). Several applications in the biomedical environment have been presented that have enabled us to visualize the energy needs of these solutions. Specific cases have been discussed, but more specifically, resources that could be used strictly in the biomedical environment. The type of energy source required to power a sensor node varies according to the final application and where the biomedical device is placed. This brings us to three approximations: (1) the possibility of utilizing discreet elements of energy storage, (2) being able to use energy resources present in the environment, and (3) being able to use energy from the human body as a resource. The use of batteries could be a limitation for the envisaged sensor nodes of the future, either on the outside of or implanted within the human body, as in Figure 9.1. The use of large batteries ensures the duration of the system, but the sensor nodes may be too large and heavy. So, smaller solutions with a high enough energy density are needed, combined with ultra-low-power electronics solutions that ensure a trade-off between the autonomy of the system and the smart functionality of the sensor, in terms of the sensor, signal processing, and communications modules. Ultra-low-power electronics for biomedical applications based on lithium–ion batteries are common for non-implantable solutions, such as BSNs, and also in implantable solutions [108, 109]. Lithium–ion batteries are divided into two types: (1) single-use

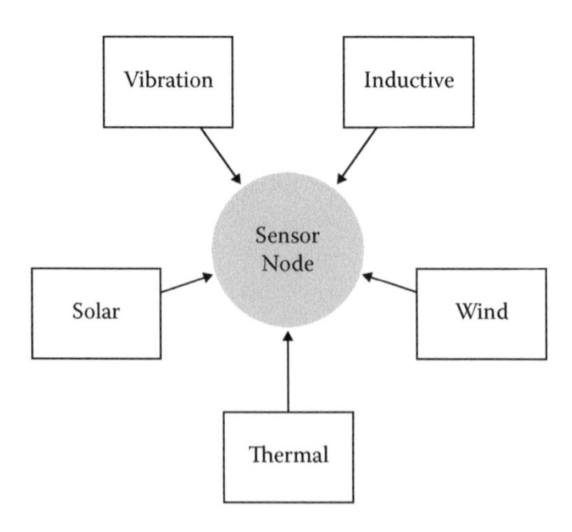

FIGURE 9.7
Combined power sources in a sensor node.

batteries, which are placed, for instance, in drug pumps, cardiac defibrillators, and pacemakers; and (2) rechargeable batteries, which are used in artificial hearts. Both types of batteries present an adequate energy density, around 1440–3600 J/cm^3, and unlike some other batteries, they do not present the memory effect; that is, they do not need to be discharged completely before a recharge phase. These batteries have a better life cycle than other types, typically 20,000 discharges and recharge cycles, but with a finite lifetime, which is typically of several years for a battery of 1 cm^3. A key aspect for these batteries is the need for battery management circuitry to ensure the range of operation, as lithium–ion batteries are extremely sensitive to overvoltage (maximum 4.2 V) and deep discharge (minimum 2 V), and to ensure high energy efficiencies. In [110], an application is reported with an average power efficiency of 89.7% and a voltage accuracy of 99.9%, conceived for biomedical applications. Other approaches are under development. Interest is especially focused on fuel cells such as the methanol fuel cell [111], but these also have their drawbacks. One is the need to replace the external reactant and the oxidant, which is analogous to the problem of recharging batteries. Although higher levels of energy are expected, based on the use of fuels such as methanol with an energy density of 17600 J/cm^3, the design issues are highly complex and proving very expensive.

Supercapacitors are another field that is being explored as an option for biomedical sensors instead of batteries, but they have a low energy density, which is a problem for systems that require a constant power source for long periods of time. In [112], a sensor is powered via a supercapacitor that is charged wirelessly, and in [113], an energy management ASIC is implemented and tested to manage supercapacitors for implants.

The next issue concerns power sources that are available in and around the human body. With reference to sources of energy present in the vicinity of the human body, and which permit the charging of implantable devices, we are able to emphasize recent works such as [114], which explores the possibility of using subcutaneous PVCs for superficial implants with a recovery capacity of a density to the power of 0.1 μW/mm^2 in conditions of strong light (sunlight). Another interesting approximation is at the level of ultrasound [115], which has advantages compared with solutions based on RF.

However, we can consider other scenarios. For instance: Is it possible to recover energy from human bodily motion? Is it possible to recover energy from the body itself? These are two questions addressed in this section. First we introduce the concept of a harvester system based on an energy harvester, an energy storage element, and power management regulation (Figure 9.8). Four types of energy harvester based on ambient sources are introduced when this system is described—that is, systems that acquire electrical energy from environmental energy sources: mechanical, solar, thermal, and RF (the details of which are beyond the scope of the present chapter). The PMU has two main roles: (1) to generate a raw DC voltage from the harvester unit, essentially acting as a conversion module, and (2) to regulate the output voltage

FIGURE 9.8
Schematic of an energy harvesting solution.

from the energy storage unit to the sensor node electronics. The energy storage unit receives energy from the conversion module and stores it.

Attention is especially focused on-body harvesting. The energy can be generated passively or actively. The natural motion of the body is one of the main research topics, where the conversion is based on mechanical (vibration) to electrical energy [20]. In this context, the power that can be harvested when a human being walks or runs has been studied at different locations, with an average of 0.5 mW/cm^3 for the hip, chest, elbow, upper arm, and head, and a maximum of 10 mW/cm^3 for the ankle and knee. The mechanical energy can be converted to electrical energy based, for instance, on electromagnetic [20], electrostatic [10], or piezoelectric [12] principles. There are several references to types of design that convert this mechanical energy to electrical energy based on the three methods of transduction just mentioned [116]. The piezoelectric approach is of special interest because high voltages are obtained for low strains, with a maximum energy density of 17.7 mJ/cm^3. Several examples are presented in the literature. In particular, in [19], piezoelectric micro-generators [117] are envisaged for drug delivery devices or dental applications. Examples of actively generated power are human power through peddling [118] or the Freeplay® or AladdinPower® rechargeable products (Table 9.3). Walking alone has been analyzed and a prototype implemented in [119], but that prototype is excessively large for our purposes. An extended analysis in terms of the placement of the harvester and the type of everyday activity is presented in [120]. Based on a commercial cantilever beam harvester from Midé Technologies, the PEH20W Volture, these different scenarios have been analyzed. The greatest level of power extracted was from a shank at the instep of the foot while the subject runs fast (28.74 μW), and the lowest level of power (0.02 μW) was also from a shank but while performing knee rehabilitation exercises sitting down. As expected, the maximum levels are obtained for activities and placements where higher amplitudes and impacts

TABLE 9.3
Examples of Human Power Generated Actively

Activity	Power generation
Finger (pushing pen)	0.3 W
Legs (cycling at 25 km/h)	100 W
Hand and arm (Freeplay)	21 W
Hand (AladdinPower)	3.6 W

are produced. An example of a wearable wireless sensor based on a kinetic energy harvester is presented in [121]. Electromagnetic transduction is used for average human motion at 0.5 Hz, with a simple architecture implemented via commercial components and a supercapacitor but not used as the final storage element of the system. It is used to transfer the charge to a smaller supply capacitor, thus improving system start-up. In [122], a MEMS piezoelectric generator is used to harvest energy from vibrations; it also uses supercapacitors as storage elements. An example of a MEMS designed for implantable devices is given in [123], where it is stated that the micro-generator is able to generate more energy per unit volume than conventional batteries—that is, an RMS power of 390 μW for 1 mm^2 of footprint area and a thickness of 500 μm, which is smaller than the volume of a typical battery in a pacemaker. The other main source of power from the human body is heat, which is limited by the Carnot efficiency, which states that the maximum power that can be recovered is in the range of 2.4–4.8 W, where other possible sources are available, such as arm motion (0.33 W), exhalation (0.40 W), or breathing (0.42 W). In the case of body heat, TEGs are used [124]. Specific designs must be considered when working with TEGs in these conditions, as the voltages generated are very small, and DC–DC converters with a specific step-up are conceived in order to boost the voltage [50]. An example of a thermogenerator is Thermo Life, which generates 60 μW/cm^2 for a temperature difference of 5°C [8]. It should also be stated that the recovery depends on the specific placement of the TEG. Placing it in the neck is not the same as placing it in the head, both of which are parts of the body that are warmer than other zones. The recovery range is 200–320 mW for the neck and 600–960 mW for the head (three times the surface area of the neck). There are other approximations in the area of the recovery of energy from the human body itself. From the concept of a fuel cell, a biogenerator for implantable devices has emerged. In some ways, the basic concept is the use of fluids in the body as a fuel source for the fuel cell, which would be an inexhaustible energy source. Interesting approaches are the use of glucose and oxygen dissolved in blood as fuel sources, as in [125] and [126]. Advanced approaches also explore a shift to the use of white blood cell capacities in biofuel cells [127] or approaches such as that in [94], where the fuel cell is based on the use of a microorganism to convert the chemical energy of glucose into electrical energy in a PBS (Phosphate Buffered Salin) solution.

9.3 Multi-Source Self-Powered Device Conception

At this point, a sensor node could be powered by following different approaches, and the choice depends on the placement of the sensor, which defines its accessibility, its size, and its weight. A battery, a supercapacitor, or a fuel cell are possible choices on the one hand. On the other, the use of some kind of EHS that recovers energy from the body, mostly based on vibration to electrical conversion or thermoelectric generation, are also envisaged. However, there are some other approaches that can also be considered depending on the placement of the sensor; these include the possibility of using environmental resources to harvest energy from for the sensor node, such as light [6, 128] or radio waves [7]. An approach that was introduced previously was RF coupling, both as a way to supply energy to the sensor node and to use the RF link for communications purposes, typically for short distances between the master and slave modules in the 13.56 MHz ISM band. For instance, at 4 MHz and a distance of 25 mm, for the subcutaneous powering of a biomedical device, the energy recovered was 5 mW [129]. A particular example of this is introduced in [130], where the rectifier module is designed to work at the 915 MHz ISM bandwidth (only in region 2). The performance of the RF source is quite small, from a typical 4 μW/cm^2 for GSM to 1 μW/cm^2 for the Wi-Fi band. For coils, typical values are lower than 1 μW/cm^2, but with as much as 1 mW for close inductive coils (a few centimeters). In the 915 MHz ISM bandwidth, at 1.1 m, the energy recovered is around 20 μW [9]. New approaches are being developed. In particular, the use of ultrasonic powering instead of RF powering is of great interest. In [131] and [115], 1 V is generated with a power capability of 21.4 nW.

At this point, the issue that we need to consider is the possibility of placing or using more than one energy harvester. This approach, combining indoor solar cells, RF coupling at the ISM 13.56 MHz band, and mechanical vibration based on piezoelectric transducers, was presented in [1] together with experimental results. The system does not rely on only one harvesting source, and the objective is to use different energy harvesters. In [132], a specific PMU was designed for a multi-source and multi-load EHS with the aim of establishing a minimum number of conversion stages and magnetic components and developing a specific algorithm. In [20], an approach to the envisaged platform architecture for a multi-harvesting WSN is presented. Another interesting example commented on in [85] combines the recovery of energy from the motion of a human being, based on PZTs, with the recovery of energy from RF signals, depending on the operating mode, but no more details are reported. In [133], a combination of vibration energy harvesting and charging circuitry for a lithium–ion battery is presented. The architecture proposed in [1, 134] is used to demonstrate the feasibility of combining different harvester sources for a 1 cm^3 multi-sensor node. This is a generic architecture envisaged for standard self-powered smart sensors that could be applied in different fields, such as the integrity of structures, surveillance, and medical devices and implants. Electromagnetic coupling with an indoor solar cell module and a piezoelectric generator was used. The envisaged platform is defined as a multi-harvesting power chip (MHPC) with a total power consumption of 160 μW. The integrated circuit is designed with 0.13 μm technology, which is a low-voltage technology up to 3.3 V. The capacity of a small system to recover a few microwatts from the energy present in the environment of the multi-sensor node, combining vibrations, light, and RF, is a proof of concept and recovers a total power of more than 1 mW. The system is initially analyzed with each different source working alone. The system recovers 360 μW from the piezoelectric generator, which is a commercial PZT (QP40W) [135] operating at 7 m/s^2 at 80 Hz. Then, two indoor solar cells [136] in conditions of 1500 lx are used, with a total harvesting power of 2.76 mW. Finally, an RF link generator (TRF7960) emitting at full power (200 mW) is used, with a distance between the base station and the antenna receiver of 25 mm, and recovers 4.5 mW. The total power expected to be recovered from the combination of the three harvested sources would be 7.8 mW, but the experimentally recovered power was 6.2 mW [1, 134]. The difference of 1.6 mW between the theoretical total and the actual amount of power that was recovered is due to variations in the effective load conditions for the light and RF modules. This suggests research is needed into the trade-off between the implementation of peak power tracking circuitry for the light and RF modules and the improvements that could be achieved, compared with the cost in silicon area and the power consumption of each module. The three main elements in the design of the MHPC ASIC (application-specific integrated circuit) are as follows: (1) A bandgap reference circuit based on peak regulation. (2) A linear dropout (LDO) regulator based on the bandgap reference circuit, which uses a PMOS (P-channel metal-oxide-semiconductor) switch and an error amplifier [1]. The switch is placed at the output stage of the simple two-stage amplifier, where MN1 and MN2 define the ground current. These elements have a power requirement of 30 μW. (3) Finally, two integrated rectifiers that must be placed in the MHPC, one for AC–DC rectification of the piezoelectric generator signal and the other one for the RF-coupled signal. The MHPC has its own control unit that plays the role of the PMU, based on the concept presented in [65]. The system combines a control switch for each of the energy harvester channels in order to be able to operate as a single and independent harvesting channel, with independent storage elements for each channel in multiple storage device (MSD) configuration or combining of all them into a single storage device (SSD). In SSD mode, the PMU controls the energy stored in the single storage element (a supercapacitor) and transfers energy when an adequate voltage is reached in the capacitor. When the voltage reaches the low value threshold (V_{min}), the system opens the charge transfer to the load until the maximum voltage level (V_{max}) is reached again. The PMU also incorporates power on reset (POR). This circuit is used to reduce the power consumption of the module during the start-up phase.

Based on this MHPC ASIC conception, a CMOS architecture for an implantable device is envisaged [137]. Nowadays, interest in nanobiosensors is increasing in the field of medical diagnosis. The development of such devices and the telemedicine environments that can be derived from them have great market potential, as has been pointed out previously. Different approaches are required for discrete, small (cm^3) devices and for implantable devices, and the performances, communications capabilities, and so on are very different. The size of the implantable device is envisaged as that of a capsule, ideally less than 4.5 cm long and 2.5 cm in diameter, following the same philosophy as some subcutaneous implantable contraceptive devices such as Norplant®, Jadelle®, and Implanon®. One proposal is an implantable event detector system or an event detector that works as an alarm. When the concentration moves out of the range of accepted values and reaches a threshold value, the alarm is activated. The proposed generic implantable architecture is presented in [134]. It is composed of three biosensor electrodes, an antenna, and electronic modules. Such a system combines different modules. There is the antenna and the AC–DC module that is used to supply energy to the device (inductive powering) and the communication setup (backscattering) based on AM modulation. Then, a low-voltage and low-power potentiostat is integrated. The biosensor is the only part of the implantable device interacting with the biological environment. It detects the desired target, generating an electrical signal. The biosensor design must be carefully selected depending on the target sample to be detected and the total size of the device. Several biosensor configurations formed by simple two, three or four electrodes can be used for single target detection and more complex array structures of microsensors can be introduced for multi-analyte detection. The antenna and its associated electronics are used for two main aspects. Firstly, it is used to supply energy to the implantable device (inductive powering) working together with an integrated AC–DC module. Secondly, the same antenna transmits the information to the external reader through the electronic communications setup (backscattering). In this scenario, one antenna is used for both power generation and communication. This reduces the antenna operation frequency to tens of megahertz due to the inductive power drop caused by the human skin. Moreover, the number of transmitted information is limited and the size of the antenna is considerable big. Another scenario considers the use of two antennas in the same implantable device, one for communication and the other for power. In this case, the communication link can be established in the hundreds of megahertz range (usually in the 400 MHz ISM band), allowing higher communication rates and reducing the size of the antenna. The second antenna is focused to power the electronics through a dedicated inductive link operating at lower frequencies than the communication antenna. In this way, each antenna can be optimized for its functionality. Then, the integrated electronics is introduced to drive the biosensor and to generate the data to be transmitted. Usually a low-voltage, low-power potentiostat circuit or similar instrumentations are used to control the implantable sensors.

9.4 Summary and Conclusions

This chapter introduces a review of state-of-the-art energy harvesting, focusing on the idea of using different types of energy sources to drive ultra-low-power electronics. In addition, it provides an overview of the most widely used MPPT algorithms to maximize the efficiency of conversion during energy extraction. As a result, combining highly efficient low-power energy recovery electronics with MPPT algorithms opens the door to develop new portable devices in very different areas: from human wearable and implantable devices to SHM and smart sensor motes. In this way, it is possible to envisage new autonomous, smart, portable instruments (or multi-instruments) easily adapted to daily human activity or industrial (technological) advances. Collecting energy from any possible environmental source or human activity has the enormous advantage of reducing the need for bulky batteries in handheld devices while furthering the miniaturization and weight reduction of portable instruments.

9.5 Acknowledgments

The authors would like to thank the support of the SMARTER Project (PCIN-2013-069), Ministerio de Economía y Competitividad, and the MINAUTO Project (TEC2016-78284-C3-3-R), Ministerio de Economía, Industria y Competitividad, Agencia Estatal de Investigación, with FEDER-UE funds.

References

1. J. Colomer-Farrarons, P. Miribel-Català, A. Saiz-Vela, J. Samitier, "A multiharvested self-powered system in a low-voltage low-power technology," *IEEE Trans. Ind. Electron.*, vol. 58, no. 9, pp. 4250–4263, 2011.
2. G.-Z. Yang, *Body Sensor Networks*. 2006.
3. A. K. Dey and D. Estrin, "Perspectives on pervasive health from some of the field's leading researchers," *IEEE Pervasive Comput.*, vol. 10, no. 2, pp. 4–7, 2011.
4. S. Mehraeen, S. Jagannathan, and K. A. Corzine, "Energy harvesting from vibration with alternate scavenging circuitry and tapered cantilever beam," *IEEE Trans. Ind. Electron.*, vol. 57, no. 3, pp. 820–830, 2010.
5. E. Carlson, K. Strunz, and B. Otis, "20mV input boost converter for thermoelectric energy harvesting," *2009 Symp. VLSI Circuits*, vol. 2, pp. 162–163, 2009.
6. A. Nasiri, S. A. Zabalawi, and G. Mandic, "Indoor power harvesting using photovoltaic cells for low-power applications," *IEEE Trans. Ind. Electron.*, vol. 56, no. 11, pp. 4502–4509, 2009.

7. T. Sogorb, J. V. Llario, J. Pelegri, R. Lajara, and J. Alberola, "Studying the feasibility of energy harvesting from broadcast RF station for WSN," in *2008 IEEE Instrumentation and Measurement Technology Conference*, 2008, pp. 1360–1363.
8. S. Roundy, D. Steingart, L. Frechette, P. Wright, and J. Rabaey, "Power sources for wireless sensor networks," *Sens. Networks*, vol. 2920, pp. 1–17, 2004.
9. E. M. Yeatman, "Energy scavenging for wireless sensor nodes," in *2nd International Workshop on Advances in Sensors and Interface (IWASI)*, 2007, vol. 90, pp. 1–4.
10. M. E. Kiziroglou, C. He, and E. M. Yeatman, "Rolling rod electrostatic microgenerator," *IEEE Trans. Ind. Electron.*, 2009, vol. 56, no. 4, pp. 1101–1108.
11. T. T. Le, J. Han, A. V. Jouanne, K. Mayaram, and T. S. Fiez, "Piezoelectric micropower generation interface circuits," *IEEE J. Solid-State Circuits*, vol. 41, no. 6, pp. 1411–1420, Jun. 2006.
12. L. Garbuio, M. Lallart, D. Guyomar, C. Richard, and D. Audigier, "Mechanical energy harvester with ultralow threshold rectification based on SSHI nonlinear technique," in *IEEE Trans. Ind. Electron.*, 2009, vol. 56, no. 4, pp. 1048–1056.
13. J. P. Curty, N. Joehl, C. Dehollain, and M. J. Declercq, "Remotely powered addressable UHF RFID integrated system," *IEEE J. Solid-State Circuits*, vol. 40, no. 11, pp. 2193–2202, 2005.
14. S. Carrara, A. Cavallini, S. Ghoreishizadeh, J. Olivo, and G. De Micheli, "Developing highly-integrated subcutaneous biochips for remote monitoring of human metabolism," in *2012 IEEE Sensors*, 2012, pp. 1–4.
15. "Thermo Life LPTG. Thermo Life Energy Corp." Online: http://www.poweredbythermolife.com. Accessed date: February 27, 2013.
16. V. Leonov, C. Van Hoof, and R. J. M. Vullers, "Thermoelectric and hybrid generators in wearable devices and clothes," in *Proceedings: 6th International Workshop on Wearable and Implantable Body Sensor Networks (BSN2009)*, 2009, pp. 195–200.
17. O. Garcia-Morchon, T. Falck, T. Heer, and K. Wehrle, "Security for pervasive medical sensor networks," in *Proceedings of the 6th Annual International Conference on Mobile and Ubiquitous Systems: Computing, Networking and Services*, 2009, pp. 1–10.
18. J. Penders, B. Gyselinckx, R. Vullers, M. De Nil, V. S. R. Nimmala, J. Van De Molengraft, F. Yazicioglu, et al., "Human++: From technology to emerging health monitoring concepts," in *Proceedings of the 5th International Workshop on Wearable and Implantable Body Sensor Networks, BSN2008, in conjunction with the 5th International Summer School and Symposium on Medical Devices and Biosensors (ISSS-MDBS) 2008*, 2008, pp. 94–98.
19. M. R. Mhetre, N. S. Nagdeo, and H. K. Abhyankar, "Micro energy harvesting for biomedical applications: A review," *3rd International Conference on Electronics Computer Technology*, vol. 3, pp. 1–5, Apr. 2011.
20. J. F. Christmann, E. Beigné, C. Condemine, and J. Willemin, "An innovative and efficient energy harvesting platform architecture for autonomous microsystems," in *Proceedings of the 8th IEEE International NEWCAS Conference (NEWCAS2010)*, 2010, pp. 173–176.
21. C. Lu, V. Raghunathan, and K. Roy, "Efficient design of micro-scale energy harvesting systems," *IEEE J. Emerg. Sel. Top. Circuits Syst.*, vol. 1, no. 3, pp. 254–266, 2011.
22. J. M. Jornet, "A joint energy harvesting and consumption model for self-powered nano-devices in nanonetworks," *IEEE Int. Conf. Commun.*, 2012, pp. 6151–6156.

23. F. Khoshnoud and C. W. De Silva, "Recent advances in MEMS sensor technology-mechanical applications," *IEEE Instrum. Meas. Mag.*, vol. 15, no. 2, pp. 14–24, 2012.
24. S. Benecke, J. Ruckschloss, N. F. Nissen, and K.-D. Lang, "Energy harvesting on its way to a reliable and green micro energy source," *Electronics Goes Green 2012+, ECG 2012 Joint International Conference and Exhibition: Proceedings*, 2012.
25. J. Lu, Y. Zhang, T. Itoh, and R. Maeda, "Design, fabrication, and integration of piezoelectric MEMS devices for applications in wireless sensor network," *Symposium on Design, Test, Integration and Packaging of MEMS/MOEMS (DTIP)*, May 2011, pp. 217–221.
26. S. Trolier-Mckinstry, F. Griggio, C. Yaeger, P. Jousse, D. Zhao, S. S. N. Bharadwaja, T. N. Jackson, et al., "Designing piezoelectric films for microelectromechanical systems," in *IEEE Trans. Ultrasonics, Ferroelectrics, Frequency Cont.*, 2011, vol. 58, no. 9, pp. 1782–1792.
27. S. W. Liu, S. W. Lye, and J. M. Miao, "Sandwich structured electrostatic/electrets parallel-plate power generator for low acceleration and low frequency vibration energy harvesting," in *IEEE 25th International Conference on Micro Electro Mechanical Systems (MEMS)*, 2012, pp. 1277–1280.
28. K. Peterson and G. A. Rincon-Mora, "High-damping energy-harvesting electrostatic CMOS charger," in *ISCAS 2012: 2012 IEEE International Symposium on Circuits and Systems*, 2012, pp. 676–679.
29. J. Martorell, "A photonic nano-structuring approach to increase energy harvesting for organic photovoltaic cells," in *14th International Conference on Transparent Optical Networks (ICTON)*, 2012, pp. 1–1.
30. C. Lu, S. P. Park, V. Raghunathan, and K. Roy, "Low-overhead maximum power point tracking for micro-scale solar energy harvesting systems," *25th International Conference on VLSI Design (VLSID)*, 2012, pp. 215–220.
31. B. Chen, "Nanomaterials for green energy: next-generation energy conversion and storage," *IEEE Nanotechnol. Mag.*, vol. 6, no. 3, pp. 4–7, 2012.
32. M. Bareiss, A. Hochmeister, G. Jegert, G. Koblmuller, U. Zschieschang, H. Klauk, B. Fabel, et al.., "Energy harvesting using nano antenna array," *11th IEEE International Conference on Nanotechnology*, pp. 218–221, Portland, OR, 2011.
33. D. J. Paul, A. Samarelli, L. Ferre Llin, J. R. Watling, Y. Zhang, J. M. R. Weaver, P. S. Dobson, et al., "Si/SiGe nanoscale engineered thermoelectric materials for energy harvesting," in *12th IEEE International Conference on Nanotechnology (IEEE-NANO)*, 2012, pp. 1–5.
34. M. Zargham and P. G. Gulak, "High-efficiency CMOS rectifier for fully integrated mW wireless power transfer," in *ISCAS 2012: 2012 IEEE International Symposium on Circuits and Systems*, 2012, pp. 2869–2872.
35. D. Maurath, P. F. Becker, D. Spreemann, and Y. Manoli, "Efficient energy harvesting with electromagnetic energy transducers using active low-voltage rectification and maximum power point tracking," *IEEE J. Solid-State Circuits*, vol. 47, no. 6, pp. 1369–1380, Jun. 2012.
36. J. Leicht, D. Maurath, and Y. Manoli, "Autonomous and self-starting efficient micro energy harvesting interface with adaptive MPPT, buffer monitoring, and voltage stabilization," in *Proceedings of the ESSCIRC*, 2012, pp. 101–104.
37. H. Le, N. Fong, and H. C. Luong, "An energy harvesting circuit for GHz on-chip antenna measurement," in *2011 IEEE International Symposium on Radio-Frequency Integration Technology*, 2011, pp. 145–148.

38. L. J. Frizzell-Makowski, R. A. Shelsby, J. Mann, and D. Scheidt, "An autonomous energy harvesting station-keeping vehicle for persistent ocean surveillance," in *OCEANS'11 MTS/IEEE KONA*, 2011, pp. 1–4.
39. A. K. Gupta and R. Saxena, "Review on widely-used MPPT techniques for PV applications," in *2016 International Conference on Innovation and Challenges in Cyber Security (ICICCS-INBUSH)*, 2016, pp. 270–273.
40. International Energy Agency (IEA), Nuclear Energy Agency (NEA), and Organization for Economic Co-operation and Development (OECD), "Projected costs of generating electricity: 2015 edition," Paris, France, 2015.
41. C. G. Popovici, S. V. Hudişteanu, T. D. Mateescu, and N.-C. Chereches, "Efficiency improvement of photovoltaic panels by using air cooled heat sinks," *Energy Procedia*, vol. 85, pp. 425–432, Jan. 2016.
42. V. Sumathi, R. Jayapragash, A. Bakshi, and P. Kumar Akella, "Solar tracking methods to maximize PV system output: A review of the methods adopted in recent decade," *Renew. Sustain. Energy Rev.*, vol. 74, pp. 130–138, Jul. 2017.
43. J. G. Njiri and D. Söffker, "State-of-the-art in wind turbine control: Trends and challenges," *Renew. Sustain. Energy Rev.*, vol. 60, pp. 377–393, Jul. 2016.
44. R. Tiwari and N. R. Babu, "Recent developments of control strategies for wind energy conversion system," *Renew. Sustain. Energy Rev.*, vol. 66, pp. 268–285, Dec. 2016.
45. X. Liu and L. A. C. Lopes, "An improved perturbation and observation maximum power point tracking algorithm for PV arrays," in *IEEE 35th Annual Power Electronics Specialists Conference (IEEE Cat. No.04CH37551)*, 2004, pp. 2005–2010.
46. A. Safari and S. Mekhilef, "Simulation and hardware implementation of incremental conductance MPPT with direct control method using Cuk converter," *IEEE Trans. Ind. Electron.*, vol. 58, no. 4, pp. 1154–1161, Apr. 2011.
47. M. K. Rajendran, S. Kansal, A. Mantha, V. Priya, Y. B. Priyamvada, and A. Dutta, "Automated environment aware nW FOCV–MPPT controller for self-powered IoT applications," in *2016 IEEE International Symposium on Circuits and Systems (ISCAS2016)*, 2016, pp. 1818–1821.
48. I. Laird and D. D. C. Lu, "Steady state reliability of maximum power point tracking algorithms used with a thermoelectric generator," in *2013 IEEE International Symposium on Circuits and Systems (ISCAS2013)*, 2013, pp. 1316–1319.
49. S. Bandyopadhyay and A. P. Chandrakasan, "Platform architecture for solar, thermal, and vibration energy combining with MPPT and single inductor," *IEEE J. Solid-State Circuits*, vol. 47, no. 9, pp. 2199–2215, Sep. 2012.
50. Y. K. Ramadass and A. P. Chandrakasan, "A battery-less thermoelectric energy harvesting interface circuit with 35 mV startup voltage," *IEEE J. Solid-State Circuits*, vol. 46, no. 1, pp. 333–341, Jan. 2011.
51. N. Phillip, O. Maganga, K. J. Burnham, M. A. Ellis, S. Robinson, J. Dunn, and C. Rouaud, "Investigation of maximum power point tracking for thermoelectric generators," *J. Electron. Mater.*, vol. 42, no. 7, pp. 1900–1906, Jul. 2013.
52. A. Paraskevas and E. Koutroulis, "A simple maximum power point tracker for thermoelectric generators," *Energy Convers. Manag.*, vol. 108, pp. 355–365, Jan. 2016.
53. H. P. Desai and H. K. Patel, "Maximum power point algorithm in PV generation: An overview," in *2007 7th International Conference on Power Electronics and Drive Systems*, 2007, pp. 624–630.

54. D. Champier, C. Favarel, J. P. Bédécarrats, T. Kousksou, and J. F. Rozis, "Prototype combined heater/thermoelectric power generator for remote applications," *J. Electron. Mater.*, vol. 42, no. 7, pp. 1888–1899, Jul. 2013.
55. Z. Zeng, X. Li, A. Bermak, C.-Y. Tsui, and W.-H. Ki, "A WLAN 2.4-GHz RF energy harvesting system with reconfigurable rectifier for wireless sensor network," in *2016 IEEE International Symposium on Circuits and Systems (ISCAS2016)*, 2016, pp. 2362–2365.
56. J. F. Dickson, "On-chip high-voltage generation in MNOS integrated circuits using an improved voltage multiplier technique," *IEEE J. Solid-State Circuits*, vol. 11, no. 3, pp. 374–378, Jun. 1976.
57. J. H. Hyun, D. S. Ha, D. Daeick Han, and C.-M. Kyung, "A solar energy harvesting system for a portable compact LED lamp," in *IECON 2015: 41st Annual Conference of the IEEE Industrial Electronics Society*, 2015, pp. 001616–001621.
58. A. Álvarez-Carulla, J. Colomer-Farrarons, J. López-Sánchez, and P. Miribel-Català, "An adaptive self-powered energy harvester strain sensing device based of mechanical vibrations for structural health monitoring applications," in *IEEE 25th International Symposium on Industrial Electronics (ISIE)*, 2016, pp. 638–644.
59. E. Sazonov, P. Pillay, H. Li, and D. Curry, "Self-powered sensors for monitoring of highway bridges," *IEEE Sens. J.*, vol. 9, no. 11, pp. 1422–1429, 2009.
60. W. Wang, N. Wang, E. Jafer, M. Hayes, B. O'Flynn, and C. O'Mathuna, "Autonomous wireless sensor network based building energy and environment monitoring system design," in *2nd Conference on Environmental Science and Information Application Technology*, 2010, pp. 367–372.
61. H. Raisigel, G. Chabanis, I. Ressejac, and M. Trouillon, "Autonomous wireless sensor node for building climate conditioning application," in *Fourth International Conference on Sensor Technologies and Applications*, 2010, pp. 68–73.
62. G. vom Boegel, F. Meyer, and M. Kemmerling, "Batteryless sensors in building automation by use of wireless energy harvesting," in *IEEE 1st International Symposium on Wireless Systems (IDAACS-SWS)*, 2012, pp. 72–77.
63. D. Zhu, S. P. Beeby, M. J. Tudor, N. M. White, and N. R. Harris, "Novel miniature airflow energy harvester for wireless sensing applications in buildings," *IEEE Sens. J.*, vol. 13, no. 2, pp. 691–700, Feb. 2013.
64. Y. Lee, S. Bang, I. Lee, Y. Kim, G. Kim, M. H. Ghaed, P. Pannuto, et al., "A modular 1 mm^3 die-stacked sensing platform with low power I^2C inter-die communication and multi-modal energy harvesting," *IEEE J. Solid-State Circuits*, vol. 48, no. 1, pp. 229–243, Jan. 2013.
65. J. Colomer-Farrarons, P. Miribel-Català, A. Saiz-Vela, M. Puig-Vidal, and J. Samitier, "Power-conditioning circuitry for a self-powered system based on micro PZT generators in a 0.13-μm low-voltage low-power technology," *IEEE Trans. Ind. Electron.*, vol. 55, no. 9, pp. 3249–3257, Sep. 2008.
66. V. G. Koutkias, I. Chouvarda, A. Triantafyllidis, A. Malousi, G. D. Giaglis, and N. Maglaveras, "A personalized framework for medication treatment management in chronic care," *IEEE Trans. Inf. Technol. Biomed.*, vol. 14, no. 2, pp. 464–472, Mar. 2010.
67. H. Zhou and K. Hou, "Pervasive cardiac monitoring system for remote continuous heart care," in *4th International Conference on Bioinformatics and Biomedical Engineering*, 2010, pp. 1–4.

68. P. Zweifel, S. Felder, and M. Meiers, "Ageing of population and health care expenditure: A red herring?," *Health Econ.*, vol. 8, no. 6, pp. 485–496, 1999.
69. M. G. Allen, "Micromachined endovascularly-implantable wireless aneurysm pressure sensors: From concept to clinic," in *13th International Conference on Solid-State Sensors, Actuators and Microsystems, 2005: Digest of Technical Papers; TRANSDUCERS '05*, vol. 1, pp. 275–278.
70. M. ElHelw, J. Pansiot, D. McIlwraith, R. Ali, B. Lo, and L. Atallah, "An integrated multi-sensing framework for pervasive healthcare monitoring," in *Proceedings of the 3rd International ICST Conference on Pervasive Computing Technologies for Healthcare*, 2009, pp. 1–7.
71. Y. Chuo, M. Marzencki, B. Hung, C. Jaggernauth, K. Tavakolian, P. Lin, and B. Kaminska, "Mechanically flexible wireless multisensor platform for human physical activity and vitals monitoring," *IEEE Trans. Biomed. Circuits Syst.*, vol. 4, no. 5, pp. 281–294, Oct. 2010.
72. S. Mandal, L. Turicchia, and R. Sarpeshkar, "A low-power, battery-free tag for body sensor networks," *IEEE Pervasive Comput.*, vol. 9, no. 1, pp. 71–77, Jan. 2010.
73. Qiang Fang, Shuenn-Yuh Lee, H. Permana, K. Ghorbani, and I. Cosic, "Developing a wireless implantable body sensor network in MICS band," *IEEE Trans. Inf. Technol. Biomed.*, vol. 15, no. 4, pp. 567–576, Jul. 2011.
74. Nenggan Zheng, Zhaohui Wu, Man Lin, and L. T. Yang, "Enhancing battery efficiency for pervasive health-monitoring systems based on electronic textiles," *IEEE Trans. Inf. Technol. Biomed.*, vol. 14, no. 2, pp. 350–359, Mar. 2010.
75. L. Ricotti, T. Assaf, A. Menciassi, and P. Dario, "A novel strategy for long-term implantable artificial pancreas," in *2011 Annual International Conference of the IEEE Engineering in Medicine and Biology Society*, 2011, pp. 2849–2853.
76. G. T. A. Kovacs, C. W. Storment, and J. M. Rosen, "Regeneration microelectrode array for peripheral nerve recording and stimulation," *IEEE Trans. Biomed. Eng.*, vol. 39, no. 9, pp. 893–902, 1992.
77. J. J. FitzGerald, N. Lago, S. Benmerah, J. Serra, C. P. Watling, R. E. Cameron, E. Tarte, et al., "A regenerative microchannel neural interface for recording from and stimulating peripheral axons in vivo," *J. Neural Eng.*, vol. 9, no. 1, p. 16010, Feb. 2012.
78. S.-Y. Lee et al., "A programmable implantable micro-stimulator SoC with wireless telemetry: Application in closed-loop endocardial stimulation for cardiac pacemaker," in *2011 IEEE International Solid-State Circuits Conference*, 2011, pp. 44–45.
79. S. Lange et al., "An AC-powered optical receiver consuming 270μW for transcutaneous 2Mb/s data transfer," in *2011 IEEE International Solid-State Circuits Conference*, 2011, pp. 304–306.
80. Y.-T. Liao, H. Yao, B. Parviz, and B. Otis, "A 3μW wirelessly powered CMOS glucose sensor for an active contact lens," in *2011 IEEE International Solid-State Circuits Conference*, 2011, pp. 38–40.
81. H.-W. Chiu et al., "Pain control on demand based on pulsed radio-frequency stimulation of the dorsal root ganglion using a batteryless implantable CMOS SoC," *IEEE Trans. Biomed. Circuits Syst.*, vol. 4, no. 6, pp. 350–359, Dec. 2010.
82. Seung Bae Lee, Hyung-Min Lee, M. Kiani, Uei-Ming Jow, and M. Ghovanloo, "An inductively powered scalable 32-channel wireless neural recording system-on-a-chip for neuroscience applications," *IEEE Trans. Biomed. Circuits Syst.*, vol. 4, no. 6, pp. 360–371, Dec. 2010.

83. P. C.-P. Chao, "Energy harvesting electronics for vibratory devices in self-powered sensors," *IEEE Sens. J.*, vol. 11, no. 12, pp. 3106–3121, Dec. 2011.
84. S. Almouahed, M. Gouriou, C. Hamitouche, E. Stindel, and C. Roux, "The use of piezoceramics as electrical energy harvesters within instrumented knee implant during walking," *IEEE/ASME Trans. Mechatronics*, vol. 16, no. 5, pp. 799–807, Oct. 2011.
85. C. Lahuec, S. Almouahed, M. Arzel, D. Gupta, C. Hamitouche, M. Jézéquel, E. Stindel, et al., "A self-powered telemetry system to estimate the postoperative instability of a knee implant," *IEEE Trans. Biomed. Eng.*, vol. 58, no. 3, pp. 822–825, Mar. 2011.
86. H. Chen et al., "Low-power circuits for the bidirectional wireless monitoring system of the orthopedic implants," *IEEE Trans. Biomed. Circuits Syst.*, vol. 3, no. 6, pp. 437–443, Dec. 2009.
87. R. F. Yazicioğlu, C. Van Hoof, and R. Puers, *Biopotential readout circuits for portable acquisition systems*, 2009, Springer Science. https://doi.org/10.1007/978-1-4020-9093-6.
88. S. Hanson, M. Seok, Y. S. Lin, Z. Y. Foo, D. Kim, Y. Lee, N. Liu, D. Sylvester, and D. Blaauw, "A low-voltage processor for sensing applications with picowatt standby mode," *IEEE J. Solid-State Circuits*, vol. 44, no. 4, pp. 1145–1155, 2009.
89. B. Zhai, S. Pant, L. Nazhandali, S. Hanson, J. Olson, A. Reeves, M. Minuth, et al., "Energy-efficient subthreshold processor design," *IEEE Trans. Very Large Scale Integr. Syst.*, vol. 17, no. 8, pp. 1127–1137, 2009.
90. Y. K. Song, D. A. Borton, S. Park, W. R. Patterson, C. W. Bull, F. Laiwalla, J. Mislow, et al., "Active microelectronic neurosensor arrays for implantable brain communication interfaces.," *IEEE Trans. Neural Syst. Rehabil. Eng.*, vol. 17, no. 4, pp. 339–45, 2009.
91. J. H. Cheong, K. L. Chan, P. B. Khannur, K. T. Tiew, and M. Je, "A 400-nW 19.5-fJ/conversion-step 8-ENOB 80-kS/s SAR ADC in 0.18-μm CMOS," *IEEE Trans. Circuits Syst. II Express Briefs*, vol. 58, no. 7, pp. 407–411, 2011.
92. P. Cong, N. Chaimanonart, W. H. Ko, and D. J. Young, "A wireless and batteryless 10-bit implantable blood pressure sensing microsystem with adaptive RF powering for real-time laboratory mice monitoring," *IEEE J. Solid-State Circuits*, vol. 44, no. 12, pp. 3631–3644, 2009.
93. N. T. Trung and P. Häfliger, "Time domain ADC for blood glucose implant," *Electron. Lett.*, vol. 47, no. 26, p. S18, 2011.
94. R. Muller, S. Gambini, and J. M. Rabaey, "A 0.013 mm2, 5 μW, DC-coupled neural signal acquisition ic with 0.5 V supply," *IEEE J. Solid-State Circuits*, vol. 47, no. 1, pp. 232–243, 2012.
95. C. D. Ezekwe and B. E. Boser, "A mode-matching ΣΔ closed-loop vibratory gyroscope readout interface with a $0.004°/s/\sqrt{Hz}$ noise floor over a 50 Hz band," *IEEE J. Solid-State Circuits*, vol. 43, no. 12, pp. 3039–3048, 2008.
96. Y. H. Liu and T. H. Lin, "A 3.5-mW 15-Mbps O-QPSK transmitter for real-time wireless medical imaging applications," in *Proceedings of the Custom Integrated Circuits Conference*, 2008, pp. 599–602.
97. S. Diao, Y. Gao, W. Toh, C. Alper, Y. Zheng, M. Je, and C. H. Heng, "A low-power, high data-rate CMOS ASK transmitter for wireless capsule endoscopy," in *2011 Defense Science Research Conference and Expo, DSR 2011*, 2011.

98. Q. Wang, K. Wolf, and D. Plettemeier, "An UWB capsule endoscope antenna design for biomedical communications," in *3rd International Symposium on Applied Sciences in Biomedical and Communication Technologies, ISABEL 2010*, 2010.
99. R. Bashirullah, "Wireless implants," *IEEE Microw. Mag.*, vol. 11, no. 7, pp. S14–S23, Dec. 2010.
100. B. Lenaerts and R. Puers, *Omnidirectional Inductive Powering for Biomedical Implants*. Dordrecht, Netherlands: Springer, 2009.
101. A. Alomainy and Y. Hao, "Modeling and characterization of biotelemetric radio channel from ingested implants considering organ contents," *IEEE Trans. Antennas Propag.*, vol. 57, no. 4, pt. 1, pp. 999–1005, 2009.
102. K. Fotopoulou and B. W. Flynn, "Wireless power transfer in loosely coupled links: Coil misalignment model," *IEEE Trans. Magn.*, vol. 47, no. 2, pt. 2, pp. 416–430, 2011.
103. J. Walk, T. Ussmueller, R. Weigel, and F. Georg, "Remote powering systems of medical implants for maintenance free healthcare applications," *Microw. Conf. (EuMC)*, 2011 *41st Eur.*, 2011.
104. J. Abouei, J. D. Brown, K. N. Plataniotis, and S. Pasupathy, "Energy efficiency and reliability in wireless biomedical implant systems," *IEEE Trans. Inf. Technol. Biomed.*, vol. 15, no. 3, pp. 456–466, May 2011.
105. M. Sawan and B. Gosselin, "CMOS circuits for biomedical implantable devices," in *VLSI Circuits for Biomedical Applications*, K. Iniewski, Ed. Norwood, MA: Artech House, 2008.
106. G. Simard, M. Sawan, and D. Massicotte, "High-speed OQPSK and efficient power transfer through inductive link for biomedical implants," in *IEEE Trans. Biomed. Circuits Syst.*, 2010, vol. 4, no. 3, pp. 192–200.
107. J. Jung, S. Zhu, P. Liu, Y.-J. E. Chen, and D. Heo, "22-pJ/bit energy-efficient 2.4-GHz implantable OOK transmitter for wireless biotelemetry systems: In vitro experiments using rat skin-mimic," *IEEE Trans. Microw. Theory Tech.*, Dec. 2010.
108. D. M. Spillman and E. S. Takeuchi, "Lithium ion batteries for medical devices," in *Fourteenth Annual Battery Conference on Applications and Advances. Proceedings of the Conference (Cat. No.99TH8371)*, 1999, pp. 203–208.
109. R. S. Rubino, H. Gan, and E. S. Takeuchi, "Implantable medical applications of lithium-ion technology," in *Seventeenth Annual Battery Conference on Applications and Advances. Proceedings of Conference (Cat. No.02TH8576)*, 2002, pp. 123–127.
110. B. Do Valle, C. T. Wentz, and R. Sarpeshkar, "An area and power-efficient analog Li-ion battery charger circuit," *IEEE Trans. Biomed. Circuits Syst.*, vol. 5, no. 2, pp. 131–137, Apr. 2011.
111. Yuming Yang, Y. C. Liang, Kui Yao, and Choon Kiat Ong, "Low-power fuel delivery with concentration regulation for micro direct methanol fuel cell," *IEEE Trans. Ind. Appl.*, vol. 47, no. 3, pp. 1470–1479, May 2011.
112. A. Pandey, F. Allos, A. P. Hu, and D. Budgett, "Integration of supercapacitors into wirelessly charged biomedical sensors," in *6th IEEE Conference on Industrial Electronics and Applications*, 2011, pp. 56–61.
113. W. Sanchez, C. Sodini, and J. L. Dawson, "An energy management IC for bio-implants using ultracapacitors for energy storage," in *2010 Symposium on VLSI Circuits*, 2010, pp. 63–64.
114. S. Ayazian and A. Hassibi, "Delivering optical power to subcutaneous implanted devices," in *2011 Annual International Conference of the IEEE Engineering in Medicine and Biology Society*, 2011, pp. 2874–2877.

115. Yong Zhu, S. O. R. Moheimani, and M. R. Yuce, "A 2-DOF MEMS Ultrasonic Energy Harvester," *IEEE Sens. J.*, vol. 11, no. 1, pp. 155–161, Jan. 2011.
116. S. Roundy, P. K. Wright, and K. S. J. Pister, "Micro-electrostatic vibration-to-electricity converters," *Microelectromech. Syst.*, vol. 2002, pp. 487–496, 2002.
117. J. Y. Zhang, H. Kuwano, Q. Wang, and Z. Cao, "Microstructure and piezoelectric properties of AlN thin films grown on stainless steel for the application of vibration energy harvesting," *Micro Nano Lett.*, vol. 7, no. 12, pp. 1170–1172, Dec. 2012.
118. T. Kazazian and A. Jansen, "Eco-design and human-powered products", in Proceedings of the Electronics Goes Green 2004, Berlin, Germany, pp. 1–6, 6–10 September 2004.
119. J. M. Donelan, V. Naing, and Q. Li, "Biomechanical energy harvesting," in *2009 IEEE Radio and Wireless Symposium*, 2009, pp. 1–4.
120. A. Olivares, J. M. Gorriz, G. Olivares, J. Ramírez, and P. Glösekötter, "A study of vibration-based energy harvesting in activities of daily living," in *Proceedings of the 4th International ICST Conference on Pervasive Computing Technologies for Healthcare*, 2010, pp. 1–4.
121. J. Olivo, D. Brunelli, and L. Benini, "A kinetic energy harvester with fast start-up for wearable body-monitoring sensors," in *Proceedings of the 4th International ICST Conference on Pervasive Computing Technologies for Healthcare*, 2010, pp.1–7.
122. H. E. Z. Abidin, A. A. Hamzah, and B. Y. Majlis, "Design of interdigital structured supercapacitor for powering biomedical devices," in *2011 IEEE Regional Symposium on Micro and Nano Electronics*, 2011, pp. 88–91.
123. J. Martinez-Quijada and S. Chowdhury, "Body-motion driven mems generator for implantable biomedical devices," in *2007 Canadian Conference on Electrical and Computer Engineering*, 2007, pp. 164–167.
124. A. Lay-Ekuakille, G. Vendramin, A. Trotta, and G. Mazzotta, "Thermoelectric generator design based on power from body heat for biomedical autonomous devices," in *2009 IEEE International Workshop on Medical Measurements and Applications*, 2009, pp. 1–4.
125. C. Ravariu, C. Ionescu-Tirgoviste, and F. Ravariu, "Glucose biofuels properties in the bloodstream, in conjunction with the beta cell electro-physiology," in *2009 International Conference on Clean Electrical Power*, 2009, pp. 124–127.
126. F. V. Stetten, S. Kerzenmacher, A. Lorenz, V. Chokkalingam, N. Miyakawa, R. Zengerle, and J. Ducree, "A one-compartment, direct glucose fuel cell for powering long-term medical implants," in *19th IEEE International Conference on Micro Electro Mechanical Systems*, 2006, pp. 934–937.
127. G. A. Justin, Yingze Zhang, Mingui Sun, and R. Sclabassi, "An investigation of the ability of white blood cells to generate electricity in biofuel cells," in *Proceedings of the IEEE 31st Annual Northeast Bioengineering Conference*, 2005, pp. 277–278.
128. A. Bertacchini, D. Dondi, L. Larcher, and P. Pavan, "Performance analysis of solar energy harvesting circuits for autonomous sensors," in *34th Annual Conference of IEEE Industrial Electronics*, 2008, pp. 2655–2660.
129. C. Sauer, M. Stanacevic, G. Cauwenberghs, and N. Thakor, "Power harvesting and telemetry in CMOS for implanted devices," *IEEE Trans. Circuits Syst. I Regul. Pap.*, vol. 52, no. 12, pp. 2605–2613, Dec. 2005.
130. X. Zhang, H. Jiang, L. Zhang, C. Zhang, Z. Wang, and X. Chen, "An energy-efficient ASIC for wireless body sensor networks in medical applications," *IEEE Trans. Biomed. Circuits Syst.*, vol. 4, no. 1, pp. 11–18, Feb. 2010.

131. Y. Zhu, S. O. R. Moheimani, and M. R. Yuce, "Ultrasonic energy transmission and conversion using a 2-D MEMS resonator," *IEEE Electron Device Lett.*, vol. 31, no. 4, pp. 374–376, Apr. 2010.
132. S. Saggini and P. Mattavelli, "Power management in multi-source multi-load energy harvesting systems," *Proceedings of IEEE 13th European Power Electronics and Applications Conference*, 2009, pp. 1–10.
133. E. O. Torres and G. A. Rincon-Mora, "Electrostatic energy-harvesting and battery-charging CMOS system prototype," *IEEE Trans. Circuits Syst. I Regul. Pap.*, vol. 56, no. 9, pp. 1938–1948, Sep. 2009.
134. J. Colomer-Farrarons and P. L. Miribel-Català, *A CMOS Self-Powered Front-End Architecture for Subcutaneous Event-Detector Devices: Three-Electrodes Amperometric Biosensor Approach*. New York: Springer, 2011.
135. "Midé engineering smart technologies." Online: https://www.mide.com. Accessed January 22, 2013.
136. "IXYS efficiency trough technology." Online: http://www.ixys.com. Accessed February 2, 2013.
137. J. Colomer-Farrarons, P. Miribel-Català, I. Rodríguez, and J. Samitier, "CMOS front-end architecture for in-vivo biomedical implantable devices," in *35th Annual Conference of IEEE Industrial Electronics*, 2009, pp. 4401–4408.

10

RFID Supporting IoT in Health and Well-Being Applications

Sari Merilampi, Johanna Virkki, Nuno Pombo, and Nuno Garcia

CONTENTS

10.1 Introduction 253
10.2 RFID Technology in Health and Well-Being Applications 254
10.2.1 Access Control 254
10.2.2 Tracking Objects and People 255
10.2.3 Identifying Patients and Preventing Adverse Effects 255
10.2.4 Identifying Medication 255
10.2.5 Sensing and Monitoring 255
10.2.6 Smart Home Applications 256
10.3 Emerging Applications: Wearable Passive RFID Tags in Body Movement Monitoring 256
10.4 RFID and Big Data Handling in Smart (Home) Environments 258
10.4.1 Case Study: Contributions of RFID to an Intelligent Health Monitoring Bracelet 260
10.5 Conclusion 261
References 262

10.1 Introduction

In the healthcare and well-being industry, technology applications tackling future challenges are gaining more and more interest. The current trend is in preventive, health-promoting activities as well as in helping people with disabilities to live independently at home instead of in care homes. As the importance of home nursing is expected to grow in general, the freedom of movement achieved by technology is especially significant. In this study, we concentrate on radio-frequency identification (RFID) technology and its applications in this context. We introduce commercial and emerging RFID-based applications in general and investigate the suitability of the technology for a new purpose: wearable sensors and smart well-being-enhancing environments. The most mature care–related RFID applications are asset and patient identification and tracking systems in care institutions. However, the technology is suitable for the remote measuring of various parameters, such as temperature, pressure, strain, movement, and moisture. These could be used to indicate various health conditions (seizures, incontinence, wound bleeding, movement, activity, etc.) and for monitoring, informing, and adjusting smart environments to enhance comfort and well-being. These are discussed further in this chapter. We will first introduce the basics of RFID technology followed by a brief review of the health and well-being-related applications of the technology in Section 10.2. We will further discuss emerging technologies in Section 10.3, where we examine the possibilities of passive RFID technology in human body movement monitoring. Section 10.4 then links the applications into a smart environment context. Finally, the possibilities are summarized in Section 10.5.

10.2 RFID Technology in Health and Well-Being Applications

In general, RFID technology is designed for the automatic identification or tracking of objects. In many ways, it is used similarly to common barcode systems. RFID systems consist of RFID tags, which are attached to the objects to be identified, an RFID reader, which is used to read the tags and communicate with the background system, which utilizes the identification data (data collecting, data processing, data storing, data sharing, data usage for further tasks). There are various different RFID systems available, but in this chapter we will mainly concentrate on passive ultra-high frequency (UHF) RFID systems, since passive tags are simple in structure, lightweight, inexpensive, easy to integrate into versatile objects, and offer virtually unlimited operational lifetime. With UHF systems, the identification is fast and there's no need to visually see the tag. It is also possible to identify moving objects and monitor them through materials such as clothing. With read ranges of several meters, UHF RFID tags enable the monitoring of people without their conscious effort. These characteristics make the technology interesting in terms of wearable and embedded smart home applications [1, 2].

The widespread use of RFID in other industries has resulted in attempts to introduce applications in the healthcare environment as well. These include the following.

10.2.1 Access Control

RFID allows selective access into facilities (with varying access rights). The technology can be used in many ways for safety purposes in elderly care as well. To name some examples, the systems may prevent people with dementia using elevators alone or accessing nurses' facilities, or trigger an alarm when a person is going to exit from a building [2, 3].

10.2.2 Tracking Objects and People

RFID can be used to track care-related equipment, such as surgery equipment and wheelchairs, or people, such as doctors and other personnel, and thus help to efficiently locate the correct resources when needed. It can also be used to guarantee that no surgery equipment is accidently left inside a patient, since all equipment is identified as being where it should be. The technology is also useful in home care–related applications. The system could raise an alarm if a person is in a certain place for too long or does not perform similar daily routines as usual. It could also verify nurses' visits, for example [2, 3].

10.2.3 Identifying Patients and Preventing Adverse Effects

RFID technology can be used for reliable patient identification systems in hospitals and care homes. The system helps to avoid errors such as giving the patient the wrong medication or double dosing. The tag may be, for example, a bracelet that links the patient, in addition to the correct patient records, to other information not strictly care-related but essential, such as certain behaviors or habits. The data are easily available and present everywhere the patient is. However, this kind of bracelet might be useful in everyday life, since the tag could contain important information such as blood type, diagnosed conditions, and so on [3, 4].

10.2.4 Identifying Medication

The medication industry is using RFID technology for anti-counterfeit and tracking purposes. This has the potential to increase the safety of users' medication because the technology helps to efficiently identify, for example, faulty product lots that need to be withdrawn from the market. In addition, the technology helps to prevent drug piracy [5].

10.2.5 Sensing and Monitoring

The technology can also be further developed and applied to wearable wireless sensing, which also has versatile possibilities in healthcare and well-being contexts. For example, moisture, strain, temperature, and movement sensors have countless applications in these areas. An example of a novel wearable movement-monitoring sensor is discussed in Section 10.3 [6–13].

10.2.6 Smart Home Applications

RFID can also be applied in home environments. The tags may be used in the same way as the industry—for example, in access controlling, tracking, and sensing. To give some examples, the technology could be applied to smart cabinets or fridges to analyze their use and to monitor their contents. RFID could also measure the environmental conditions. Smart locking systems are also available. The house key can be an RFID implant or key-like device. Access rights can be managed from software. RFID can also be part of wider smart home systems and indoor tracking. This is discussed further in Section 10.4 [7, 12, 13, 14].

10.3 Emerging Applications: Wearable Passive RFID Tags in Body Movement Monitoring

Passive RFID tags are composed only of an antenna and a small integrated circuit (IC) component that includes the specific tag ID. Thanks to the energy-efficient mechanism of digitally modulated signal backscattering utilized in tag-to-reader wireless communication in UHF RFID technology, the tags can be read from distances of several meters.

These RFID tags are promising candidates to work as energy-autonomous wireless sensors. We can identify two main types of passive RFID sensor tags:

1. RFID tags with a traditional sensor attached as part of the tag or incorporated into the tag's IC.
2. RFID tags in which the sensing ability is integrated into the tag structure.

In the former type, the RFID tag is typically used in power supplies and data transfer, and in the latter, the tag performs the sensor function itself [15]. Novel RFID ICs have I/O ports for external sensors and there are also ICs available with integrated sensors, such as a temperature sensor, pressure sensor, and moisture sensor. The selection of integrated sensors is still very limited, but the use of external sensors substantially widens the sensing possibilities. However, external sensors may require battery-assisted tags to be used for providing enough supply power [3].

In the latter type, antenna-based sensing integrates sensing capabilities in passive RFID tags with a minimal increase in the overall complexity and power consumption of the tag. These types of simple tags can be also integrated into clothes for wireless wearable sensors with low complexity and cost [11, 16–18].

It is possible to establish maintenance-free sensors, without onboard sensors or other electronics, by using passive UHF RFID tags as the sensing elements [15, 19–22]. Clothing-integrated passive RFID-based strain sensors, especially, have versatile possibilities all around us [15, 20–21]. The antenna stretching effect can be observed in the strength of the backscattered signal received from the tag. Due to this, these wearable strain sensors will allow physiotherapy exercises to be remotely monitored as well as posture and flexibility during normal everyday tasks. Also, they can be used to provide real-time feedback to neurological patients undergoing motor rehabilitation.

Example passive UHF RFID strain sensors are presented in Figure 10.1, where we show three different stretchable tag antennas that are cut from stretchable conductive textiles, embroidered with conductive yarn on a stretchable fabric, and 3D-printed with stretchable conductive ink on a stretchable fabric. As all these stretchable tags are textile based, they can be easily integrated into clothing.

In addition to strain sensors, passive RFID technology has a lot more to offer body movement monitoring. Variations of backscattered signal strength from on-body passive UHF RFID tags can provide information about the subject's state—for example, if the person is standing or moving [23]. Further, both 2D and 3D kinematics can be detected by RFID tags attached to the body [24, 25]. The inclination of the tags was estimated

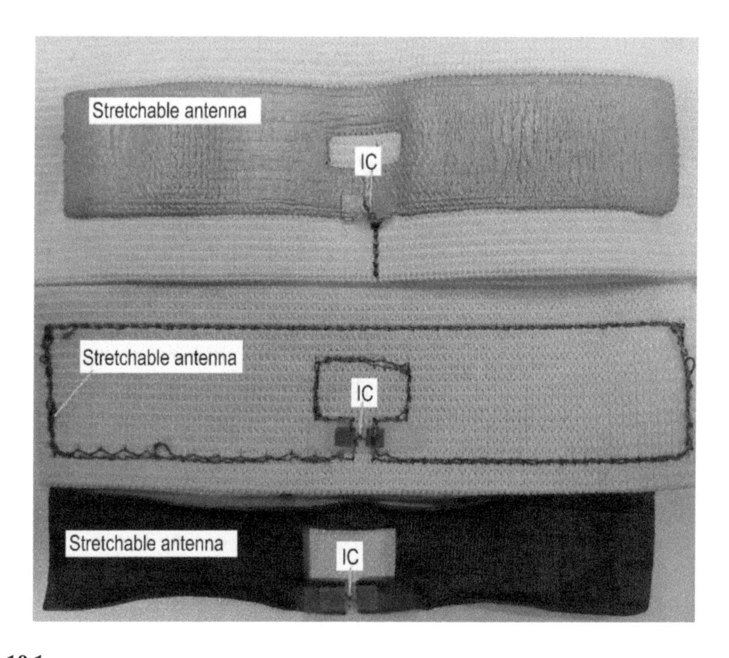

FIGURE 10.1
Textile-based passive UHF RFID strain sensor tags: 3D-printed (top), embroidered (middle), and cut from conductive textile (bottom).

based on the polarization of the tags' backscattered signals. Finally, a passive UHF RFID system, consisting of commercial tags attached on body, has been demonstrated to classify periodic movements and to recognize single gestures occurring within random sequences [26]. In this passive RFID system, the training procedure played a key role in accuracy, and each participating person was first given personal training. Based on the current state of the art as well as the active research in this area, passive RFID tags offer great future possibilities for movement monitoring. As these tags can be integrated into clothing, this technology can act as if it were invisible.

For monitoring purposes, RFID readers can be embedded, for example, in hospital beds and doors, or they can be used as handheld devices. The data can be queried at predefined intervals and transmitted to a background system for automatic processing or only on demand for the benefit of the user.

These RFID-based body movement–monitoring systems can give special groups, such as disabled people with different limitations, more opportunities for independent living and motivational therapy, thus significantly improving their quality of life.

10.4 RFID and Big Data Handling in Smart (Home) Environments

Smart home environments challenge for social computing solutions based on assisted living and ambient intelligence together to provide care services not only in patients' or elderly persons' homes but also in medical and occupational environments. Thus, all the intervenient should be asked to periodically interact with the system so as to either obtain healthcare and well-being information such as medication or active aging or clinical guidance. The collected data include activities and medication reminders, objective measurements of physiological parameters, and feedback based on observed patterns, questionnaires, and scores. In line with this, wearables, sensors, and ubiquitous devices are the cornerstones of smart environment solutions. In this scenario, RFID technology is desirable for the localization and tracking of people and objects. A generic RFID consisted of antennas connected to RFID readers. These antennas manage the obtained information from the sensor tag and send it to the remote processing center. In the smart home context, the RFID interconnects several sensor tags (either passive or active); it is therefore possible to integrate them into computerized systems focused on different purposes, such as security, monitoring, healthcare, and well-being. However, the indoor environment presents challenges for RFID design and implementation, such as the localization. In fact, there are several approaches based on different techniques, such as distance estimation, proximity, and scene analysis.

- Distance estimation: This technique is also known as *triangulation*, since it determines the target location based on triangles, which means it is supported by three different reference points. In fact, the position of tagged objects is estimated by measuring their distances from these multiple reference points.
- Scene analysis: This technique relies on the collection of the features, also known as *fingerprints*, of a scene in order to estimate the location of tagged objects. The estimation is based on the closest deductive location based on signal strength, which is susceptible to error due to the environment, or pattern recognition and machine learning algorithms.
- Proximity: This technique relies on the relative position in an extensive grid of antennas. When a single antenna detects a tagged object, this means the object is in the antenna's coverage area. On the contrary, when multiple antennas detect the same tagged object, the position is obtained by the strongest signal, representing the proximity to a specific antenna. In addition, the accuracy is directly related to the density of the antennas in the environment.

In the last few years, several authors have addressed different solutions for RFID in smart home environments, chiefly based on location and tracking, as summarized in Table 10.1. The inclusion criteria are as follows: (1) generic and/or health and well-being RFID initiatives that can be applied in the smart home environment, (2) studies presenting clearly stated results and consequently discussion, and (3) studies written in the English language.

The fundamentals of localization using tag references was introduced in the Landmarc system proposed by [27]. The Landmarc system relies on signal reception to estimate the location. Firstly, the system computes the signal strength vector to determine the power of the signal of each tagged object. Secondly, the signal is classified using an eight-level scale that varies depending on the signal strength and consequently its distance, in which the shortest distance is classified as 1 and the longest is classified as 8. Inspired by this study, [28–30] proposed several improvements aimed at, for example, avoiding interference or mean location errors, thus contributing to the improvement of the overall accuracy.

Moreover, [31] develop the concept of the *floor surface* using a grid of RFID tags, whereas [32] propose to determine indoor localization based on a probabilistic measurement model. On the contrary, several authors [33–38] propose solutions based on algorithms aimed at providing reliable solutions by means of accurate outcomes. On the other hand, this approach should be implemented with some caution—namely, to avoid increasing either costs or

TABLE 10.1

Studies Related to RFID-Based Location and Tracking in Smart Home Contexts

Study	Main Advantages	Main Limitations
[27]	Cost-effective solution Less infrastructure required Accuracy	Complexity implementation
[28]	Reduces computational burden compared with [27] Flexible localization method	Cost-effective solution Complexity implementation
[29]	Reduces computational burden compared with [27]	Cost-effective solution Complexity implementation
[30]	Cost-effective solution	Complexity implementation Scalability
[31]	Location based on grids Virtual mapping	Cost-effective solution Complexity implementation
[32]	Map learning approach	Cost-effective solution Complexity implementation
[33]	Accuracy Flexibility	Cost-effective solution Complexity implementation
[34]	Accuracy	Scalability
[35]	Cost-effective solution Accuracy	System was not tested in a large-scale experimental space
[36]	Reduces energy consumption	Cost-effective solution Complexity implementation
[37]	Accuracy	Cost-effective solution Complexity implementation
[38]	Accuracy	Cost-effective solution
[39]	Less infrastructure required	Adaptability to different scenarios

complexity. Finally, [39] propose a system to support people with disabilities based on fixed location tags on the floor.

Although the studies related to smart home contexts are mainly concerned with location and tracking, RFID could be used as part of other smart home or remote care systems. A case study on the subject is presented in the next section.

10.4.1 Case Study: Contributions of RFID to an Intelligent Health Monitoring Bracelet

The next few decades are expected to see an increasingly large number of elderly people in the most advanced societies, such as those in Europe, Oceania, and North America [40]. As an ever higher percentage of the elderly are living alone, it challenges families in terms of emergencies that may lead to morbidity or eventually to death. Nevertheless, for every emergency situation requiring a faster and more proficient approach, the existing solutions are merely based on wrist bracelets, emergency calls, or pendant devices with an

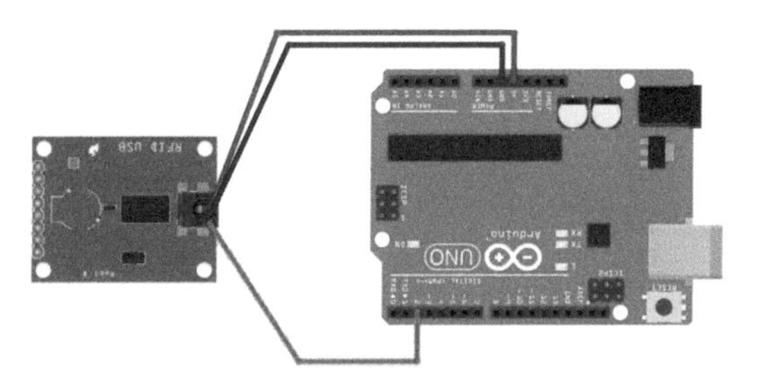

FIGURE 10.2
Arduino and RFID reader.

emergency button. In addition, these solutions are dependent on the individual, which raises several limitations. On the one hand, they require the individual to be conscious—in order to press the emergency button, for example. On the other hand, they are dependent on a successful remote phone call and/or text message. In line with this, there are promising fully automated solutions that are able to recognize either human behavior or health conditions and to contact the emergency services whenever needed. Thus, the proposed solution encompasses an high-tech bracelet capable of collecting vital signs and recognizing daily human activities, aimed at providing a fully automated health monitoring and emergency system. This universal device is designed to support individuals living alone or being remotely tracked due to, for example, a chronic disease. The intelligent bracelet has an RFID reader, thus being capable to detect false positive situations. For example, if the individual suddenly sits down, it may causes a similar G-force to falling on the floor, therefore triggering an emergency. Thus, to disable the alarm, the individual just needs to put his or her bracelet within reachable distance of the RFID reader. The implementation scheme uses an Arduino board and a USB RFID reader (Figure 10.2).

The early-stage prototype, as depicted in Figure 10.3, includes an Arduino board on which five different sensors are connected: GPS, an accelerometer, and pulse, temperature, and skin contact detectors.

10.5 Conclusion

RFID has various possibilities in well-being contexts. The technology already has applications such as access control, tracking, and the identification of equipment, medication, and people. However, the technology also enables

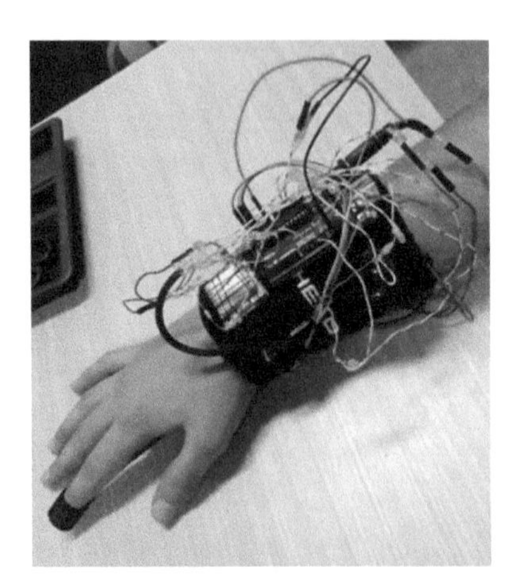

FIGURE 10.3
Intelligent bracelet prototype.

the invisible, comfortable, and wearable monitoring of various health-related indicators. These solutions could be implemented in smart environments as part of more complex systems or they could be used as stand-alone systems. However, most of the applications are still in the developmental phase. Multi-disciplinary development is needed to harness the potential of RFID technology in health- and care-related applications.

References

1. D. Dobkin. 2008. *The RF in RFID: Passive UHF RFID in Practice.* Amsterdam, the Netherlands: Elsevier.
2. A. Want. 2006. An introduction to RFID technology. *IEEE Pervasive Comput.*, 5(1): 25–33.
3. S. Merilampi. 2016. RFID-based sensing in health and care: Technological research for a better future. *Proceedings of the Third Research Forum of Changzhou University and Satakunta University of Applied Sciences*, Pori, Finland, March 9, 2016, 46–49.
4. S. Ajami, A. Rajabzadeh. 2013. Radio frequency identification (RFID) technology and patient safety. *J. Res. Med. Sci.*, 18(9): 809–813.
5. C. Swedberg. 2015. Smart septa system uses RFID to authenticate medications. *RFID.* Online: http://www.rfidjournal.com/articles/pdf?12649 (retrieved October 2016).
6. S. Merilampi, A. Sirkka, K. Iniewski. 2016. Introduction to smart eHealth and eCare technologies. In J. Virkki, S. Merilampi, S. Arrayo eds., *The Possibilities and Challenges of RFID-Based Passive Wireless Components in Healthcare Applications.* Boca Raton, FL: CRC Press.

7. A.A. Babar, S. Manzari L. Sydänheimo, A.Z. Elsherbeni, L. Ukkonen. 2012. Passive UHF RFID tag for heat sensing applications. *IEEE Trans. Antennas Propag.*, 60(9): 4056–4064.
8. S. Merilampi, H. He, L. Sydänheimo, L. Ukkonen, J. Virkki. 2016. The possibilities of passive UHF RFID textile tags as comfortable wearable sweat rate sensors. *37th Progress in Electromagnetic Research Symposium (PIERS)*, Shanghai, China, August, 8–11, 2016, 3984–3987.
9. F. Long X. Zhang T. Björninen J. Virkki, Y.-C. Chan, L. Ukkonen. 2015. Implementation and wireless readout of passive UHF RFID strain sensor tags based on electro-textile antennas. *European Conference on Antennas and Propagation*, Lisbon, Portugal, April 13–17, 2015, 1–5.
10. S. Merilampi, T. Björninen, L. Ukkonen, P. Ruuskanen, L. Sydänheimo. 2011. Embedded wireless strain sensor based on Printed RFID tag. *Sensor Rev.*, 31(1): 32–40.
11. C. Occhiuzzi, S. Cippitelli, G. Marrocco. 2010. Modeling, design and experimentation of wearable RFID sensor tag. *IEEE Trans. Antennas Propag.*, 58(8): 2490–2498.
12. Q. Qiao, L. Zhang, F. Yang, Z. Yue, A.Z. Elsherbeni. 2013. UHF RFID temperature sensor tag using novel HDPE-BST composite material. *IEEE International Symposium on Antennas and Propagation & USNC/URSI National Radio Science Meeting*, July 7–13, Orlando, FL, 2313–2314.
13. J. Virtanen, L. Ukkonen, T. Björninen, A.Z. Elsherbeni, L. Sydänheimo. 2011. Inkjet-printed humidty sensor for passive UHF RFID systems. *IEEE Trans. Instrum. Meas.*, 60(8): 2768–2777.
14. P. Sayer. 2017. Digiwell's implanted NFC chip lets you open doors with a wave of your hand. PC World. Online: https://www.pcworld.com/article/3043524/security/digiwell-will-teach-you-an-nfc-trick-your-old-dog-may-already-know.html (retrieved November 30, 2017).
15. S. Merilampi, T. Björninen, L. Sydänheimo, L. Ukkonen. 2012. Passive UHF RFID strain sensor tag for detecting limb movement. *Int. J. Smart Sensing Intell. Syst.*, 5(2), 315–328.
16. Lemey, S., Declercq, F., and Rogier, H. 2014. Textile antennas as hybrid energy-harvesting platforms. *Proceedings of the IEEE, 102*(11), 1833–1857.
17. C. Occhiuzzia, C. Vallese, S. Amendola, S. Manzari, G. Marrocco. 2014. NIGHT-care: A passive RFID system for remote monitoring and control of overnight living environment. *Procedia Comput. Sci.*, 32, 190–197.
18. O. O. Rakibet, C. V. Rumens, J. C. Batchelor, S. J. Holder. 2014. Epidermal passive RFID strain sensor for assisted technologies. *IEEE Antennas Wirel. Propag. Lett.*, 13, 814–817.
19. G. Marrocco. 2010. Pervasive electromagnetics: Sensing paradigms by passive RFID technology. *IEEE Wirel. Commun.*, 17(6), 10–17.
20. S. Merilampi, T. Björninen, L. Ukkonen, P. Ruuskanen, L. Sydänheimo. 2011. Embedded wireless strain sensor based on printed RFID tag. *Sensor Rev.*, 31(1), 32–40.
21. F. Long, X. Zhang, T. Björninen, J. Virkki, L. Sydänheimo, Y.C. Chan, L. Ukkonen. 2015. Implementation and wireless readout of passive UHF RFID strain sensor tags based on electro-textile antennas, *EuCAP 2015*, Lisbon, Portugal, April 13–17, 2015, 1–5.

22. D. Shuaib, S. Merilampi, L. Ukkonen, J. Virkki. 2017. The possibilities of embroidered passive UHF RFID textile tags as wearable moisture sensors. *SeGAH 2017,* Perth, WA, April 2–4, 2017, 1–5.
23. S. Manzari, C. Occhiuzzi, G. Marrocco. 2012. Feasibility of body-centric systems using passive textile RFID tags. *IEEE Antennas Propagat. Mag.,* 54(4), 49–62.
24. R. Krigslund, S. Dosen, P. Popovski, J. Dideriksen, G. F. Pedersen, D. Farina. 2013. A novel technology for motion capture using passive UHF RFID tags. *IEEE Trans. Biomed. Eng.,* 60(5), 1453–1457.
25. R. Krigslund, P. Popovski, G. F. Pedersen. 2013. 3D gesture recognition using passive RFID tags. *IEEE APS/URSI 2013,* Orlando, FL, July 7–13, 2013, 2307–2308.
26. S. Amendola, L. Bianchi, G. Marrocco. 2015. Movement detection of human body segments: Passive radio-frequency identification and machine-learning technologies. *IEEE Antennas Wirel. Propag. Mag.,* 57(3), 23–37.
27. L. M. Ni, Y. Liu, Y. C. Lau, A. P. Patil. 2004. LANDMARC: Indoor location sensing using active RFID. *Wireless Networks,* 10: 701–710. doi:10.1023/B:WINE.0000044029.06344.dd.
28. K.-L. Sue, C.-H. Tsai, M.-H. Lin. 2006. FLEXOR: A flexible localization scheme based on RFID. In: *Information Networking: Advances in Data Communications and Wireless Networks; International Conference,* I. Chong, K. Kawahara, eds., ICOIN 2006, Sendai, Japan, January 16–19, 2006. Revised selected papers, Berlin: Springer, 306–16.
29. G.-Y. Jin, X.-Y Lu, M.-S. Park. 2006. An indoor localization mechanism using active RFID tag. *IEEE International Conference on Sensor Networks, Ubiquitous, and Trustworthy Computing (SUTC'06),* 1, 4, 4. doi:10.1109/SUTC.2006.1636157.
30. Y. Zhao, Y. Liu, L. M. Ni. 2007. VIRE: Active RFID-based localization using virtual reference elimination. *2007 International Conference on Parallel Processing (ICPP 2007),* Xi'an, China, Sept 10–14, 2007, 56–56. doi:10.1109/ICPP.2007.84.
31. R. Tesoriero, J. A. Gallud, M. D. Lozano, V. M. R. Penichet. 2009. Tracking autonomous entities using RFID technology. *IEEE Trans. Consum. Electron,* 55: 650–655. doi:10.1109/TCE.2009.5174435.
32. D. Hahnel, W. Burgard, D. Fox, K. Fishkin, M. Philipose. 2004. Mapping and localization with RFID technology: Robotics and automation. *2004 IEEE International Conference on Robotic and Automation Proceedings (ICRA '04),* 2004, vol. 1, New Orleans, LA, April 26-May 1, 1015–1020. doi:10.1109/ROBOT.2004.1307283.
33. B. Jachimczyk, D. Dziak, W. J. Kulesza. 2014. RFID: Hybrid scene analysis–neural network system for 3D indoor positioning optimal system arrangement approach. *2014 IEEE International Instrumentation and Measurement Technology Conference (I2MTC) Proceedings,* Montevideo, Uruguay, May 12–15, 2014, 191–196. doi:10.1109/I2MTC.2014.6860732.
34. D. Fortin-Simard, K. Bouchard, S. Gaboury, B. Bouchard, A. Bouzouane. 2012. Accurate passive RFID localization system for smart homes. *IEEE 3rd International Conference on Networked Embedded Systems for Every Application (NESEA),* Liverpool, UK, Dec 13–14, 2012, 1–8. doi:10.1109/NESEA.2012.6474010.
35. Z. Xiong, Z. Song, A. Scalera, E. Ferrera, F. Sottile, P. Brizzi, R. Tomasi, M. A. Spirito. 2013. Hybrid WSN and RFID indoor positioning and tracking system. *EURASIP J. Embed. Syst.,* 6. doi:10.1186/1687-3963-2013-6.

36. R. D'Errico, M. Bottazzi, F. Natali, E. Savioli, S. Bartoletti, A. Conti, D. Dardari, et al. 2012. An UWB-UHF semi-passive RFID system for localization and tracking applications. *2012 IEEE International Conference on RFID-Technologies and Applications (RFID-TA)*, Nice, France, Nov 5–7, 2012, 18–23. doi:10.1109/RFID-TA.2012.6404509.
37. P. Vorst, S. Schneegans, B. Yang, A. Zell. 2008. Self-localization with RFID snapshots in densely tagged environments. *2008 IEEE/RSJ International Conference on Intelligent Robots and Systems*, 1353–1358. doi:10.1109/IROS.2008.4650715.
38. D. Joho, C. Plagemann, and W. Burgard. 2009. Modeling RFID signal strength and tag detection for localization and mapping. *2009 IEEE International Conference on Robotics and Automation*, Kobe, Japan, May 12–17, 2009, 3160–3165. doi:10.1109/ROBOT.2009.5152372.
39. S. Park, S. Hashimoto. 2010. An intelligent localization algorithm using read time of RFID system. *Adv. Eng. Inform.*, 24: 490–497. doi:10.1016/j.aei.2010.05.001.
40. United Nations. 2015. World population ageing. New York: United Nations.

11

Low-Power Biosensor Design Techniques Based on Information Theoretic Principles

Nicole McFarlane

CONTENTS

11.1 Introduction..267
11.2 Noise and Information Rates in Amplifiers..268
11.3 A First Order Model for the Information Rate of an Amplifier..........269
11.4 Metrics for Tradeoffs in Power and Noise..273
11.5 Tradeoffs in a Simple Amplifier Design: An Example........................274
11.6 Information Rate of a Simple Amplifier Model....................................278
11.7 Conclusion..280
References...280

11.1 Introduction

Mixed Signal CMOS technology has become a popular research area for integrated biosensing applications. However, while modern CMOS processes, through the fulfillment of Moore's Law, realize decreasing minimum sizing, this is accompanied by a lessening power supply. Further, the inherent physical noise still remains the same. This trend leads to poor signal to noise ratios and dynamic range performance being significant challenges to sensitive and accurate low power biosensing.

The application of information theory to circuits has been introduced to model various topologies. These topologies include chopper stabilized amplifiers, active pixel sensors, and single photon avalanche diodes [1–6]. By using circuit design methodologies based on information theory it is possible to create mixed signal systems that can operate at lower power, while efficiently transmitting information in the presence of high intrinsic physical and environmental noise. The methodology, and its implications, in this chapter were previously presented as part of [7].

11.2 Noise and Information Rates in Amplifiers

Noise, from a variety of sources, distorts signals of interest, setting a lower bound on the minimum detectable signal (Fig. 1). Additionally, biosignals tend to be extremely small, and hover around the levels of intrinsic system noise. The intrinsic transistor noise sources include thermal noise, a white band noise source with a constant value spectral density, and flicker noise which varies inversely with the frequency, gate current noise, and shot noise. The flicker noise and thermal noise sources are typically dominant, particularly at the lower frequencies where biosignals are found. The output noise spectral density for the thermal and flicker noise current sources is,

$$S_{I_d} = \gamma 4KTg_m + \frac{K_f I_d^{A_f}}{f^{E_f} C_{ox} L_{\mathrm{eff}}^2} \tag{11.1}$$

where K_f, A_f, and E_f are parameters are dependent on the fabrication process [8–12]. The constant γ depends on the region of operation. In strong inversion it has a value of 2/3 while in weak inversion it has a value of 1/2. The other intrinsic noise sources, gate leakage current nosie and shot noise, are typically non-dominant sources of noise for MOSFETs. p-MOSFETs typically have

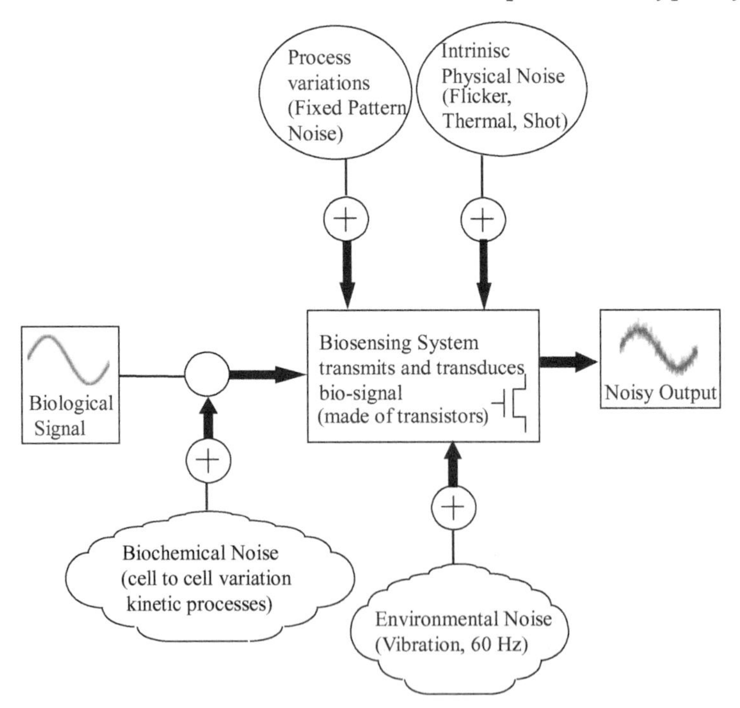

FIGURE 11.1
Overview of possible noise sources in biosensing microsystems.

lower noise than n-MOSFETs in a given process. The output current noise is tranformed to voltage noise through the equivalent output impedance [13–15],

$$S_{outv} = S_{Id}Z_{out2}. \tag{11.2}$$

For any mixed signal, multi-stage system, including amplifiers, the voltage noise at the input is more important to consider than the output noise. The noise at the input node of any stage is determined from the gain of the output signal with respect to the input signal. For multiple stages, the noise is divided by the gain for each stage until the input node is reached. Process parasitics, including the gate-to-source and gate-to-drain capacitances (C_{gd} and C_{gs}), may need to be considered to accurately account for frequency reach. Since each device is assumed to have an independent noise generator, the total noise density is simply the sum of all the noise sources. Thus, the total input noise at the input of the amplifier is determined by developing the reach of each noise source on the output divided by the frequency dependent gain of the amplifier. A method for extracting and measuring noise characteristics of transistors is shown in Figure 11.2 [16, 17]. Amplifier noise is measured at the output directly using the spectrum analyzer and a voltage buffering circuit to prevent the analyzer from loading the amplifier. The voltage buffer and amplifier should be supplied using a noiseless power source such as a battery with voltage regulation. The measured experimental noise for an NMOS transistor in a standard CMOS process is shown in Figure 11.3 [7].

11.3 A First Order Model for the Information Rate of an Amplifier

For differential amplifiers, the noise referred to the input terminals are dominated by the initial stage, and specifically by the input differential pair. The voltage noise source at the input of the amplifier can be represented as,

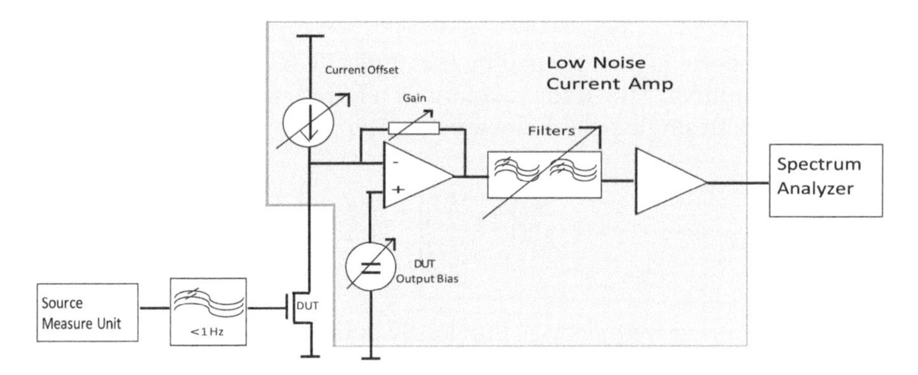

FIGURE 11.2
Setup for noise parameters, K_f, and A_f, extraction. DUT is the Device Under Test.

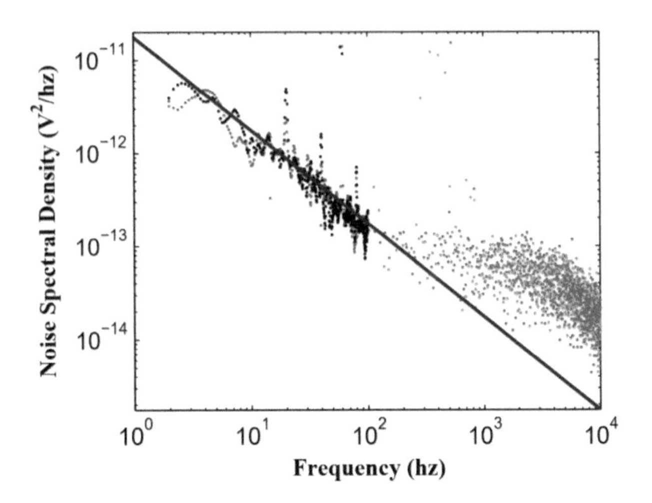

FIGURE 11.3
Flicker (1/f) noise profile of an NMOS transistor with $Kf=10^{-26}$. (Multiple lines reflect different measurement ranges during the same experiment.) [7].

$$S_{V_g} = \frac{4KT\gamma}{g_m} + \frac{K_f}{WLC_{ox} f g_m^2} \tag{11.3}$$

More generically, following [18], this has the form,

$$S_{V_g} = S_o\left(1+\frac{f_k}{f}\right) \tag{11.4}$$

where:

- S_0 is the amount of frequency independent white noise from thermal sources, and
- f_k is the corner frequency defined as the point where the frequency at which the thermal noise is equal to the flicker noise.

The output noise is then shaped by the, typically low pass, transfer function of the amplifier. This leads to an input referred noise for a generic low pass system with single pole characteristics, [19],

$$S_{n_{in}} = \frac{S_0}{A_0^2}\left(1+\frac{f_k}{f}\right)\left(1+\left(\frac{f}{f_c}\right)^2\right) \tag{11.5}$$

where:

- f_k is the corner frequency as previously defined,
- S_0 is the noise from thermal sources,
- A_0 is the midband gain of the system, and
- f_c is the bandwidth of the system.

These parameters are functions of the physical makeup, including the fabrication process, of the transistors. This includes the noise parameters for flicker noise (K_f, A_f, and, E_f) the transconductance, g_m, the input differential pair transistors aspect ratio, $W=L$, the output resistance, g_o, and the bias current from the tail current source, I.

In biological applications, for example in electrophysiological experiments, the input biological signal presented to the amplifier is extremely weak. The assumption is made that the noise sources are Gaussian in nature. Following classic information theory, the amplifier can be appropriately modeled as a Gaussian communication channel. The solution to maximizing the information efficiency of this channel is well known as the water-filling technique where the input signal is spectrally diffused over frequency locations or bins where the noise is at a minimum [20, 21]. This may be mathematically expressed as,

$$C = \int_{f_1}^{f_2} \log_2\left(\frac{\nu}{S_{n_{in}}(f)}\right) df \tag{11.6}$$

ν is the power spectral density of the input signal power plus the spectral density of the flicker and thermal noise signals for the system's bandwidth.

$\Delta f = f_2 - f_1$ is a fixed number, and the input signal power is defined as,

$$\begin{aligned} p &= \int_{f_1}^{f_2} \left(\nu - S_{n_{in}}(f)\right) df \\ v &= S_{n_{in}}(f_1) = S_{n_{in}}(f_2) \end{aligned} \tag{11.7}$$

Assuming an ideal, single pole, low pass transfer function for the filter the information rate is,

$$\nu = S_{n_{in}}(f_1) \text{ and } f_2 = f_c = f_{3dB}$$

$$\begin{aligned} I = \frac{1}{\ln 2}\Bigg[& 2(f_c - f_1) + f_c \ln \frac{\left(1+\frac{f_k}{f_1}\right)\left(1+\frac{f_1^2}{f_c^2}\right)}{\left(1+\frac{f_k}{f_c}\right)^2} \\ & + f_k \ln \frac{f_1 + f_k}{f_c + f_k} - 2f_c\left(\frac{\pi}{4} - \tan^{-1}\frac{f_1^2}{f_c^2}\right)\Bigg] \end{aligned} \tag{11.8}$$

with the signal power being related to the bandwidth by

$$P_{\text{sig}} = \frac{S_0}{A_0^2}\left[\frac{2}{3}\frac{f_2^3 - f_1^3}{f_c^2} + \frac{f_k}{2f_c^2}\left(f_2^2 - f_1^2\right) - f_k 1n \frac{f_1}{f_2}\right] \tag{11.9}$$

$$0=\left(1+\frac{f_k}{f_2}\right)\left(1+\frac{f_2^2}{f_c^2}\right)-\left(1+\frac{f_k}{f_1}\right)\left(1+\frac{f_1^2}{f_c^2}\right) \tag{11.10}$$

As defined previously, in water-filling, the power of the input signal is first allocated to frequency bins where the thermal + flicker noise power is at a minimum, before being allocated to frequencies where these noise sources are higher. In this manner, as the signal power increases, so does the information rate. Assuming the flicker-thermal corner frequency lies within the bandwidth, the first two quantities rely only on the frequency where the noise corner occurs and system bandwidth. Increasing the bandwidth or the frequency of the noise corner results in an information rate increase. Typically,

$$f_1 << f_c \text{ and } f_2 \approx f_c$$

Therefore as the bandwidth, f_c increases, $\tan^{-1}(f_2/fc)$ remains almost at a fixed value, while $\tan^{-1}(f_1/f_c)$ linearly increases. For a fixed noise corner frequency, f_k, larger bandwidth systems will have greater information rates. As the bandwidth increases, $f_c \gg f_{1,2}$, the last two terms cancel. With lowered corner frequencies, $f_c < f_{1,2}$, the sum approaches $2(f_2 - f_1)$.

Most systems will have multiple poles and zeroes, for example, most two stage amplifiers will have at least two dominant poles and a zero. Other poles and zeroes are generally high enough to be neglected for the amplifier performance. The information rate can be generalized for a system with poles and zeroes as,

$$\begin{aligned} C=\frac{1}{\ln 2}\Bigg[&2\left(f_2-f_1\right)+f_k\ln\frac{f_1+f_k}{f_2+f_k} \\ &-\sum_{i=1}^{n}2f_{pi}\left(\tan^{-1}\frac{f_2}{f_{pi}}-\tan^{-1}\frac{f_1}{f_{pi}}\right) \\ &+\sum_{j=1}^{m}2f_{zj}\left(\tan^{-1}\frac{f_2}{f_{zj}}-\tan^{-1}\frac{f_1}{f_{zj}}\right)\Bigg] \end{aligned} \tag{11.11}$$

where $f_{1,2}$ are determined by the shape of the spectrum for the power density of the noise sources and the quantity of available signal power. Clearly, higher signal powers will require higher bandwidth (subject to the same of the spectral density function). Each additional pole decreases the information rate, while zeroes increase the information rate. A decrease in information capacity will be dominated by the dominant pole, while any increase will be dominated by the first zero. From this result, signal power should be placed at frequencies above the dominant pole where any amplification is typically minimal. This, this requires careful design strategies to optimize signal power, noise power, and power supply using information rate and typical application design constraints.

The input referred noise is lowered with decreasing bias currents due to reduced bandwidth and increased midband gain. The dominant pole location is a function of the bias current, thus lowered bias current also influences the frequency location of the noise minimum. Thus, the water-filling frequency allocation of signal power may vary significantly with the bias current. If the bandwidth is about the same or smaller than the frequency of the thermal-flicker noise intersection, the frequency location of the minimum value of noise may increase.

11.4 Metrics for Tradeoffs in Power and Noise

Many other metrics allow for the characterization of the interplay noise of a system and the available power specifications. Specifically, the noise efficiency factor or NEF is a popular metric for characterizing noise and power of a system. It compares the noise of the given system to an ideal bipolar transistor. The ideal bipolar is assumed to have no base resistance and only have thermal noise sources [22]. Given some frequency bandwidth, f, bipolar collector current, I_c, and total current draw from the power supplies of the amplifier, I_{tot}, the NEF is given by,

$$\mathrm{NEF} = \sqrt{\Delta f \frac{\pi}{2} \frac{4kTV_T}{I_c}} \sqrt{\frac{2I_{tot}}{\pi 4kTV_T \Delta f}} \tag{11.12}$$

where

V_T is the thermal voltage, 0.025 V. A less noise and power efficient amplifier will have a greater NEF value. Improved system noise performance is implied by having a lowered noise efficiency factor.

Another metric known as the bit energy, *BE*, measures the energy cost of using the system with optimized amplifier performance [23]. This is measured by the power consumed by the amplifier, P_{amp}, and maximum information rate or information capacity, *I*, such that

$$BE = \frac{P_{amp}}{I} \tag{11.13}$$

Lower bit energy implies a more efficient amplifier in terms of cost to utilize the system and performance. That is the cost of maximizing amplifier performance will be minimized.

The total power in an amplifier is a function of the power supplies, V_{dd} and V_{ss}, as well as the total current for the amplifier and any biasing an start-up circuitry,

$$P_{amp} = I_{tot}\left(V_{dd} - V_{ss}\right) \tag{11.14}$$

The power increases as the current increases. However, the dominant pole also increases with the current. This leads to a less efficient operation, as the maximum information transfer rate decreases and the cost of using the system increases. The noise efficiency factor is related to the bit energy as they both include the total current and noise. However, the bit energy takes the available power resources explicitly into account in its formulation, making it a more accurate measure.

11.5 Tradeoffs in a Simple Amplifier Design: an Example

A simple operational transconductance amplifier (OTA) is shown in Figure 11.4. The major design parameters and constraints for an amplifier using the EKV model are the bandwidth, mid band gain, and intersection of the flicker and thermal spectral density plots. In the EKV model these are all formulated in terms of the inversion coefficient, IC, is [24]

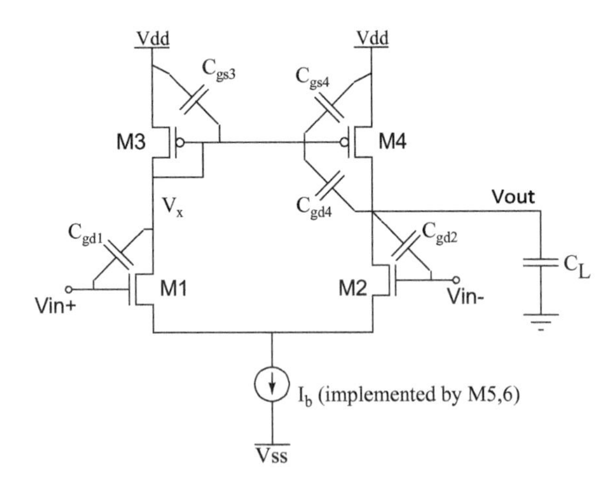

FIGURE 11.4
Simple differential operational transconductance amplifier explicitly showing the parasitic capacitances between the terminals of thetransistor [7].

$$A_0 = \frac{\kappa}{U_T}\frac{1-e^{-\sqrt{IC}}}{\sqrt{IC}}V_{A2,4}$$

$$S_0 = \frac{4kT\gamma}{\dfrac{\kappa I_D}{U_T}\dfrac{1-e^{-\sqrt{IC}}}{\sqrt{IC}}}$$

$$f_k = \frac{K_f}{WLC_{ox}4kT\dfrac{\kappa}{U_T}\dfrac{1-e^{-\sqrt{IC}}}{\sqrt{IC}}}$$

$$f_{3dB} = \frac{ICI_0W/L}{2\pi V_{A2,4}C_{out}}$$

$$V_{A2,4} = \frac{V_{A2}V_{A4}}{V_{A4}+V_{A2}}$$

where the inversion coefficient is defined as the ratio of the drain current to the reverse saturation current, $I_d = I_s$. $W = L$ is the transistor aspect ratio, and V_A is the early voltage (reciprocal of the transistor channel length modulation parameter). $U_T = V_T$ is the thermal voltage, 0.025 V, C_{ox} is the capacitance of the gate oxide, I_o is the characteristic current, and κ is the slope in the subthreshold. The simplified model of uses a constant of 1/2 for transistors operating in above threshold, and a value of 2/3 for transistors operating in strong inversion. Over weak, moderate, and strong inversion regions, a more complete model is [24],

$$\gamma = \frac{1}{1+IC}\left(\frac{1}{2}+\frac{2}{3}IC\right) \tag{11.15}$$

The design parameters to be determined are the bias tail current, typically implemented as a current mirror source with M_5 and M_6, and transistor aspect ratios. The midband gain at lowered frequencies is traditionally written as,

$$A_0 = g_{m1}\left(r_{o2}\,||\,r_{o4}\right) = \sqrt{\frac{\mu C_{ox}W/L}{I_{\text{bias}}}}\frac{1}{\lambda_2+\lambda_4} \tag{11.16}$$

in saturation. The MOSFET square law equations are assumed and is the inverse of the early voltage. The dividing line that separates the weak, moderate and strong inversion lie at inversion coefficient values of 0.1, 1, and 10.

The minimum inputs occurs due to the tail current transistor entering into triode (assuming strong inversion)

$$V_{I\min} \geq \sqrt{\frac{I_B}{\beta_1}} + V_{th1} + \sqrt{\frac{2I_B}{\beta_5}} + V_{SS} \tag{11.17}$$

The maximum input signal occurs when M_4 enters triode and eventually turns off, so that,

$$V_{G1} = V_{DD} - \sqrt{\frac{I_b}{\mu C_{ox} W / L}} - V_{th3} + V_{th1} \tag{11.18}$$

The common mode gain is a function of the tail current output resistance,

$$|A_c| = \frac{1}{2 g_{m4} r_{o5}} = \frac{1}{\frac{4\kappa 1 - e^{-\sqrt{IC}}}{U_T \sqrt{IC}}} \tag{11.19}$$

and the common mode rejection ratio (CMRR) is given by $\left|\frac{A_v}{A_c}\right|$. The slew rate increases with increasing bias current and lowered load capacitances. The load capacitance consists of parasitic capacitances and any explicit capacitances at the output node. Increased current increases both the thermal and flicker noise levels. These noise levels decrease with transistor area. Thus, increasing inversion coefficient increases the noise as well as the thermal and flicker noise corner frequency, system bandwidth, and cost of using the system (bit energy). Gain decreases with decreasing inversion coefficient. However, the system information rate shows a complex response to design parameters [1].

For the amplifier, ignoring process variations means $g_{m1} = g_{m2}$ and $g_{m3} = g_{m4}$. The input referred noise, assuming no frequency effect, is [25],

$$v_{eq}^2 = \frac{i_{n1}^2}{g_{m1}^2} + \frac{i_{n2}^2}{g_{m1}^2} + \left(\frac{g_{m3}}{g_{m1}}\right)^2 \left(\frac{i_{n3}^2}{g_{m3}^2} + \frac{i_{n4}^2}{g_{m3}^2}\right) \tag{11.20}$$

Figure 11.4 shows the parasitic capacitances which accept the transfer function. The capacitances of the NMOS and PMOS pair, from the gate to source and from the gate to the drain, are assumed to be equal, the sources of differential pair, M1 and M2, are considered to be at virtual ground, and since the bias current adds equally to both sides, it is not considered to add to the system noise. The system transfer function from input to output nodes, and from the gate of the current mirror active load to the input is,

$$H(f) = \frac{-1/2\left(sC_{gd1} - g_{m1}\right)\left(sC_{gd1} + sC_1 + 1/R_1 + g_{m4}\right)}{\left(sC_{\text{out}} + 1/R_{\text{out}}\right)\left(sC_{gd1} + 1/R_1 + sC_1 + sC_{gd4}\right) - sC_{gd4}\left(sC_{gd4} - g_{m4}\right)}$$

$$\frac{V_x}{V_{\text{in}}} = \frac{1/2\left(sC_{gd1} - g_{m1}\right)\left(sC_{dg4} + sC_{\text{out}} + 1/R_{\text{out}}\right)}{sC_{gd4}\left(sC_{gd4} - g_{m4}\right) - \left(sC_{\text{out}} + 1/R_{\text{out}}\right)\left(sC_{gd1} + 1/R_1 + sC_1 + sC_{gd4}\right)}$$

The poles and zeroes of the amplifier, assuming C_{gd4} is small, are,

$$z_1 = \frac{g_{m1}}{C_{gd2}}$$
$$z_2 = \frac{2g_{m4}}{C_1 + C_{gd1}}$$
$$p_1 = \frac{1}{R_0\left(C_L + C_{gd2}\right)} \tag{11.21}$$
$$p_2 = \frac{g_{m4}}{C_1 + C_{gd1}}$$

where:

$$C_1 = C_{gs3} + C_{gs4}$$
$$C_{out} = C_L + C_{gd2} + C_{gd4}$$
$$R_1 = r_{o1} \| r_{o3} \| 1/g_{m3} \tag{11.22}$$
$$R_{out} = r_{o2} \| r_{o4}$$

Given that $i = g_m v_{gs}$, the noise voltage is reflected back to the gate of the transistor as i_n^2/g_m^2 at low frequencies. At mid frequencies, the gate-to-drain capacitance may not necessarily be ignored, and the noise voltage at the gate is $i^2{}_n/(g_m + sC_{gd})^2$.

The width of a transistor may be taken out of the design space by considering instead that the bias current, inversion level, and transistor length are known [26, 27]. Bias current and length of the transistor both change linearly with the inversion level. Both aspect ratio, $W = L$, and area, $W \times L$, impact the system, and should be explored first after which the effects of length and bias current are explored [26].

Increased inversion coefficient, that is moving from weak to moderate to strong inversion, affects the information rate and energy cost of using the system (bit energy). The stronger the inversion level, the better the information rate, implying that above threshold operation will give a cleaner transfer of the input signal to the measured output signal. The cost of using the amplifier is lowered at the lower inversion levels. Since the input differential pair should contribute the most noise, a generalization can be made of the design specifications as functions of the transistor length and region of operation (Table 11.1).

TABLE 11.1

Summary of Trends with Design Parameters

Parameter	IC ↑	L ↑
A_o	↓	↑
f_k	↑	↓
S_o	↑	↑
A_c	↑	↓
Bit energy	↑	↓
Information rate	↑	↓

11.6 Information Rate of a Simple Amplifier Model

Capacitive feedback with an operational transconductance amplier (OTA) is a popular architecture for weak biosignal aquisistion [3, 28, 29]. The design has been widely implemented in various CMOS processes. A wide range OTA is used as the OTA for the system shown in Figure 11.5. For a typical design in a 500nm process the input differential pair transistors have W = 35m, L = 2.1m, C2 = 200fF, and C1 = 20pF. For a 130 nm process the input differential pair have W = 24m, L = 1.2m, C2 = 98.3fF, and C1 = 10pF. The power supply goes from 5V to 1.8V between the two processes. The psuedo resistors have a large resistance value. The noise spectrum referred to the input terminals of the OTA is a function of the input capacitance (C_{in}) and the output noise (S_{OTA}), as is [28],

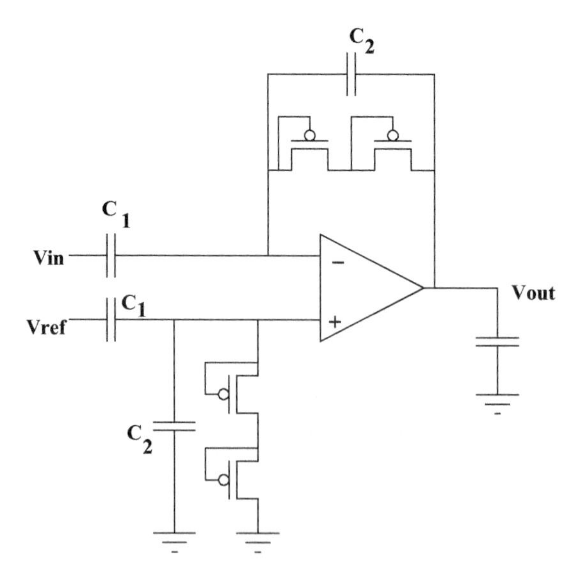

FIGURE 11.5
Frequently used capacitive feedback amplifier topology used for biosignal acquisition [28].

$$S_{amp} = \left(\frac{C_1 + C_2 + C_{in}}{C_1} \right)^2 S_{OTA} \tag{11.23}$$

Figure 11.6 shows the measured noise at the output, frequency dependent magnitude of the transfer function, spectral density of the noise referred to the input terminals, information rate, and energy cost of using the amplifier [7]. Due to a wider minimal noise spectral density bandwidth, the larger process shows a higher information rate than the shorter channel process. There is a slight decrease in voltage between the processes, however, it is unable to compensate for the noise driven information rate measure. With

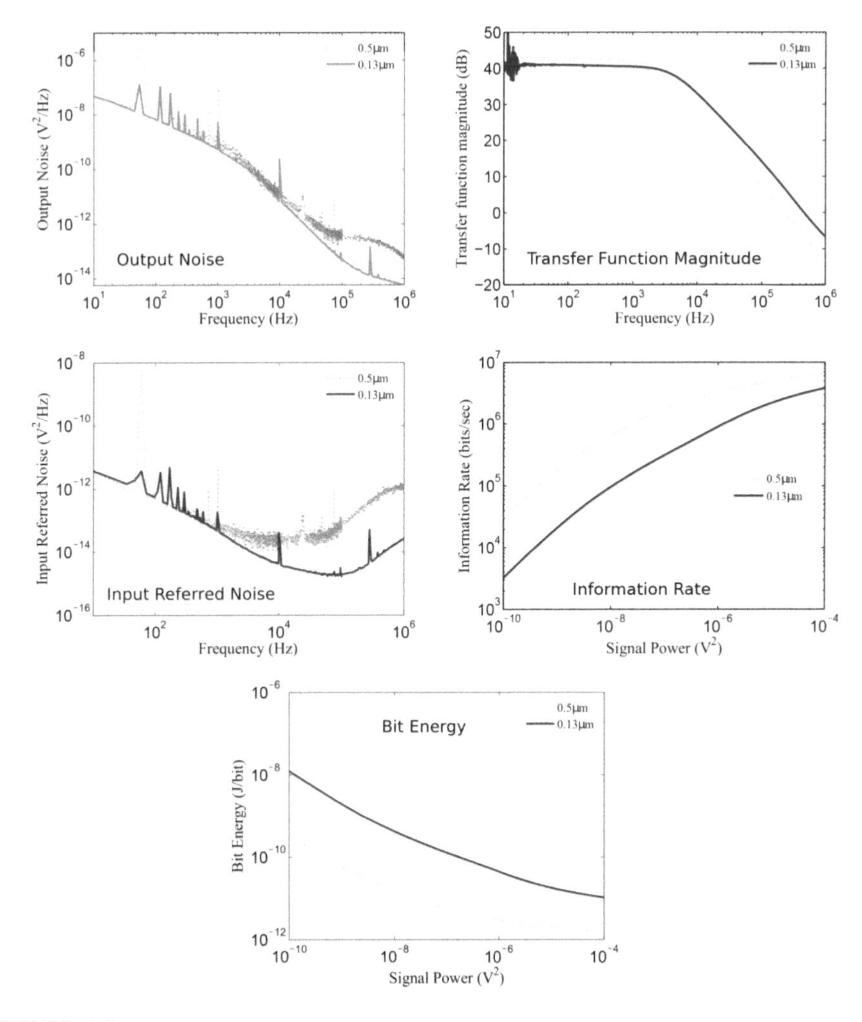

FIGURE 11.6
Comparison of larger and submicron process output voltage noise, input-referred noise, transfer function, information rates, and bit energy [7].

the fundamental noise sources actually increasing with decreasing process lengths, this result shows that the expected improvement in the energy costs (due to lowered voltage supplies) of sub-micron and deep sub-micron processes does not materialize.

11.7 Conclusion

The input referred thermal and flicker noise spectral density for an amplifier used for weak biological signals along with the information rate and energy cost of using the amplifier was theoretically calculated. The information rate and bit energy was derived from treating the circuits as communication channels according to information theory principles. This has been incorporated into an inversion coefficient based design methodology. The use of information rate to optimize amplifiers suggests optimum operation occurs at frequencies where gain is not maximum. The information derived from this methodology has the potential to be incorporated into the mixed signal design flow and can be particularly important biological applications, where weak signals need to be detected in the presence of fixed system noise.

References

1. N. McFarlane and P. Abshire, "Comparative analysis of information rates of simple amplifier topologies," *IEEE International Symposium of Circuits and Systems*, Rio de Janeiro, May 2011, pp. 785–788.
2. M. Loganathan, S. Malhotra, and P. Abshire, "Information capacity and power efficiency in operational transconductance amplifier," *IEEE International Symposium on Circuits and Systems*, Vancouver, BC, vol. 1, pp. 193–196, May 2004.
3. N. M. Nelson and P. A. Abshire, "An information theoretic approach to optimal amplifier operation," *IEEE, Midwest Symposium on Circuits and Systems*, Covington, KY August 2005, vol. 1, pp. 13–16.
4. N. Nelson and P. Abshire, "Chopper modulation improves OTA information transmission," *IEEE International Symposium on Circuits and Systems*, New Orleans, LA, May 2007, pp. 2275–2278.
5. D. Sander, N. Nelson, and P. Abshire, "Integration time optimization for integrating photosensors," *IEEE International Symposium on Circuits and Systems*, Seattle, WA, June 2008, pp. 2354–2357.
6. J. Gu, M. H. U. Habib, and N. McFarlane, "Perimeter-gated single-photon avalanche diodes: An information theoretic assessment," *IEEE Photonics Technology Letters*, vol. 28, no. 6, pp. 701–704, 2016.
7. N. McFarlane. "Information power efficiency tradeoffs in mixed signal CMOS circuits," Ph.D. dissertation, University of Maryland, College Park, MD, 2010.

8. M. J. Kirton and M. J. Uren, "Noise in solid-state microstructures: A new perspective on individual defects, interface states and low-frequency (1/f) noise," *Advances in Physics*, vol. 38, no. 4, pp. 367–468, 1989.
9. I. Bloom and Y. Nemirovsky, "1/f noise reduction of metal-oxide-semiconductor transistors by cycling from inversion to accumulation," *Applied Physics Letters*, vol. 58, pp. 1664–1666, 1991.
10. E. A. M. Klumperink, S. L. J. Gierink, A. P. Van der Wel, and B. Nauta, "Reduction MOSFET 1/f noise and power consumption by switched biasing," *IEEE Journal of Solid State Circuits*, vol. 35, no. 7, pp. 994–1001, 2000.
11. Y. Isobe, K. Hara, D. Navarro, Y. Takeda, T. Ezaki, and M. Miura-Mattausch, "Shot noise modelling in metal oxide semiconductor field effect transistors under sub threshold condition," *IEICE Transactions on Electronics*, vol. E90-C, no. 4, pp. 885–894, 2007.
12. L. Callegaro, "Unified derivation of johnson and shot noise expressions," *American Journal of Physics*, vol. 74, pp. 438–440, May 2006.
13. S.-C. Liu, J. Kramer, G. Indiveri, T. Delbruck, and R. Douglas, *Analog VLSI: Circuits and Principles*. Cambridge, MA: MIT Press, 2002.
14. C. Jakobson, I. Bloom, and Y. Nemirovsky, "1/f noise in CMOS transistors for analog applications from subthreshold to saturation," *Solid State Electronics*, vol. 42, pp. 1807–1817, 1988.
15. Y. Nemirovsky, I. Brouk, and C. G. Jakobson, "1/f noise in CMOS transistors for analog applications," *IEEE Transactions on Electron Devices*, vol. 48, no. 5, pp. 212–218, 2001.
16. R. Tinti, F. Sischka, and C. Morton, "Proposed system solution for 1/f noise parameter extraction." Online: http://www.agilent.com. Accessed 12 June 2018.
17. A. Blaum, O. Pilloud, G. Scalea, J. Victory, and F. Sischka, "A new robust on-wafer 1/f noise measurement and characterization system," *Proceedings of the 2001 International Conference on Microelectronic Test Structures*, Kobe, Japan, 2001, pp. 125–130.
18. C. C. Enz and G. C. Temes, "Circuit techniques for reducing the effects of op-amp imperfections: Autozeroing, correlated double sampling, and chopper stabilization," *Proceedings of the IEEE*, November 1996, vol. 84, no. 11, pp. 1584–1614.
19. C. C. Enz, E. A. Vittoz, and F. Krummenacher, "A CMOS chopper amplifier," *IEEE Journal of Solid-State Circuits*, vol. 22, no. 3, pp. 335–342, 1987.
20. T. M. Cover and J. A. Thomas, *Elements of Information Theory*. New York: John Wiley, 1991.
21. P. A. Abshire, "Implicit energy cost of feedback in noisy channels," in *Proceedings of the 41st IEEE Conference on Decision and Control (CDC02)*, Las Vegas Nevada, December 2002, vol. 3, pp. 3217–3222.
22. M. S. J. Steyaert, W. M. C. Sansen, and C. Zhongyuan, "A micropower low-noise monolithic instrumentation amplifier for medical purposes," *IEEE Journal Solid State Circuits*, vol. 22, no. 6, pp. 1163–1168, 1987.
23. A. G. Andreou, An information theoretic framework for comparing the bit energy of signal representations at the circuit level. *Low-Voltage/Low Power Integrated Circuits and Systems*. IEEE Press, Hoboken, NJ, 1999, pp. 519–540.
24. C. C. Enz, F. Krummenacher, and E. A. Vittoz, "An analytical MOS transistor model valid in all regions of operation and dedicated to low-voltage and low-current applications," *Analog Integrated Circuits Signal Processing*, vol. 8, no. 1, pp. 83–114, 1995.

25. P. R. Gray, P. J. Hurst, S. H. Lewis, and R. G. Meyer, *Analysis and Design of Analog Integrated Circuits*. New York: John Wiley, 2001.
26. D. Binkley, B. Blalock, and J. Rochelle, "Optimizing drain current, inversion level, and channel length in analog CMOS design," *Analog Integrated Circuits and Signal Processing*, vol. 47, pp. 137–163, 2006.
27. D. Binkley, C. Hopper, S. Tucker, B. Moss, J. Rochelle, and D. Foty, "A CAD methodology for optimizing transistor current and sizing in analog CMOS design," *IEEE Transaction on Computer-Aided Design*, vol. 22, pp. 225–237, 2003.
28. R. R. Harrison and C. Charles, "A low-power low-noise CMOS amplifier for neural recording applications," *IEEE Journal of Solid State Circuits*, vol. 38, no. 6, pp. 958–965, 2003.
29. S. B. Prakash, N. M. Nelson, A. M. Haas, V. Jeng, P. Abshire, M. Urdaneta, and E. Smela, "Biolabs-on-a-chip: Monitoring cells using CMOS biosensors," *IEEE/NLM Life Science Systems and Applications Workshop*, Bethesda, MD, July 2006, pp. 1–2.

12

Modern Application Potential of Miniature Chemical Sensors

Paul Hichwa and Cristina E. Davis

CONTENTS

12.1 Introduction 283
12.2 Modern Applications of Miniature Chemical Sensors 285
12.2.1 Medical Applications 285
12.2.2 Military and Security Applications 287
12.2.3 Agriculture Applications 288
12.3 Examples of Miniature Chemical Sensors 288
12.3.1 Ion Mobility Spectrometry 288
12.3.2 Metal Oxide Sensors 290
12.3.3 Micro-GC Detectors 291
12.3.4 E-Nose Detectors 292
12.4 Data Analysis for Chemical Detection and Prediction 293
12.5 Future Direction Trends for the Field 294
References 296

12.1 Introduction

Advancements in gas-phase chemical detection and analysis are opening exciting new pathways for impactful devices. From novel medical applications to aiding security and optimizing agriculture, portable chemical sensing platforms, in particular, are poised to play a crucial role in breakthrough developments. Areas such as disease diagnostics and health monitoring using non-invasive breath analysis have specifically gained increasing interest in the last few decades [1–10]. Additionally, the real-time detection of trace levels of explosives and toxic chemical species, quality control in the food industry, and monitoring air quality are other active areas of research. Many candidate technologies are manufactured using microelectromechanical systems (MEMS) processes.

While these areas are growing in the research setting, robust chemical detection and analysis has not yet reached widespread commercialization in a portable form. Reliable commercial chemical analysis has long been dominated by laboratory benchtop platforms that are bulky and expensive, require resource-intensive operation, and are limited to laboratory-localized analysis by highly skilled personnel. Benchtop-based experimental platforms for chemical analysis include nuclear magnetic resonance (NMR), infrared spectroscopy (IR), liquid chromatography (LC), and the "gold standard" for gas-phase chemical analysis, gas chromatography–mass spectrometry (GC-MS) [11, 12]. These standard laboratory benchtop technologies have proven to be highly accurate, sensitive, selective, and robust. Furthermore, the complex data obtained from these laboratory instruments lends itself to sophisticated analytical methods and the construction of chemical databases such as the National Institute of Standards and Technology Chemistry WebBook developed over the past few decades (https://webbook.nist.gov/chemistry). Based on the growing application potential of chemical sensing combined with the high cost and lead time associated with the traditional "standard" techniques, the miniaturization of chemical sensors and the incorporation of high-confidence analysis into portable device platforms has recently gained momentum and will likely open the door to an even greater array of new applications.

The goal of bringing chemical analysis to the field via portable platforms is not new. In 1979, Stanford University researchers invented a microfabricated GC system using a 1.5 m long separating capillary column and a thermal conductivity detector (TCD) on a compact silicon wafer, all enclosed in an anodic bonded glass wafer [13]. This was among the first of many research groups to begin extending the then traditional integrated circuit manufacturing methods to realize and optimize a variety of miniature chemical sensors [14]. This was followed over the years by other examples, and in the mid-1990s, Sandia National Laboratory created its own MicroChemLab program in the hope of producing battery-powered microanalytical systems for gas-phase analytes [15]. These are just two examples, and other platforms also helped expand this emerging field.

The drive to miniaturize technologies stems from the advantages of portable systems. Aside from the obvious gain in mobility that shrinking equipment size brings, major advantages also include the significant reduction in operating costs; the simplification of a previously labor-intensive, multi-step sampling-to-analysis procedure; potential reduction in analyte loss during sampling, storage, and transportation to the laboratory; and the reduction of errors due to time variations during sampling, which can confound outcomes. In some cases, miniaturization allows a chemical measurement that could not easily be taken on a larger-scale instrument, and sometimes lower power requirements are enabled by MEMS technology. In short, in-field measurements have become more affordable and increase the options for when and where chemical analysis can be performed. As more chemical sensors and components scale to nano-size, the feasibility of robust mobile analytical systems is rapidly becoming a reality.

Portable chemical sensor development is not without its challenges. Developing suitable chemical sensors necessitates a multi-disciplinary approach for design and engineering, including MEMS design, microfabrication process development, systems integration and controls design, and real-time streaming data analysis. Ultimately, commercial chemical sensors and analyzers must be fast, selective, sufficiently sensitive, and robust across a multitude of environments and sampling procedures. The complexity of chemical sensing, in part, arises from the unavoidable nature of on-site device use, where a multitude of inputs must be tracked, prioritized, and processed. Gas-phase analysis adds the further challenge of simply handling and sampling air. Optimal gas flow, pressure, humidity, and temperature can all influence a chemical sensor's selectivity, sensitivity, and robustness. Furthermore, in some application areas, standardization and normalization procedures for sample collection, preconcentration, and biochemical analysis methods are critical. For example, breath analysis may vary based on recent food intake, the time in between sampling to analysis, and other factors. As applications demand more data in order to make useful suggestions/decisions, portable chemical sensors will need to become more diverse in their selectivity, particularly in biological use cases. In addition to the normal requirements of a functional chemical sensor, portable platforms demand low-power requirements, exhibit minimal space consumption, and have low supporting electronics infrastructure. Before mobile chemical sensor devices can contribute in these new application areas, research and device performance issues must first address these concerns.

The main purpose of this chapter is to outline modern applications for miniaturized gas-phase chemical instrumentation systems as well as provide several examples of microfabricated MEMS devices and data analysis methods presented in the literature. Next, we will introduce some of the challenges that portable systems face and identify a few future directions and emerging trends in the field. While miniaturized chemical sensor–based analysis systems have yet to reach the commercial market in a meaningful way, the numerous emerging applications are likely to spur new research and development.

12.2 Modern Applications of Miniature Chemical Sensors

12.2.1 Medical Applications

Chemical analysis is already standard practice in the medical field. Blood tests, toxicology, endocrinology, and urinalysis all routinely measure and analyze urine, blood, and/or tissue samples in the clinic. With the medical community constantly seeking non-invasive means for diagnosis and health monitoring, *breath analysis* has become a powerful potential option. A subset of diagnostically relevant metabolites from the human body are exhaled and emanate from our pulmonary system. By monitoring those chemicals as they leave the body, we can potentially diagnose or monitor health conditions or diseases. The benefits of this non-invasive technique may include relief from patient discomfort during sampling, potential increases in patient compliance, and potential increases in early disease detection. There is also the potential for repeat measurements that can be done frequently because they are so non-invasive. Breath analysis not only has the potential to offer relief from difficulties associated with standard invasive diagnostic techniques in adults, it could also open the door to easier monitoring and testing for pediatric, neonatal, and geriatric patients.

The most abundant chemicals in exhaled breath are carbon dioxide, carbon monoxide, nitric oxide, hydrogen, and methane [16]. Profiles of these chemicals can lead to insight about a person's health including asthma, cystic fibrosis, and malabsorption of substances such as lactose and fructose [16, 17]. Similarly, breath measurements can aid in measuring for energy expenditure, detect whether alcohol is present in the blood, aid the diagnosis of diabetes by measuring ketones, and help identify asthma by monitoring exhaled nitric oxide levels [17]. Other metabolites are observed in the breath, and they have equally important diagnostic value. Metabolite monitoring in breath for health applications is an increasingly popular area of research. Metabolites are small molecules produced by the body as intermediates or end products of metabolism. Sensing and analyzing exhaled metabolites, often in the form of volatile organic compounds (VOCs), can provide insights into the biochemical reactions happening in the body and ultimately detect changes in patient physiology, metabolism, and overall health. Current research often attempts to correlate disease or disorders with particular VOCs and develop key biomarker profiles.

Many experiments have shown that VOCs can be detected and used to analyze the health of patients, including in respiratory medicine [1–3, 18] (e.g., distinguishing chronic obstructive pulmonary disease (COPD) patients [19–23]), diabetes monitoring [5, 24], oncology (e.g., lung cancer [25–29] and breast cancer [4, 30]), portable devices measuring acetone levels in breath [31], and asthma [3]. Similarly, the detection and analysis of VOCs in the headspace of fecal matter is another potential non-invasive means for detecting disease [32, 33]. As more is uncovered about the importance and impact that the microbiome has on health, sensing VOCs in fecal matter may become more prominent.

One approach is to detect a specific well-defined VOC in the presence of interfering gaseous chemical species using a highly selective detector. While this is straightforward, it is very difficult to identify the critical VOCs needed to diagnose a disease [10]. Using pattern recognition and semi-selective sensor techniques is an alternative [10], but this approach struggles with the specificity and selectivity of the diagnostic response. Bringing these approaches and research to portable gas analyzers increases the options for the non-invasive monitoring of patient health in hospitals or even in patients' homes. In vitro diagnostics, drug screening, and asthma monitoring could become available as portable hardware devices shrink in size and reliability. They may even be wearable in the future. However, modern clinical criteria are stringent and include low cost, low energy, easy repeatability at specific time intervals, and must have little to no impact on the day-to-day activities of the patients. Regulatory standards surrounding the wearables and the *quantified self* are yet to be determined, which also can also lead to uncertainty in this area of development. Finally, more studies are needed to establish the standards for these types of devices and the breath sampling requirements that will be needed in this field.

12.2.2 Military and Security Applications

Rapid chemical detection is an important tool in military and security environments. From detecting chemical warfare agents (CWAs) in the field to monitoring for explosive materials in public spaces, substantial interest is focused on bringing chemical detection to smaller portable platforms. Due to background air contaminants potentially obscuring the detection of dangerous chemicals, suitable detection requires a combination of reliable selective and highly sensitive sensors coupled with an on-site data processing system.

One of the most successful technologies employed for trace-level explosive detection has been ion mobility spectrometry (IMS). Chemicals such as trinitrotoluene (TNT), product ions for TNT, and some fragmented product ions from nitroglycerin have favorable properties for IMS detection—namely, they form relatively stable, uncomplicated negative gas-phase ions [34]. This has led to several early commercial instruments based on IMS technology being used in security screening checkpoints, such as in modern airports today. These have included the PD-5® and Plastec® (Graseby Dynamics acquired by Smiths Detection), Ionscan® (Barringer Research, Ltd), and Orion® (Intelligent Detection Systems) among others [34].

Other notable technologies that have focused on explosive detection include differential mobility spectrometry (DMS) sensors [35], microfabricated GC systems [36], microcantilevers [37, 38], portable optical systems [39–41], and metal oxide–based sensors [42, 43]. Some of these sensors have been deployed and sold commercially, such as DMS sensors. However, many others have been limited to research settings.

Detecting threats from CWAs and toxic industrial chemicals (TICs) is another area of interest crossing into both military and civilian environments. These applications require the fast detection of extremely low concentrations of specific chemicals. CWAs include nerve gases such as sarin, soman, tabun and VX agents; blistering agents such as sulfur mustard, nitrogen mustards, and lewisite; and choking agents or pulmonary intoxicants such as phosgene. Here, IMS is also a main technique for the on-site analysis of CWAs. Sarin and mustard gas have been demonstrated to have characteristic DMS spectra as well and are commercially relevant. The ions of these chemical compounds fragment at specific electric field intensities, leading to the characteristic spectra and subsequent detection of sarin and mustard gas [44]. Other methods that have been employed include surface acoustic wave (SAW), electrochemical, and spectrophotometric sensors [45, 46]. However, SAW sensors have the drawback of being moisture sensitive and spectrophotometric techniques often require relatively high concentrations for detection.

12.2.3 Agriculture Applications

As the demand for food supply increases with the rise in the global population, another area that could greatly benefit from on-site chemical analysis is the agriculture and food sector. Environmental monitoring, plant health, precision agriculture, and food quality are a few examples where chemical sensors could play a large role. Volatile analysis is showing promising results that could aid with plant disease diagnosis [47, 48]. This is critical for halting epidemics and even eventually reducing the reliance on pesticides. Food quality assurance is another major industry interested in direct, non-invasive, real-time quality assurance. In some cases, gas-phase chemical detection can accomplish this. Several studies have already assessed the use of gas chromatography and IMS in food and drink quality and purity [49–52]. In addition, electronic noses (e-noses) are also emerging as potential tools for food quality analysis [53–55], although they currently have issues with reproducibility and calibration issues outside lab environments.

12.3 Examples of Miniature Chemical Sensors

Chemical sensors come in a variety of forms, all based on different underlying physics and chemical reaction theories. A few of the major types of chemical sensors include optical sensors, electrochemical devices (which take a chemical input and convert it to an electrical signal using a potentiometric or amperometric detector), chemoresistive (e.g., metal oxide) devices, and resonant devices (e.g., SAW sensors) [56]. In this chapter, we will focus on ion mobility devices, metal oxide–based sensors, and chemical analysis systems, including micro-gas chromatography technologies and e-nose devices.

12.3.1 Ion Mobility Spectrometry

One technology already deployed in portable platforms for chemical analytics is IMS. In the most general terms, IMS is an analytical technique based on ion movement in gases. The first IMS devices were built in the 1970s and focused on military applications [57–59]. Since then, commercial devices have emerged and been popularized for screening items for trace explosive materials in airports [34]. IMS is also used in CWA detection in military settings [45]. While ion mobility techniques are not necessarily new, their applications to medicine and environmental monitoring has recently gained momentum [60, 61]. Such applications include air quality monitoring [62, 63] and medical and biological research [64, 65].

IMS involves the ionization of a gas sample and a subsequent analysis of the resulting ions. First, the gas is introduced to the device using a carrier gas and passed through an ionization chamber. Here, the analytes within the gas sample become ionized through collisions with the ionized carrier gas. Next, the ions move into a drift region via a shutter window. In this drift region, the ionized molecules gain velocity and an external electric field, as well as collisions with neutral counter-flowing drift-gas, separate the ions. This drift-gas also ensures neutral molecules do not enter the drift region. Finally, at the end of the drift tube, detecting electrodes (faraday plates) pick up signals as ions collide with the electrodes themselves. There are typically separate electrodes for both positive and negative ions.

Relatively low detection limits, simple architecture and low cost compared with other benchtop chemical sensing methods, make ion mobility technology attractive for use in portable instruments. However, in some cases selectivity can restrict particular applications [45]. Humidity and temperature can also influence the outcome of IMS data, adding layers of complexity to the data analysis steps. Finally, chemical interference in highly complex or contaminated environments can create challenges during the interpretation of IMS results. In order to overcome some of these challenges, researchers have often combined IMS with other chemical separation and detection techniques, including GC [66], pyrolysis-GC [67], metal oxide gas sensors [68], microfabricated preconcentrators [69], and multi-capillary columns [70–72].

Other forms of ion mobility, such as high-field asymmetric ion mobility spectrometry (FAIMS) and DMS have also shown promise for chemical sensing in portable systems [73, 74]. FAIMS and DMS are very similar to IMS in that they focus on ion mobility as a separation principle; however, these techniques use an asymmetric waveform electric field alternating between high and low strength applied across two parallel plates instead of a cylindrical tube in the drift region. In DMS, a low-strength DC compensation or *steering* voltage can then be adjusted or swept through a range, selecting particular ions to pass into the detecting electrode region. Compared with IMS, this creates gas-phase separation based on ion mobility in low and high electric fields and offers the advantage of gas analysis at atmospheric pressure. In 2001, researchers developed a MEMS-based DMS, showing the potential for portable chemical detection platforms to be based around this technology [75]. These devices have been effectively miniaturized and commercially matured for many defense and non-defense applications. However, most

DMS systems are limited in extended-use portability by the high power demands necessary for DMS devices. Another challenge with DMS is categorizing complex data from biological samples, but this can generally be tackled using advanced signal processing and algorithms [76]. DMS applications have been used both to identify single chemicals for specific applications [77, 78] and to detect complex signatures and multiple chemicals in biological applications.

12.3.2 Metal Oxide Sensors

One of the most common types of miniature chemical sensors is the metal oxide gas sensor. Already widely used in portable chemical detection systems for commercial and industrial use, metal oxide gas sensors are solid-state gas detectors based on semiconducting materials. Often classified as a chemoresistor, the conductivity of the sensor is a function of analyte concentration. Because of the low cost, ease of production, availability in MEMS design allowing for arrays of unique sensors, and a simple electronic measurement interfaces [79, 80], this type of chemical sensor is relatively easy to integrate into portable system. In fact, metal oxide–based carbon monoxide and combustible gas sensors have been commercially available (both in thick film or pellet based) since the 1970s for use in automotive monitoring, air quality monitoring, and other toxic gas detection applications [81]. Perhaps one of the most common and best understood oxide-based sensors is SnO_2, a conductive metal oxide known to vary as a function of specific gases including oxygen, hydrogen, carbon monoxide, carbon dioxide, and nitrogen dioxide. Other metal oxide films commonly used are tin oxide and zinc oxide.

At the core, metal oxide–based sensors rely on electrical impedance changing as a function of gas composition and concentration. The basic idea behind these sensors involves gas molecules of a certain species being adsorbed onto the surface of a resistive material. In other words, the surface holds the gas molecules as a thin film. Electrons from a resistive material then become trapped in the adsorbed gas species, causing a reduction in conductivity [82]. Resistive materials are typically polymers doped with metal oxides. A variety of film material choices and various geometries, thicknesses, and configurations allow for many design options. By changing the crystal structure and morphology, dopants, contact geometry, and the temperature of operation of an oxide-based sensor, the selectivity can be tuned. Specific chemical detection can then occur via arrays of differently doped sensors and using different modes of operation, such as pulsing temperature [83]. Most chemical detection cases then require some sort of pattern recognition during the data analysis step, and this category of sensors struggles with operational issues in complex chemical environments.

Generally, the structure of a metal oxide sensor is comprised of three parts. The first is the sensing film, where the reaction with the gas species occurs. The second is a pair of electrodes used to measure the resistance changes of the sensing film. The final part is typically a heater used to attain an operational or optimal reaction temperature for the device. The manufacture of metal oxide devices is relatively uncomplicated in that there are two general structures for micromachining using metal oxide as the sensing material: closed membrane and suspended membrane [84]. While most of the operations involved in manufacturing these are relatively standard in the MEMS world, careful material selection and device design is needed based on the application and operating environment. Additionally, there are frequent issues with performance and drift in some devices.

These types of chemical sensors typically have moderate sensitivity, selectivity, and robustness but often require significant power [56]. Because these sensors tend to be relatively noisy in ambient environments, this also creates challenges during design and optimization for on-site applications. Other challenges with impedance-based chemical sensors include identifying single gases, reliable gas concentration quantification, complex gas mixture analysis, and humidity and temperature sensitivity. While the manufacture and production typically involve simple micromachining procedures for a single sensor, the next challenge is in the design of integrated MEMS sample preparation and sample delivery to the metal oxide sensor [85].

12.3.3 Micro-GC Detectors

GC-MS remains the "gold standard" for laboratory gas chemical analysis, offering high sensitivity and selectivity. Because GC is an established method for detecting and assessing the concentration of VOCs in the environment [86], bringing this technology to the micro-scale for portable systems has been a goal since the 1970s. Micro-GCs remain a relatively active area of research. Typical for laboratory-based standard GC analysis, gas samples are collected in the field through a trap system such as sorbent tubes, stored and/or transported as necessary, then analyzed by trained technicians or scientists. The miniaturization of a GC system could potentially keep the benefits of a selective and sensitive analysis system but remove the trapping, storage, and transportation steps.

In a GC system, a gas sample flows through a long, serpentine channel, also called the *column*. During this time the chemical species within the gas separate based on interactions with a mobile phase and a stationary phase of the column. The mobile phase is usually an inert carrier gas, such as helium or nitrogen, which carries the sample through the column. The stationary phase is a filling within the column. The chemical species within the sample elute out of the column at different times based on the respective chemical properties and interaction within the column. The gas flow rate, column length, and temperature of the column can all influence the retention time and the subsequent outcome of the analysis. Traditionally, these systems are large and benchtop, although some suitcase-sized instruments are now being developed commercially.

While some research prototypes have focused on connecting arrays of microsensors to a GC in a portable platform [87], others have focused on miniaturizing the GC itself [86, 88–94]. These recent prototypes seem to remain application specific, focusing on innovations in GC architecture in order to improve selectivity and sensitivity for particular applications. While there are many devices using GC on a portable platform, these are still relatively bulky and expensive [95–98]. Related research in this area includes the development of MEMS-based gas preconcentrators, placed before the GC and other chemical analysis systems in order to amplify the chemical signal [99, 100]. As the design of reliable MEMS-based GC systems continues to improve, the optimization of the GC system architecture will allow for application-specific devices to be created.

12.3.4 E-Nose Detectors

E-nose devices have long been touted as a real-time gas-phase chemical detection platform. The concept of e-noses aspires to mimic the behavior of the human olfactory system by including multiple selective surfaces tiled as an array in the device. Typically, this means that an e-nose device consists of an array of gas-phase chemical sensors connected to a processing unit containing a pattern recognition algorithm [101]. Each element of the array may only have partial selectivity to specific chemicals of interest. The first developments of the e-nose technology date back to the 1960s, when researchers sought to measure signals thought to come from the interactions of volatiles with polarized microelectrodes [102]. In the 1980s, this concept was combined with a classification algorithm to attempt to interpret data outputs [103–105].

The types of sensors used in e-nose devices vary from chemoresistors, such as metal oxide sensors; acoustic-based sensors, such as piezoelectric crystals and SAW devices; electrochemical-based potentiometric or amperometric sensors; and optical sensors. While most of the sensors will only have partial specificity and/or sensitivity, the array architecture attempts to detect a variety of chemicals within a potentially complex gas mixture. However, this means that there are inherent disadvantages tied to the e-nose array architecture. Limitations arise when several chemical sensors optimally operate at different device parameters or settings (including temperature and humidity). Furthermore, there are limitations on physical configuration and sensitivity levels, depending on the chosen sensors in the array. This poses a challenge in creating usable, commercially viable e-nose chemical sensors for real-world applications.

Another challenging aspect involves the data analysis side of e-nose devices. In order to attempt a prediction of the chemicals present in a sample, many known and labeled gas samples are typically needed to train a predictive model before the e-nose can detect chemicals in real time. This poses a challenging problem when an e-nose encounters a new gas sample mixture.

Pattern recognition approaches are being attempted to create semi-supervised predictive methods for e-nose devices [106, 107], but the fundamental limitations of the device architecture remain.

Attempts have been made to use e-noses in environmental monitoring [108, 109], food detection, and medical applications [110], and they have been proposed for explosive detection [111, 112]. While they sometimes work in a laboratory setting, there are few (if any) commercial success stories of this category of devices. They can suffer from drift and calibration issues, and they also are frequently challenged by complex mixtures.

12.4 Data Analysis for Chemical Detection and Prediction

Complimentary to the chemical sensing hardware, real-time and on-site chemical detection and analysis systems need sophisticated data analysis processing. This component is critical to any reliable chemical prediction or detection system. In most cases, the overarching question for chemical detection is what chemical compounds are present and in what concentrations. Nowadays, the general strategy to accomplish this revolves around creating statistical prediction models, often borrowing heavily from machine learning methods, a branch of artificial intelligence focused on automatically learning complex patterns. For the purpose of this chapter, machine learning, pattern recognition, and statistical learning methods will be used interchangeably. Since computational and statistical analysis is a diverse and vibrant area of research, here we will only focus on introducing a few popular analysis techniques used in chemical prediction.

In general, pattern recognition uses two types of methods. The first is supervised learning and the second is unsupervised learning. Broadly speaking, supervised learning uses known input data and output classifications or regression to train a statistical learning model that can then be used to predict an outcome/output from the input data. Unsupervised learning, on the other hand, strives to find unseen or unnoticed patterns in a dataset and then automatically cluster data into similar patterns. A typical workflow largely remains the same independent of the technique used. First, signals from the sensor are formatted and preprocessed. This step can incorporate de-noising, smoothing, and feature/event detection. High noise levels in the chemical composition of some samples can be a problem for miniaturized chemical sensors, especially for miniature chemoresistors [81]; this step is critical for reliable outcomes. Visualization of the data is often used in parallel to ensure the quality of collected samples as well as identify potential outliers. Next, characteristic features are selected or computationally extracted from each observation event. This step can include deconvolution of large datasets into fewer, more critical data points. Feature selection and extraction also has high impact on the outcome and successful predictability of a model, since the selected or extracted features are used to train a supervised or unsupervised predictive model. Finding a suitable model for a particular application usually takes several iterations. Finally, the trained model can be used onboard a portable platform to predict the output of a sample in the field.

One popular computation and mathematical technique in chemical analysis and detection is principal component analysis (PCA). This technique can be used to reduce the dimensionality of a dataset and find linearly independent variables; in this case, they are called the *principal components*. This is particularly useful for complex gas mixtures or biological-based datasets with high dimensionality. When highly dimensional datasets are used to train a statistical learning method, it can lead to *overfitting*, a phenomenon where a model is too closely fit to a specific set of data points. In real-world applications, "uninteresting" compounds are present and usually in high concentrations. This can interfere with the analysis of metabolites/analytes of actual interest. For example, in 2010, Westhoff used PCA on IMS samples to pinpoint a single analyte out of 104 features that created separation between healthy individuals and COPD patients [22]. Similarly, PCA is popular to use with DMS data. A recent study used PCA combined with hierarchical cluster analysis (HCA) for GC-DMS data to classify biodiesel and fuel blends [113]. E-noses have also used PCA [101]. PCA can also be combined with other statistical learning methods such as genetic algorithms used on GC-DMS data [114].

There are a myriad of statistical learning methods that have been used in chemical detection and analysis, each with its own underlying theories. For example, discriminant analysis, partial least squares, learning vectors, support vector machines, and back-propagation neural network methods have been used [101] [115, 116]. However, suitable predictive models that are fast, accurate, and robust must be chosen based on the combination of sensors used and the desired application. In 2012, Phillips used 12 different learning methods to analyze VOC data in COPD and healthy individuals, exhibiting the difficulty in finding optimal techniques in order to help accurately diagnose patients [20]. Needless to say, data analysis is a non-trivial component integral to the successful deployment of a portable chemical device.

12.5 Future Direction Trends for the Field

Despite substantial progress in the research and design of miniaturized components for chemical detection systems, there are several challenges in developing holistic, reliable, and robust solutions for real-time on-site chemical sensing applications. A few prominent ones include engineering optimal flow and pressure dynamics, heat transfer design and optimization, material selection for integrated sensor systems, and the microfabrication complexities involved with this.

Compared with other forms of chemical sampling and sensing, optimal flow and pressure is often critical to gas-phase microsensor performance. For example, adsorption and desorption pressure and flow parameters for components such as preconcentrators, metal oxide–based sensors, and GC systems greatly influence calibration curves as well as overall device sensitivity and selectivity limits. Because of the complex nature of gas flow in channels, research will need to focus on benchmarking microsensor performance and standardizing optimal gas flow dynamics and requirements. Furthermore, gas pumps and valves have become a limiting factor for device size. Research into minimizing pump use through the optimization of control design and even the incorporation of micropumps into the microsensor chip design may be areas for development in the future.

Another challenging aspect revolves around the innate necessity to optimize temperature and humidity for desired chemical reactions within sensor components. As the miniaturization of sensor components continues, the tasks of chip packaging and system integration become complex when a chemical analysis system demands specific temperatures throughout different components for effective functionality. While solutions for heat transfer can be borrowed and adapted from the integrated circuit industry, chemical sensors might require higher levels of stringency, necessitating significant resources be devoted to heat transfer design.

Finally, the choice of materials for chemical microsensors is critical. Typically, integrated circuits use aluminum, polysilicon, and copper; however, for electrochemical microsensors, the conductors must often be gold or platinum in order to obtain sufficiently low levels of detection [81]. Further, microchannels and other gas containment components of a chemical sensor must be inert in order to avoid contamination between gas sampling and ensure effective device reusability. In addition to material restrictions, many chemical sensors require high aspect ratios in the *z*-direction. This is not typical in integrated circuit fabrication, creating complexity in design and microfabrication processes for microsensors at scale. The lack of appropriate chemical electronic fabrication tools in traditional integrated circuit fabrication facilities could be a bottleneck in the widespread manufacture of portable chemical analysis and detection systems.

In closing, the field of miniature chemical sensor systems is rapidly expanding. Some commercial successes are already available, and further research will likely only accelerate the development and utility of these systems. Also, as the concept of the Internet of Things (IoT) continues to grow as a systems operations concept, we envision distributed networks of these chemical sensors throughout industry segments. They are likely to be valuable to both end users and service providers. The future looks very bright for these tools.

References

1. Cohen-Kaminsky, S., et al., A proof of concept for the detection and classification of pulmonary arterial hypertension through breath analysis with a sensor array. *American Journal of Respiratory and Critical Care Medicine*, 2013. **188**(6): pp. 756–759.
2. Allers, M., et al., Measurement of exhaled volatile organic compounds from patients with chronic obstructive pulmonary disease (COPD) using closed gas loop GC-IMS and GC-APCI-MS. *Journal of Breath Research*, 2016. **10**(2): p. 026004.
3. Bos, L.D.J., et al., Exhaled breath metabolomics as a noninvasive diagnostic tool for acute respiratory distress syndrome. *European Respiratory Journal*, 2014. **44**(1): p. 188.
4. Phillips, M., et al., Volatile markers of breast cancer in the breath. *The Breast Journal*, 2003. **9**(3): pp. 184–191.
5. Phillips, M., et al., Increased breath biomarkers of oxidative stress in diabetes mellitus. *Clinica Chimica Acta*, 2004. **344**(1): pp. 189–194.
6. Phillips, M., et al., Volatile organic compounds in breath as markers of lung cancer: A cross-sectional study. *The Lancet*, 1999. **353**(9168): pp. 1930–1933.
7. Poli, D., et al., Exhaled volatile organic compounds in patients with non-small cell lung cancer: cross sectional and nested short-term follow-up study. *Respiratory Research*, 2005. **6**(1): p. 71.
8. Bajtarevic, A., et al., Noninvasive detection of lung cancer by analysis of exhaled breath. *BMC Cancer*, 2009. **9**(1): p. 348.
9. Westhoff, M., et al., Ion mobility spectrometry for the detection of volatile organic compounds in exhaled breath of patients with lung cancer: results of a pilot study. *Thorax*, 2009. **64**(9): p. 744.
10. Nakhleh, M.K., et al., Diagnosis and classification of 17 diseases from 1404 subjects via pattern analysis of exhaled molecules. *ACS Nano*, 2017. **11**(1): pp. 112–125.
11. Katajamaa, M., and M. Orešič, Data processing for mass spectrometry-based metabolomics. *Journal of Chromatography A*, 2007. **1158**(1): pp. 318–328.
12. DeHaven, C.D., et al., Organization of GC/MS and LC/MS metabolomics data into chemical libraries. *Journal of Cheminformatics*, 2010. **2**(1): p. 9.
13. Terry, S.C., J.H. Jerman, and J.B. Angell, A gas chromatographic air analyzer fabricated on a silicon wafer. *IEEE Transactions on Electron Devices*, 1979. **26**(12): pp. 1880–1886.
14. Wise, K.D., Integrated sensors, MEMS, and microsystems: reflections on a fantastic voyage. *Sensors and Actuators A: Physical*, 2007. **136**(1): pp. 39–50.
15. Lewis, P.R., et al., Recent advancements in the gas-phase MicroChemLab. *IEEE Sensors Journal*, 2006. **6**(3): pp. 784–795.
16. Zhou, M., Y. Liu, and Y. Duan, Breath biomarkers in diagnosis of pulmonary diseases. *Clinica Chimica Acta*, 2012. **413**(21): pp. 1770–1780.
17. Gardner, W.J., and A.T. Vincent, Electronic noses for well-being: Breath analysis and energy expenditure. *Sensors*, 2016. **16**(7).
18. Basanta, M., et al., Non-invasive metabolomic analysis of breath using differential mobility spectrometry in patients with chronic obstructive pulmonary disease and healthy smokers. *Analyst*, 2010. **135**(2): pp. 315–320.

19. Christiansen, A., et al., A systematic review of breath analysis and detection of volatile organic compounds in COPD. *Journal of Breath Research*, 2016. **10**(3): p. 034002.
20. Phillips, C.O., et al., Machine learning methods on exhaled volatile organic compounds for distinguishing COPD patients from healthy controls. *Journal of Breath Research*, 2012. **6**(3): p. 036003.
21. Van Berkel, J.J.B.N., et al., A profile of volatile organic compounds in breath discriminates COPD patients from controls. *Respiratory Medicine*, 2010. **104**(4): pp. 557–563.
22. Westhoff, M., et al., Differentiation of chronic obstructive pulmonary disease (COPD) including lung cancer from healthy control group by breath analysis using ion mobility spectrometry. *International Journal for Ion Mobility Spectrometry*, 2010. **13**(3): pp. 131–139.
23. Cazzola, M., et al., Analysis of exhaled breath fingerprints and volatile organic compounds in COPD. *COPD Research and Practice*, 2015. **1**(1): p. 7.
24. Walton, C., et al., The use of a portable breath analysis device in monitoring type 1 diabetes patients in a hypoglycaemic clamp: Validation with SIFT-MS data. *Journal of Breath Research*, 2014. **8**(3): p. 037108.
25. Rudnicka, J., et al., Determination of volatile organic compounds as potential markers of lung cancer by gas chromatography–mass spectrometry versus trained dogs. *Sensors and Actuators B: Chemical*, 2014. **202**: pp. 615–621.
26. Corradi, M., et al., Exhaled breath analysis in suspected cases of non-small-cell lung cancer: A cross-sectional study. *Journal of Breath Research*, 2015. **9**(2): p. 027101.
27. Machado, R.F., et al., Detection of lung cancer by sensor array analyses of exhaled breath. *American Journal of Respiratory and Critical Care Medicine*, 2005. **171**(11): pp. 1286–1291.
28. Xing, C., et al., A study of an electronic nose for detection of lung cancer based on a virtual SAW gas sensors array and imaging recognition method. *Measurement Science and Technology*, 2005. **16**(8): p. 1535.
29. Phillips, M., et al., Detection of lung cancer using weighted digital analysis of breath biomarkers. *Clinica Chimica Acta*, 2008. **393**: pp. 76–84.
30. Lavra, L., et al., Investigation of VOCs associated with different characteristics of breast cancer cells. *Scientific Reports*, 2015. **5**: p. 13246.
31. Blaikie, T.P.J., et al., Portable device for measuring breath acetone based on sample preconcentration and cavity enhanced spectroscopy. *Analytical Chemistry*, 2016. **88**(22): pp. 11016–11021.
32. van Gaal, N., et al., Faecal volatile organic compound analysis using field asymmetric ion mobility spectrometry: non-invasive diagnostics in pediatric inflammatory bowel disease. *Journal of Breath Research*, 2017. **12**(1): p. 016006.
33. Saptalena, L.G., A. Kuklya, and U. Telgheder, Fast detection of coliform bacteria by means of gas chromatography–differential mobility spectrometry. *Analytical and Bioanalytical Chemistry*, 2016. **408**(14): pp. 3715–3725.
34. Ewing, R.G., et al., A critical review of ion mobility spectrometry for the detection of explosives and explosive related compounds. *Talanta*, 2001. **54**(3): pp. 515–29.
35. Pavlačka, M., et al., Analysis of explosives using differential mobility spectrometry. *International Journal for Ion Mobility Spectrometry*, 2016. **19**(1): pp. 31–39.

36. Collin, W.R., et al., Microfabricated gas chromatograph for rapid, trace-level determinations of gas-phase explosive marker compounds. *Analytical Chemistry*, 2014. **86**(1): pp. 655–663.
37. García-Romeo, D., et al., Portable low-power electronic interface for explosive detection using microcantilevers. *Sensors and Actuators B: Chemical*, 2014. **200**: pp. 31–38.
38. Iritia, M.P.P., et al., Explosives detection by array of Si LaTeX-cantilevers coated with titanosilicate-type nanoporous materials. *IEEE Sensors Journal*, 2016. **16**(10): pp. 3435–3443.
39. Chen, N., et al., Portable and reliable surface-enhanced Raman scattering silicon chip for signal-on detection of trace trinitrotoluene explosive in real systems. *Analytical Chemistry*, 2017. **89**(9): pp. 5072–5078.
40. Gillanders, R.N., I.D.W. Samuel, and G.A. Turnbull, A low-cost, portable optical explosive-vapour sensor. *Sensors and Actuators B: Chemical*, 2017. **245**: pp. 334–340.
41. Zandieh, O., and S. Kim, Sensitive and selective detection of adsorbed explosive molecules using opto-calorimetric infrared spectroscopy and micro-differential thermal analysis. *Sensors and Actuators B: Chemical*, 2016. **231**: pp. 393–398.
42. Firtat, B., et al., Miniaturised MOX based sensors for pollutant and explosive gases detection. *Sensors and Actuators B: Chemical*, 2017. **249**: pp. 647–655.
43. Guo, L., et al., Sensitive, real-time and anti-interfering detection of nitro-explosive vapors realized by ZnO/rGO core/shell micro-Schottky junction. *Sensors and Actuators B: Chemical*, 2017. **239**: pp. 286–294.
44. Maziejuk, M., et al., Fragmentation of molecular ions in differential mobility spectrometry as a method for identification of chemical warfare agents. *Talanta*, 2015. **144**: pp. 1201–1206.
45. Mäkinen, M.A., O.A. Anttalainen, and M.E.T. Sillanpää, Ion mobility spectrometry and its applications in detection of chemical warfare agents. *Analytical Chemistry*, 2010. **82**(23): pp. 9594–9600.
46. Kim, P., et al., Towards the development of a portable device for the monitoring of gaseous toxic industrial chemicals based on a chemical sensor array. *Sensors and Actuators B: Chemical*, 2008. **134**(1): pp. 307–312.
47. Pasamontes, A., et al., *Citrus tristeza virus* infection in sweet orange trees and a mandarin × tangor cross alters low molecular weight metabolites assessed using gas chromatography mass spectrometry (GC/MS). *Metabolomics*, 2016. **12**(3): p. 41.
48. McCartney, M.M., et al., Coupling a branch enclosure with differential mobility spectrometry to isolate and measure plant volatiles in contained greenhouse settings. *Talanta*, 2016. **146**: pp. 148–154.
49. Plutowska, B. and W. Wardencki, Application of gas chromatography–olfactometry (GC–O) in analysis and quality assessment of alcoholic beverages: A review. *Food Chemistry*, 2008. **107**(1): pp. 449–463.
50. Hajslova, J., T. Cajka, and L. Vaclavik, Challenging applications offered by direct analysis in real time (DART) in food-quality and safety analysis. *TrAC Trends in Analytical Chemistry*, 2011. **30**(2): pp. 204–218.
51. Lehotay, S.J., and J. Hajšlová, Application of gas chromatography in food analysis. *TrAC Trends in Analytical Chemistry*, 2002. **21**(9): pp. 686–697.
52. Vautz, W., et al., Ion mobility spectrometry for food quality and safety. *Food Additives & Contaminants*, 2006. **23**(11): pp. 1064–1073.

53. Barié, N., M. Bücking, and M. Rapp, A novel electronic nose based on miniaturized SAW sensor arrays coupled with SPME enhanced headspace-analysis and its use for rapid determination of volatile organic compounds in food quality monitoring. *Sensors and Actuators B: Chemical*, 2006. **114**(1): pp. 482–488.
54. Winquist, F., et al., Performance of an electronic nose for quality estimation of ground meat. *Measurement Science and Technology*, 1993. **4**(12): p. 1493.
55. Tothill, I.E., Biosensors developments and potential applications in the agricultural diagnosis sector. *Computers and Electronics in Agriculture*, 2001. **30**(1): pp. 205–218.
56. Kim, I.-D., A. Rothschild, and H.L. Tuller, Advances and new directions in gas-sensing devices. *Acta Materialia*, 2013. **61**(3): pp. 974–1000.
57. Baumbach, J.I., and G.A. Eiceman, Ion mobility spectrometry: arriving on site and moving beyond a low profile. *Applied Spectroscopy*, 1999. **53**(9): pp. 338A–355A.
58. Hill, H.H., Jr., W.F. Siems, R.H. St Louis, D.G. McMinn, Ion mobility Spectrometry. *Analytical Chemistry*, 1990. **62**: pp. 1201A–1209A.
59. Mäkinen, M., M. Nousiainen, and M. Sillanpää, Ion spectrometric detection technologies for ultra-traces of explosives: A review. *Mass Spectrometry Reviews*, 2011. **30**(5): pp. 940–973.
60. Cumeras, R., et al., Review on ion mobility spectrometry, Part 1: Current instrumentation. *Analyst*, 2015. **140**(5): pp. 1376–1390.
61. Hauschild, A.-C., et al., Computational methods for metabolomic data analysis of ion mobility spectrometry data: Reviewing the state of the art. *Metabolites*, 2012. **2**(4): pp. 733–755.
62. Palmer, P.T., and T.F. Limero, Mass spectrometry in the U.S. space program: Past, present, and future. *Journal of the American Society for Mass Spectrometry*, 2001. **12**(6): pp. 656–675.
63. Eiceman, G.A., et al., Monitoring volatile organic compounds in ambient air inside and outside buildings with the use of a radio-frequency-based ion-mobility analyzer with a micromachined drift tube. *Field Analytical Chemistry & Technology*, 2000. **4**(6): pp. 297–308.
64. Baumbach, J.I., and M. Westhoff, Ion mobility spectrometry to detect lung cancer and airway infections. *Spectroscopy Europe*, 2006. **18**(6): pp. 22–27.
65. Lokhnauth, J.K., and N.H. Snow, Solid phase micro-extraction coupled with ion mobility spectrometry for the analysis of ephedrine in urine. *Journal of Separation Science*, 2005. **28**(7): pp. 612–618.
66. Snyder, A.P., et al., Portable hand-held gas chromatography/ion mobility spectrometry device. *Analytical Chemistry*, 1993. **65**(3): pp. 299–306.
67. Snyder, A.P., et al., Detection of the picolinic acid biomarker in Bacillus spores using a potentially field-portable pyrolysis–gas chromatography–ion mobility spectrometry system. *Field Analytical Chemistry & Technology*, 1996. **1**(1): pp. 49–59.
68. Utriainen, M., E. Kärpänoja, and H. Paakkanen, Combining miniaturized ion mobility spectrometer and metal oxide gas sensor for the fast detection of toxic chemical vapors. *Sensors and Actuators B: Chemical*, 2003. **93**(1): pp. 17–24.
69. Martin, M., et al., Microfabricated vapor preconcentrator for portable ion mobility spectroscopy. *Sensors and Actuators B: Chemical*, 2007. **126**(2): pp. 447–454.
70. Ruzsanyi, V., et al., Detection of human metabolites using multi-capillary columns coupled to ion mobility spectrometers. *Journal of Chromatography A*, 2005. **1084**(1): pp. 145–151.

71. Maddula, S., et al., Detection of volatile metabolites of *Escherichia coli* by multi capillary column coupled ion mobility spectrometry. *Analytical and Bioanalytical Chemistry*, 2009. **394**(3): pp. 791–800.
72. Bödeker, B., W. Vautz, and J.I. Baumbach, Peak comparison in MCC/IMS-data: Searching for potential biomarkers in human breath data. *International Journal for Ion Mobility Spectrometry*, 2008. **11**(1): pp. 89–93.
73. Krebs, M.D., et al., Detection of biological and chemical agents using differential mobility spectrometry (DMS) technology. *IEEE Sensors Journal*, 2005. **5**(4): pp. 696–703.
74. Kolakowski, B.M., and Z. Mester, Review of applications of high-field asymmetric waveform ion mobility spectrometry (FAIMS) and differential mobility spectrometry (DMS). *Analyst*, 2007. **132**(9): pp. 842–864.
75. Miller, R.A., et al., A MEMS radio-frequency ion mobility spectrometer for chemical vapor detection. *Sensors and Actuators A: Physical*, 2001. **91**(3): pp. 301–312.
76. Peirano, D.J., A. Pasamontes, and C.E. Davis, Supervised semi-automated data analysis software for gas chromatography/differential mobility spectrometry (GC/DMS) metabolomics applications. *International Journal for Ion Mobility Spectrometry*, 2016. **19**(2): pp. 155–166.
77. Davis, C.E., et al., Analysis of volatile and non-volatile biomarkers in human breath using differential mobility spectrometry (DMS). *IEEE Sensors Journal*, 2010. **10**(1): pp. 114–122.
78. Krylov, E.V., et al., Selection and generation of waveforms for differential mobility spectrometry. *Review of Scientific Instruments*, 2010. **81**(2): p. 024101.
79. Wolfrum, E.J., et al., Metal oxide sensor arrays for the detection, differentiation, and quantification of volatile organic compounds at sub-parts-per-million concentration levels. *Sensors and Actuators B: Chemical*, 2006. **115**(1): pp. 322–329.
80. Tomchenko, A.A., et al., Semiconducting metal oxide sensor array for the selective detection of combustion gases. *Sensors and Actuators B: Chemical*, 2003. **93**(1): pp. 126–134.
81. Wilson, D.M., et al., Chemical sensors for portable, handheld field instruments. *IEEE Sensors Journal*, 2001. **1**(4): pp. 256–274.
82. Sun, Y.-F., et al., Metal oxide nanostructures and their gas sensing properties: A review. *Sensors*, 2012. **12**(3): pp. 2610–2631.
83. Barsan, N., M. Schweizer-Berberich, and W. Göpel, Fundamental and practical aspects in the design of nanoscaled SnO_2 gas sensors: A status report. *Fresenius' Journal of Analytical Chemistry*, 1999. **365**(4): pp. 287–304.
84. Simon, I., et al., Micromachined metal oxide gas sensors: Opportunities to improve sensor performance. *Sensors and Actuators B: Chemical*, 2001. **73**(1): pp. 1–26.
85. Jack, W.J., Microelectromechanical systems (MEMS): Fabrication, design and applications. *Smart Materials and Structures*, 2001. **10**(6): pp. 1115.
86. Garg, A., et al., Zebra GC: A mini gas chromatography system for trace-level determination of hazardous air pollutants. *Sensors and Actuators B: Chemical*, 2015. **212**: pp. 145–154.
87. Lu, C.-J., et al., Portable gas chromatograph with tunable retention and sensor array detection for determination of complex vapor mixtures. *Analytical Chemistry*, 2003. **75**(6): pp. 1400–1409.

88. Iglesias, R.A., et al., Hybrid separation and detection device for analysis of benzene, toluene, ethylbenzene, and xylenes in complex samples. *Analytical Chemistry*, 2009. **81**(21): pp. 8930–8935.
89. Zampolli, S., et al., Real-time monitoring of sub-ppb concentrations of aromatic volatiles with a MEMS-enabled miniaturized gas-chromatograph. *Sensors and Actuators B: Chemical*, 2009. **141**(1): pp. 322–328.
90. Lu, C.-J., et al., First-generation hybrid MEMS gas chromatograph. *Lab on a Chip*, 2005. **5**(10): pp. 1123–1131.
91. Zampolli, S., et al., Selectivity enhancement of metal oxide gas sensors using a micromachined gas chromatographic column. *Sensors and Actuators B: Chemical*, 2005. **105**(2): pp. 400–406.
92. Sun, J.H., et al., A micro gas chromatography column with a micro thermal conductivity detector for volatile organic compound analysis. *Review of Scientific Instruments*, 2013. **84**(2): p. 025001.
93. Nasreddine, R., et al., Development of a novel portable miniaturized GC for near real-time low level detection of BTEX. *Sensors and Actuators B: Chemical*, 2016. **224**: pp. 159–169.
94. Agah, M., et al., High-speed MEMS-based gas chromatography. *Journal of Microelectromechanical Systems*, 2006. **15**(5): pp. 1371–1378.
95. 2017PID analyzers. Available from: http://www.hnu.com (cited 2017).
96. 2017Inficon. Available from: http://www.photovac.com (cited 2017).
97. Rae Systems, Connected Intelligent Gas Detection Systems. Available from: http://www.raesystems.com (cited 2017).
98. 2017Electronic Sensor Technology. Available from: http://www.estcal.com (cited 2017).
99. James, F., et al., Development of a MEMS preconcentrator for micro-gas chromatography analyses. *Procedia Engineering*, 2014. **87**: p. 500–503.
100. McCartney, M.M., et al., An easy to manufacture micro gas preconcentrator for chemical sensing applications. *ACS Sensors*, 2017. **2**(8): pp. 1167–1174.
101. Gardner, J.W., and P.N. Bartlett, A brief history of electronic noses. *Sensors and Actuators B: Chemical*, 1994. **18**(1): pp. 210–211.
102. Wilkens, W.F., and J.D. Hartman, An electronic analog for the olfactory processesa. *Journal of Food Science*, 1964. **29**(3): pp. 372–378.
103. K. Persaud, G.H.D., Analysis of discrimination mechanisms of the mammalian olfactory system using a model nose. *Nature*, 1982. **299**: pp. 352–355.
104. A. Ikegami, M.K., Olfactory detection using integrated sensors, in *Proceedings of the 3rd International Conference on Solid-State Sensors and Actuators*, 1985, Philadelphia, PA, pp. 136–139.
105. Kaneyasu, A.I., H. Arima, and S. Iwanga, Smell identification using a thick-film hybrid gas sensor, in *IEEE Transactions on Components, Hybrids Manufacturing Technology*, 1987. **10**(2): pp. 267–273.
106. Jian, Y., et al., A novel extreme learning machine classification model for e-nose application based on the multiple Kernel approach. *Sensors*, 2017. **17**(6): p. 1434.
107. Huang, T., et al., A novel semi-supervised method of electronic nose for indoor pollution detection trained by M-S4VMs. *Sensors*, 2016. **16**(9): p. 1462.
108. Ameer, Q., and S.B. Adeloju, Polypyrrole-based electronic noses for environmental and industrial analysis. *Sensors and Actuators B: Chemical*, 2005. **106**(2): pp. 541–552.

109. Lamagna, A., et al., The use of an electronic nose to characterize emissions from a highly polluted river. *Sensors and Actuators B: Chemical*, 2008. **131**(1): pp. 121–124.
110. Leopold, H.J., et al., Factors influencing continuous breath signal in intubated and mechanically-ventilated intensive care unit patients measured by an electronic nose. *Sensors*, 2016. **16**(8): p. 1337.
111. Wang, J., Electrochemical sensing of explosives. *Electroanalysis*, 2007. **19**(4): pp. 415–423.
112. Gardner, J.W., Review of conventional electronic noses and their possible application to the detection of explosives, in *Electronic Noses & Sensors for the Detection of Explosives*, J.W. Gardner and J. Yinon, eds. Dordrecht, the Netherlands: Springer, 2004, pp. 1–28.
113. Pasupuleti, D., G.A. Eiceman, and K.M. Pierce, Classification of biodiesel and fuel blends using gas chromatography–differential mobility spectrometry with cluster analysis and isolation of C18:3 me by dual ion filtering. *Talanta*, 2016. **155**: pp. 278–288.
114. Eiceman, G.A., et al., Pattern recognition analysis of differential mobility spectra with classification by chemical family. *Analytica Chimica Acta*, 2006. **579**(1): pp. 1–10.
115. Haddi, Z., et al., A portable electronic nose system for the identification of cannabis-based drugs. *Sensors and Actuators B: Chemical*, 2011. **155**(2): pp. 456–463.
116. Papadopoulou, O.S., et al., Sensory and microbiological quality assessment of beef fillets using a portable electronic nose in tandem with support vector machine analysis. *Food Research International*, 2013. **50**(1): pp. 241–249.

13

Optical Flow Sensing and Its Precision Vertical Landing Applications

Mohammad K. Al-Sharman, Murad Qasaimeh, Bara J. Emran, Mohammad A. Jaradat, and Mohammad Amin Al-Jarrah

CONTENTS

13.1 Introduction 304
13.2 Small-Scale Flybarless Helicopter Modeling 306
13.3 Survey of OF Sensing Techniques 307
  13.3.1 OF Computation 308
  13.3.2 OF Algorithms 310
    13.3.2.1 Block-Matching Methods 310
    13.3.2.2 Differential Methods 311
    13.3.2.3 Feature-Based Methods 311
    13.3.2.4 Image Interpolation Methods 311
    13.3.2.5 Energy-Based Methods 311
    13.3.2.6 Phase-Based Methods 312
    13.3.2.7 Fractional-Order Operator–Based Method 312
    13.3.2.8 Fusion-Based Methods 312
  13.3.3 Evaluation Methodology and Results 312
  13.3.4 Survey of OF-Based Navigation Approaches 314
  13.3.5 Specification Comparison of Different OF Sensors 315
13.4 OF Sensor Model 317
  13.4.1 Intelligent OF Sensor Fusion Design 319
13.5 Simulation Results 323
  13.5.1 Position Estimation 323
  13.5.2 Velocity Estimation 323
13.6 Conclusion 325
References 327

13.1 Introduction

The use of miniature unmanned vertical takeoff and landing (VTOL) vehicles has been extensively increasing in several civil and military applications due to their high agility and capability in performing missions close to the ground [1–4]. Accomplishing such missions successfully requires high-quality and accurately updated measurements from the onboard sensors. For example, the commonly used GPS/INS solution becomes inaccurate during vertical terminal-phase landing. Therefore, in precision landing applications, more precise sensors (i.e., vision sensors) have to be integrated with the existent GPS/INS solution to provide an accurate measurement augmentation.

Vision-based landing systems have been extensively developed in recent decades due to the high update rate and accuracy of their vision sensors. Several published articles have utilized vision sensors to perform landing on a predefined location [5, 6] or to identify the most suitable landing location [7]. For instance, in [5], a visual landing system was introduced for unmanned helicopters. The study proposed a computationally heavy landing system for an AVATAR helicopter equipped with a CCD camera for vision feedback and a high-cost Novatel RT-20 DGPS for position estimation. Although precise landing with 47 cm average position error was achieved, the proposed high-cost vision landing system is still computationally heavy and ill-suited for smaller-payload helicopters. In a similar study [6], vision-based landing with 54 cm maximum position error was performed with the Yamaha-RMAX helicopter using a set of expensive sensors. Besides the computational burden of the proposed landing algorithm, a landing pad with special patterns is required to achieve precise pose estimation. In their research, Meingast, Geyer, and Sastry [8] resolved the problem of landing a helicopter in unidentified areas using vision-based terrain recovery. This solution was also computationally complex, and a Yamaha R-50 helicopter with a 20 kg payload was used to carry the equipped sensors. Furthermore, Sharp, Shakernia, and Sastry [9] used vision sensor feedback to provide real-time attitude and position estimates relative to a patterned landing pad. Recently, the attitude and position estimation of a quadrotor was proposed using an optical flow (OF) algorithm in [10], and an auto-landing with 30 cm position error was achieved.

The ADNS series of OF sensors are utilized in unmanned aerial vision applications because of their precise capture of the surface's motion [11]. The ADNS-2620 optical mouse sensor has a speed of 3000 frames per second (fps), a resolution of 400 counts per inch (cpi), and an accurate motion of 12 inches per second (ips) [12]. On the other hand, the ADNS-3080 sensor has a higher speed and a better resolution compared with the ADNS-2620 [13]. The sensor's speed is up to 6400 fps, its resolution is up to 1600 cpi, and it can detect an accurate motion of 12 ips. However, ADNS OF sensors have a major shortcoming: a certain amount of light must be enveloping the captured surface in order for them to perform well, which makes them inappropriate for indoor or low-light applications. Honegger [14] has removed the light constraint and has recently introduced a special type of OF CMOS camera called the PX4FLOW sensor camera. This camera has the ability to perform efficiently in outdoor, indoor, and low-light conditions. PX4FLOW sensor cameras have been used to estimate the translational velocities x and y during the indoor and outdoor hovering of a quadrotor. The PX4FLOW kit is shown in Figure 13.1.

The PX4FLOW kit is equipped with three main components: a CMOS camera, an ultrasonic sensor, and a 3D gyro sensor. Having an ultrasonic sensor is important for obtaining measurements of the vertical distance to the ground, whereas the 3D gyro sensor is important for providing the onboard processor with the unmanned aerial vehicle (UAV)'s 3D rotations. By knowing the vertical distance and the 3D rotations, the PX4FLOW flow readings are compensated and converted into the metric scale. The captured frames of the PX4FLOW are processed by a high-performance ARM Cortex M4 processor, which works at a rate of 250 fps. Moreover, the PX4FLOW kit has M12 lenses with a 21° field of view and an IR block covered with an Aptina MT9V034 imager. Based on the PX4FLOW specifications and performance, its sensor was selected to be used for the OF sensor modeling experiment [4]. The modeling experiment tests the OF sensor's performance at four different heights to analyze the sensitivity of the sensor to height variations.

Rigorous state estimation techniques are fundamental for the VTOL vehicle to boost its maneuverability during an autonomous mission [15]. However, obtaining accurate state estimates of the vehicle is challenging due to problems associated with the low-cost onboard sensors (i.e., large drifts, immense noise, and measurement biases). Commercial off-the-shelf (COTS) sensors with such errors are still preferred in VTOL UAVs because of their

FIGURE 13.1
PX4FLOW smart optical sensor [14].

low power consumption, light weight, and compact size. Applying an effective sensor fusion algorithm to the COTS measurements can provide accurate state estimates. For example, in [16], a sensor fusion architecture between the GPS/INS and the Kinematic OF model was introduced for velocity and position estimation. In [15, 17, 3], high-accuracy helicopter attitude and flapping state estimates are achieved using a flybarless helicopter platform.

The robustness of the vehicle's state estimations in the fusion filter is linked to the accuracy of the measurement noise statistics and a priori information of the process, which are practically demonstrated by R and Q matrices, respectively [18, 19]. When designing a VTOL estimator, assigning appropriate statistical R and Q matrices is considered challenging; inaccurate a priori information degrades the performance of the state estimator, which might result in the divergence of the filter. Therefore, different adaptation techniques have been introduced to tackle the problem of imperfect a priori information [20, 19]. The innovation-based adaptive estimation (IAE) and multiple-model adaptive estimation (MMAE) techniques are commonly used to adapt the statistical R and Q matrices of the Kalman filter (KF) [18]. The IAE technique provides an adaptation of the R and Q matrices based on an evaluation of the discrepancy between the actual and theoretical covariance of the filter's innovation sequence. In [19, 21, 22], the IAE adaptation approach proved its ability to compensate for the lack of precise measurement noise statistics in the KF. Recently, the application of fuzzy logic rules to adjust for the values of the statistical matrices has been studied in a great deal of published research work. The fuzzy-adapted KF showed better performance in rejecting measurement noise and estimating navigational states accurately [23, 24].

This chapter reviews the available OF sensing techniques and their applications in UAV navigation and control. It also introduces an intelligent OF sensor fusion technique for terminal-phase landing, which is also presented in this work. The chapter is outlined as follows. Section 13.2 is a shortened introduction to a dynamic model of the flybarless helicopter. Section 13.3 surveys OF algorithms and their applications in the field of UAV control and estimation. Section 13.4 presents a quantitative evaluation of different OF algorithms. The intelligent OF sensor fusion design is presented in Section 13.5. The simulation results of the intelligent state estimation algorithm are demonstrated in Section 13.6. Finally, the proposed work is concluded in Section 13.7.

13.2 Small-Scale Flybarless Helicopter Modeling

This section briefly details the small-scale flybarless helicopter model. The nonlinear model of the Maxi Joker 3 helicopter is used in this study (Figure 13.2). Elaborated details on modeling the Maxi Joker 3 helicopter

FIGURE 13.2
Maxi Joker 3 helicopter

using the top-down modeling approach can be found in [4, 15]. This approach models the helicopter dynamics in four major blocks. The actuator dynamics, the flapping and thrust dynamics, the force and torques, and the rigid body equations are modeled in a Matlab (R) Simulink environment. The modeling of the flybarless helicopter is thoroughly illustrated in [4].

13.3 Survey of OF Sensing Techniques

Various types of OF sensors, ranging from simple optical mouse sensors to more complex stereo camera setups, have been studied in the literature. Moreover, different algorithms have been proposed to improve the estimation of optical and motion fields from a sequence of image frames. Deciding on the use of OF algorithms or sensors is a crucial task for the success of any autonomous application of UAVs. Recently, OF-aided navigation and positioning solutions have been used to enhance GPS-based systems for UAVs. OF sensors have made this possible because they can estimate position changes by tracking the movements in sequential images and converting pixel changes into global positional changes. In this section, the focus is on surveying the most recent algorithmic advances achieved using single OF sensors to enhance GPS-based navigation systems. The goal is to assess how the use of different algorithms can yield different results in terms of the quality of the estimated position measurements and execution time.

This section gives a brief overview of OF sensing concepts and how these concepts can be used to measure positional changes in terms of real-world distances. It also presents a summary of existing OF algorithms used in UAV navigation and positioning applications. In addition, a detailed comparison between different OF algorithms is made here. It compares the well-known algorithms, analyzes the results obtained to date, and draws a number of conclusions from them. The algorithms were chosen to focus on the most efficient OF algorithms used in OF sensors. Finally, the section summarizes state-of-the-art sensors and approaches used to provide accurate OF estimates.

13.3.1 OF Computation

OF computation is about detecting and tracking points across every two consecutive image frames to estimate the velocity and position changes on a plane parallel to the ground. The described method assumes that the camera or OF sensor is mounted on the UAV's body and is aimed toward the ground. Thus, the OF algorithms take two images as their input and generate a set of vectors that describes the amount of pixel displacement and its direction. There are two different kinds of OF outputs: sparse OF and dense OF. Sparse OF is computed for certain specified pixels in the whole image, while dense OF is computed across all pixels. Although dense techniques are slower to compute, they provide more accurate motion estimations.

The OF estimation problem can be described as follows. Given a set of points or pixels in the first frame, the algorithm tries to locate the same points in the second frame to compute pixel displacements. This displacement can be used later to estimate the real-world traveled distance per moved pixel. Most of the current algorithms depend on brightness constancy and spatial smoothness assumption. It is assumed that pixels keep the same grayscale intensity between two frames, given a high frame rate. The spatial smoothness assumption is that neighboring pixels have the same motion because they belong to the same surface.

After locating the two corresponding sets of points, vectors can be computed. The source and destination points can be represented by Equations 13.1 and 13.2:

$$p_1[t] = \left(p_x(t) \;\; p_y(t)\right)^T \tag{13.1}$$

$$p_2[t+dt] = \left(p_x(t+dt) \;\; P_y(t+dt)\right)^T \tag{13.2}$$

where:

$p_1(t)$ is the original pixel coordinate in the image plane
$p_2(t + dt)$ is the same pixels after one frame, as shown in Figure 13.3

These points represent a projection of the points in the 3D camera reference frame P into the image frame. This is based on $p = f \cdot P / Z$, where f is focal length and Z is the distance from the ground. To derive equations that relate the velocity of points in the 3D scene to the 2D motion field, we need to examine the velocity vectors induced by the camera motion in 3D space and to project these vectors onto the image plane. The displacement of 3D world points P can be measured by Equation 13.3, where R and T represent the 3×3 rotation matrix and 3×1 translation matrix, respectively.

$$\text{3D Displacement} = R \times P + T - P \tag{13.3}$$

Under small-angle approximation due to the small period of time between frames (typically, the time between two frames = 1/30 s), the 3D displacement becomes 3D velocity, as shown in Equation 13.4:

$$V = -T - \omega \times P \tag{13.4}$$

where:

T is the translational component of the motion
ω is the angular velocity

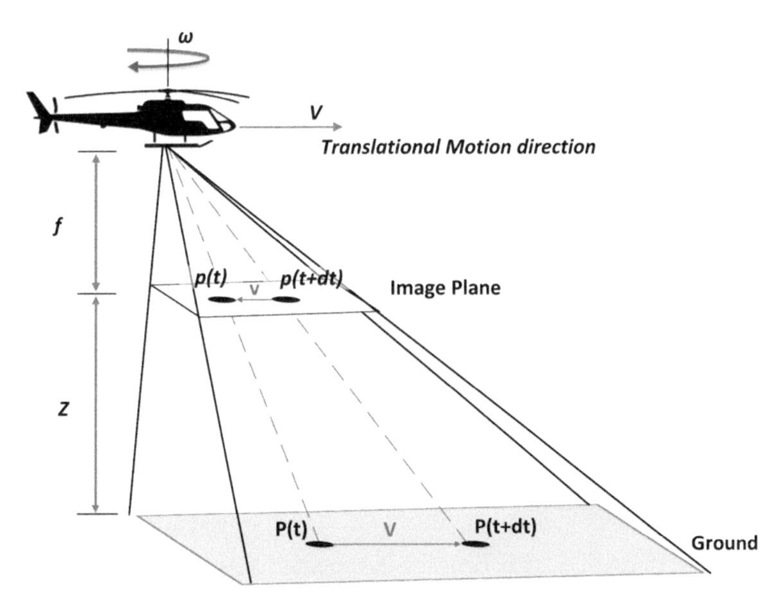

FIGURE 13.3
Basic OF scheme.

To compute the project vectors from 3D space to the 2D image plane, the 2D velocity of p is first derived as follows:

$$\frac{d}{dt}(p) = v = \frac{d}{dt}\left(f.\frac{P}{Z}\right) = f\frac{(ZV - V_Z P)}{Z^2} = f.\frac{V}{Z} - \frac{p.V_z}{Z} \tag{13.5}$$

Substituting the 3D velocity in Equation 13.4 gives Equations 13.6 and 13.7:

$$v_x = \frac{(T_z x - T_x f)}{Z} - \omega_y f + \omega_z y + \frac{(\omega_x xy - \omega_y x^2)}{f} \tag{13.6}$$

$$v_y = \frac{(T_z y - T_y f)}{Z} - \omega_x f + \omega_z x + \frac{(\omega_x y^2 - \omega_y xy)}{f} \tag{13.7}$$

13.3.2 OF Algorithms

Several OF algorithms have been proposed in the literature. These algorithms are commonly classified into one of eight major categories: block-matching methods, differential methods, image interpolation methods, feature-based methods, energy-based methods, phase-based methods, fusion-based methods, and fractional-order operator–based methods. This section provides a theoretical background of these algorithms and highlights the major strengths and limitations of each method.

13.3.2.1 Block-Matching Methods

Block-matching methods use two frames to compute a dense OF. These methods scan every pixel in the first frame and place a window of size $B \times B$ over the image, with the current pixel as the middle point. Then, the algorithm searches the next frame with a given search area to find the best match by measuring the similarity between the two blocks through the sum of absolute differences (SAD), which is usually used to measure similarity. The position with the lowest SAD value is returned as the new position of the current pixel. Computing the similarities between two blocks of size $B \times B$ in two image frames, f_k and f_{k-1}, can also be described as minimizing in the following function:

$$SAD = \sum_{i=0}^{B-1}\sum_{j=0}^{B-1}\left|f_k(x+i, y+i) - f_{k-1}(x+i+u, y+j+v)\right| \tag{13.8}$$

To generate dense OF, this process is repeated for every pixel in the first image frame. This method is time-consuming; nonetheless, it gives an accurate OF estimation.

13.3.2.2 **Differential Methods**

Differential methods, also known as *gradient-based methods,* are those methods that use spatial and temporal partial derivatives to estimate image flow. The Lucas–Kanade method is one of the most commonly used gradient-based OF methods. It compares local gradients of image intensity in space and time to estimate local translational motion. Then it finds global flow vectors that are consistent with the neighboring spatial and temporal gradients. This class of methods has received special interest due to their simplicity, computational efficiency, and robustness to noise. However, they generate sparse OFs that reduce their accuracy compared with other methods.

13.3.2.3 **Feature-Based Methods**

Feature-based OF estimation methods compute pixel displacements by detecting important points in frames called *features,* such as edges, corners, blobs, and so on, and tracking them as they move from one frame to the next. The algorithms consist of two stages: (1) detecting a list of features from two or more consecutive frames using detectors such as SIFT, Harris, the FAST corner detector, or any other local feature detection algorithm; (2) matching the two sets of features based on a similarity measurement to compute a set of motion vectors. Feature-based methods can be categorized into pixel-accurate OF methods and subpixel-accurate OF methods. The accuracy of pixel-accurate methods is limited to pixel-discrete displacement, whereas subpixel-accurate methods use linearized OF constraints to improve accuracy.

13.3.2.4 **Image Interpolation Methods**

This method was developed by Srinivasan [25] to measure OF through the use of single-stage non-iterative processes. It interpolates the position of the moving images relative to a set of reference images. This method is robust to noise and can be used for low computational applications. It does not need to detect and track image features between frames, as with feature-based methods, or to calculate the spatial and temporal derivatives of images, as with gradient-based methods.

13.3.2.5 **Energy-Based Methods**

This class of OF estimation techniques is based on the output energy of velocity-tuned filters. It extracts local energies using filters such as Gabor energy filters for every voxel in the image. Ideally, for a single translational motion, the local energies measured by a set of filters will be concentrated on a plane in the frequency domain. To estimate a certain movement velocity, the method compares the measured energies with known theoretical energy responses. These methods solve the optimization problem of fitting a plane to the estimated energy responses, using methods such as linear estimation or total least-square estimation.

13.3.2.6 Phase-Based Methods

Phase-based methods use spatio-temporal filters to compute phase contours and track them with respect to the sequence of image-related time. In these methods, zero crossing, phase gradient, or phase correlation can be used to compute the OF vectors. Furthermore, these methods give good approximation to the motion field due to their resilience to noise, occlusions, and other defects. However, they have some limitations; for instance, they might yield ambiguous results with several peaks for periodic images, such as an image of chessboard.

13.3.2.7 Fractional-Order Operator–Based Method

Two types of generalizations of the differentiations have been proposed in variational OF models. The first type generalizes the differential of OF from the integer order into the fractional order [26], whereas the second type deals with higher-order differentiations [27]. The fractional Riemann–Liouville derivative is the most widely used method of solving OF estimation problems. It can be used to define less regular functions and can be discretized through Grünwald–Letnikov approximation.

13.3.2.8 Fusion-Based Methods

Fusion-based methods [28] divide the input image frame into discrete triangles to estimate occluded regions using one of the optimization algorithms. Moreover, they use inertial estimates that provide multiple motion estimates. These two estimates are then combined using a classifier-based fusion scheme to improve the OF estimation. These methods estimate occlusions and use additional temporal information that improves the OF accuracy in difficult cases—for example, complex motions, large displacements, and difficult imaging conditions.

13.3.3 Evaluation Methodology and Results

This section presents a quantitative evaluation of different OF algorithms. The evaluation method, benchmark datasets, and experimental results are collected from a set of available survey papers in the literature [29–33] to give a brief glimpse into the performance of these algorithms. The evaluation metrics, typically used to compare different algorithms, are (1) error metrics, which include average angular error, flow end point error, residual

RMS error, and gradient-normalized RMS error, (2) statistical metrics, including robustness measures and percentile-based accuracy measures, and (3) region mask error measures.

The benchmark datasets used in these survey papers cover complex real scenes taking into consideration different kinds of artifacts of real sensors, such as motion blur and noise. The image sequences in these datasets also contain substantial motion discontinuities and non-rigid motion. These datasets are seen to be considerably challenging and are used to evaluate the progress in improving the performance of OF algorithms. Each benchmark dataset consists of a set of image sequences paired with dense, subpixel-accurate ground truth flow fields. The ground truth OF is measured by applying a fine pattern of fluorescent paints to the surfaces in the scene to be tracked with UV lighting as the scene/sensor is moving. A computer-controlled motion stage is used to move the scene in very small steps. At each step, two pairs of images and the ground truth are collected, and then the scene is moved by a small amount to generate an image sequence.

The performances of three algorithms has been investigated in [29]—namely, SAD, Lucas–Kanade, and Gunnar–Farneback. These algorithms belong to the categories of block-matching methods, differential methods, and interpolation methods, respectively. They extracted the image sequences by placing a window on a large image and then moving the window over an image. The pixels inside the window are saved as a new frame, and the ground truth associated with it is also saved. Three different images from different types of ground surfaces (grass, a grey indoor floor with small white dots, and a floor with a red carpet) were used.

The execution time of these algorithms was evaluated by running the experiment through 50 frames on an Intel i5-4300 U-based system. The average computation times are shown in Table 13.1. It shows the computation time in milliseconds for five different binning factors (1, 2, 4, 8, and 16). Binning factors represent the ratio at which the image pixels are combined; a binning factor of 4 means four pixels are combined into one pixel, which thereby reduces the total number of pixels by a factor of 4. The table shows that the Lucas–Kanade and Farneback algorithms outperform the SAD matching algorithm in terms of execution time. It also shows that Farneback is faster than Lucas–Kanade when the binning factor is at least 2.

TABLE 13.1

Computation Times in Milliseconds

	Binning Factor				
Algorithm	**1**	**2**	**4**	**8**	**16**
Lucas–Kanade	327.7	70.0	22.8	5.2	1.2
Farneback	384.8	53.0	12.3	3.1	0.7
SAD Matching	–	–	500.0	60.0	3.0

To compare the quality of estimated OF, the error value of the accumulated distance between sequential middle points is measured and compared with the ground truth. Table 13.2 shows the results on three different types of surfaces at different binning factors. It shows that all algorithms performed well on the grass dataset. On the other hand, on the indoor dataset, which is a more challenging set, block matching performs well with binning factors 4 and 8, while Farneback performs well on all binning factors.

Addressing the accuracy-versus-efficiency trade-offs of OF algorithms is essential for choosing the right algorithm. In [34], the authors present an experimental study on the accuracy-efficiency of different OF algorithms. They computed a 2D accuracy–efficiency (AE) curve to characterize the performance of the algorithms. In this curve, angular error (x-axis) is used as an accuracy metric, and throughput (y-axis; the number of output frames per unit time) is used as a metric for performance. The distance from the origin to the algorithm point represents the AE curve of the algorithm. Some algorithms were very accurate but considerably slow: the Fleet–Jepson, Horn–Shunc, and Bober algorithms, for instance. Other algorithms were fast but had a considerably low accuracy, such as the Camus and Liu algorithms. Nevertheless, other algorithms showed a relatively good accuracy and execution time: for example, the Uras and Lucas algorithms.

Previous research has focused on exploring one category of algorithms to find the optimal parameters and configuration that give the best performance. The best filtering approach for the Lucas–Kanade pyramidal OF algorithm was studied in [35]. The researchers tested the filtering approach on the input images and on all resized input images for pyramidal OF. Table 13.3 shows the average angular error and the average end point error for six different filters. It indicates that the Gaussian smoothing filter performs better than the other filters listed.

13.3.4 Survey of OF-Based Navigation Approaches

Due to recent advances in the performance of OF algorithms and the efficiency of OF sensors, researchers have begun to use OF to support the positioning and navigation of UAVs [36]. This section presents a survey of the latest contributions in this area. OF systems have been used on UAVs for different navigation and positioning tasks. These tasks include (1) velocity

TABLE 13.2

Quality Measurements of the Algorithms on the Different Types of Surfaces at Different Binning Factors

	Grass Surface			Dotted Floor			Carpet Surface		
Algorithm	4 × 4	8 × 8	16 × 16	4 × 4	8 × 8	16 × 16	4 × 4	8 × 8	16 × 16
Lucas–Kanade	0.17	0.18	0.47	0.31	0.63	2.07	0.21	0.72	2.42
Farneback	0.17	0.16	0.14	0.40	0.56	1.8	0.27	0.28	1.10
SAD Matching	0.65	0.24	0.19	5.41	2.34	7.70	0.17	0.72	2.42

TABLE 13.3

Comparison between Filtering Methods [35]

	Filtering Applied to Input Images		Filtering Applied to All Resized Images	
Filter	Average Angular Error (degrees)	Average Endpoint Error	Average Angular Error (degrees)	Average Endpoint Error
Gaussian Smooth	9.809	1.156	8.947	0.948
Median	12.276	1.351	11.961	1.307
LOG	20.604	2.273	17.296	2.005
Mean	9.955	1.170	9.028	0.947
High Boost	16.146	1.893	15.978	1.903
Laplacian	20.863	2.688	18.054	2.084
Noise Removal	10.291	1.141	9.645	1.035
Bilateral	13.909	1.353	13.909	1.353

estimation, (2) state estimation, (3) obstacle avoidance, (4) heading angle estimation, (5) motion estimation, and (6) vertical landing.

Table 13.4 presents a list of example applications of OF systems in navigation and positioning tasks. For each work, it shows the navigation tasks and the platform used for validation. It also specifies the algorithm used with a brief description. A number of observations can be made from Table 13.4. First, OF navigation algorithms have been used extensively with VTOL UAVs for indoor applications, with few studies done on fixed-wing UAVs. Second, more than one type of OF algorithm has been used to estimate optical fields based on the needs of the application. While some algorithms offer acceptable accuracy but with very simple implementation and low computational requirements, other algorithms give the highest accuracy possible at the expense of high computational complexity.

13.3.5 Specification Comparison of Different OF Sensors

Choosing the right OF sensor for a specific application is not a trivial task. It requires consideration of the limited onboard memory, the available computational power budget on the UAV, and the high amount of visual data that needs to be processed. Other constraints involve issues such as the interface type for a microprocessor, resolution, field of view, and weight. This section presents a specification comparison of different OF sensors. It analyzes ten different sensor systems, as shown in Table 13.5.

Optical mouse sensors, such as ADNS-2610 and ADNS-3080, are small and inexpensive and have a very high sampling rate of 1000 fps and higher. They are typically suitable for micro-UAVs. However, the software for these sensors is not open source and is not extendable. The result of these sensors is

TABLE 13.4

List of Different OF-Aided Navigation and Position Approaches

Navigation Tasks	Platform	OF Sensor	Algorithm	Method
Velocity Estimation [37]	Helicopter	CMOS	Image interpolation	Integration of GPS, INS, and CMOS sensor has been used for velocity and attitude estimation.
Velocity Estimation [38]	8-Rotor VTOL	Kingfishers Camera	Lucas–Kanade	The OF measurements have been used to estimate the translational speed and to keep the aircraft at a desired altitude.
Velocity Estimation [14]	Quadrotor	PX4FLOW	SAD Block matching	The researchers evaluated the velocity estimation to ground truth during the hovering flight of a quadrotor in indoor and outdoor environments.
Heading Angle [39]	Quadrotor	Firefly MV Camera	Phase correlation	OF measurements have been used to estimate the heading of a small fixed-pitch four-rotor helicopter.
Obstacle Avoidance [40]	MAV	TAOS TSL3301	Image interpolation	Only optical sensors have been used for obstacle avoidance within textured indoor environments.
Obstacle Avoidance [41]	Fixed-wing MAVs	DNS-2610	NA	The researchers used a laser range finder and OF sensors to detect obstacles and terrain.
Landing on a Moving Platform [42]	Quadrotor	Embedded Camera	Lucas–Kanade	OF measurements have been used to enable hover and landing control on a moving platform.
Landing on a Fixed Target [4]	Helicopter	PX4FLOW	SAD block matching	OF velocities were fused with GPS/INS measurements to obtain accurate position and velocity estimates during terminal-phase landing.
Motion Estimation [43]	Fixed wing	GoPro Hero I	SIFT Feature	Two rotational and two translational motions have been tested using OF estimates generated by the SIFT feature–based method.

TABLE 13.5
Comparison between OF Sensors

Sensor Name	Resolution	Dimensions	Weight	Frame Rate	Field of View
ADNS-2610	18 × 18 pix.	25 × 30 mm	15 g	1500 Hz	6.5°
		35 × 30 mm	23 g		2.5°
		50 × 30 mm	23 g		1.2°
ADNS-3080 [44]	30 × 30 pix.	10 × 10 mm	20 g	6400 fps	11°
CentEye TinyTam [45]	16 × 16 pix.	7.0 × 7.0 mm	125 mg	20 Hz	N/A
CentEye Stonyman	112 × 112 pix.	2.8 × 2.8 mm	N/A	N/A	N/A
CentEye Hawksbill	136 × 136 pix.	2.3 × 32.7 mm	N/A	N/A	N/A
OV7740 [46]	320 × 240 pix.	6.5 × 6.5 × 4 mm	N/A	120 Hz	50°
OV7670 [47]	640 × 480 pix.	30.5 × 30.5 mm	12 g	60 fps	25°
CMUCam 4 [48]	640 × 400 pix.	54.0 × 50.0 mm	27 g	50 Hz	75°
PX4FLOW [4]	188 × 120 pix.	45.5 × 35 mm	N/A	250/500 Hz	21°
	120 × 32 pix.	45.5 × 35 mm	N/A	250/500 Hz	180°
GoPro Hero [49]	1920 × 1080 pix.	42 × 60 × 30 mm	94 g	29.97 Hz	127°
	1280 × 720 pix.	42 × 60 × 30 mm	94 g	59.94 Hz	170°

also sensitive to dust and dirt and low-light conditions, which makes them inappropriate for indoor or low-light applications. A series of OF sensors has been developed by CentEye Inc. (CentEye TinyTam, CentEye Stonyman, and CentEye Hawksbill). These sensors have an asynchronous interface and flexible pixel window downsampling, but their fixed lenses and low resolution mean far objects will be blurred.

OmniVision sensors (OV7740 and OV7670) improve image quality and produce clean and fully stable color images by eliminating common lighting noise sources, such as smearing and fixed-pattern noise. Nevertheless, they have limited microprocessor memory and QVGA resolution, which is used instead of VGA. The PX4FLOW smart sensor was developed by ETH for navigation applications. It provides onboard OF estimates after subpixel refinement, rotation compensation, and metric scaling. GoPro Hero cameras have also been used in UAVs due to their high resolution, small size, and light weight.

13.4 OF Sensor Model

This section introduces the OF sensor modeling experiment for terminal-phase vertical-landing purposes. In [4], an OF modeling experiment was conducted under various simulated helicopter-landing scenarios.

A pendulum test stand was designed to obtain a dynamic model that characterizes the performance of the OF sensor. As shown in Figure 13.4, a 10,000 ppr incremental encoder is connected to the rotating axle of the pendulum to obtain accurate angular velocity measurements of the moving pendulum. The PX4FLOW sensor is looking downward to capture the apparent motion of the landing pattern scene [4] and compute the OFs in the metric scale.

A system identification process was applied to identify the OF sensor's dynamics. The dynamics are experimentally represented by picking a second-order model that has the following characteristics: a natural frequency of 13.8528 rad/s and a 0.5353 damping ratio.

After acquiring the OF model, the model is validated at different altitudes to ensure its robustness to height variations. Figure 13.5 displays the real-time validation scheme of the obtained OF model. The OF model's velocity is compared with both the real PX4FLOW velocity and the input encoder velocity [4]. From Figure 13.6, it is seen that the performance of the OF model shows significant coincidence with the performance of the real PX4FLOW

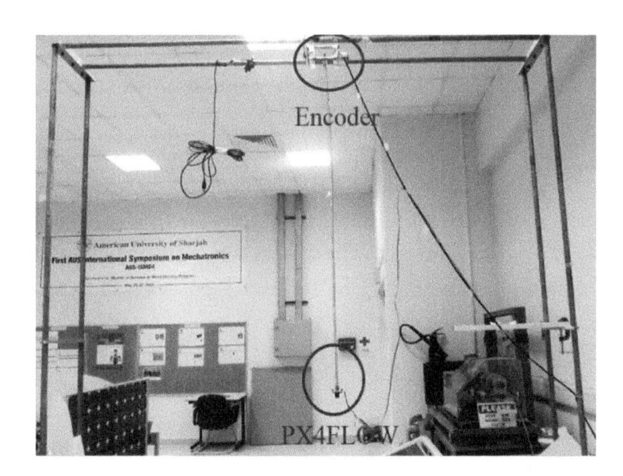

FIGURE 13.4
Pendulum test stand [4].

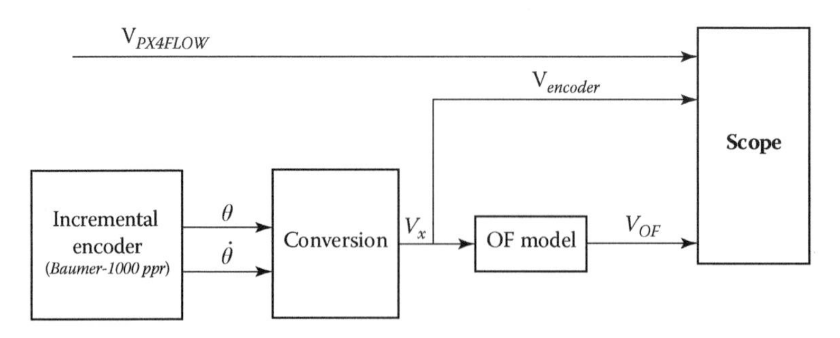

FIGURE 13.5
Validation scheme of the OF model.

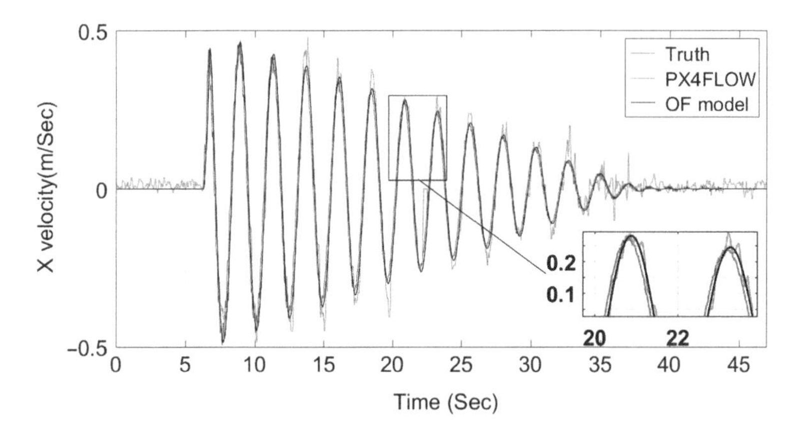

FIGURE 13.6
OF model validation.

sensor. As computed in Table 13.6, the identified OF model has small mean errors and small SD errors at various heights in comparison with the actual reading of the PX4FLOW flow sensor.

13.4.1 Intelligent OF Sensor Fusion Design

The off-the-shelf GPS/INS position measurements are inaccurate due to GPS/INS measurement characteristics, such as the quality of the GPS receiver and IMU, IMU bias error, multipath errors, and the number of satellites in view [50]. The resulting estimation error impedes precision landing for miniature helicopters. Hence, the OF sensor is used to provide an accurate augmentation to the GPS/INS unit. In this section, the position measurements of GPS/INS measurements are fused with the velocity (m/s) measurements of the OF sensor using an intelligent adaptive fuzzy Kalman filter (AFKF) estimation algorithm (Figure 13.7).

The KF is not famed only for its ability to reject the sensor's noise but also for its extensive use as an ideal minimum mean square error (MMSE) state

TABLE 13.6
Mean Error between the Output of the Validated OF Model and the Actual Reading at Each Height

Height (cm)	Mean Error (m/sec)	Standard Deviation Error (m/sec)
60	0.0064	0.0354
80	0.0186	0.0642
100	0.0613	0.1815
150	0.0950	0.2815

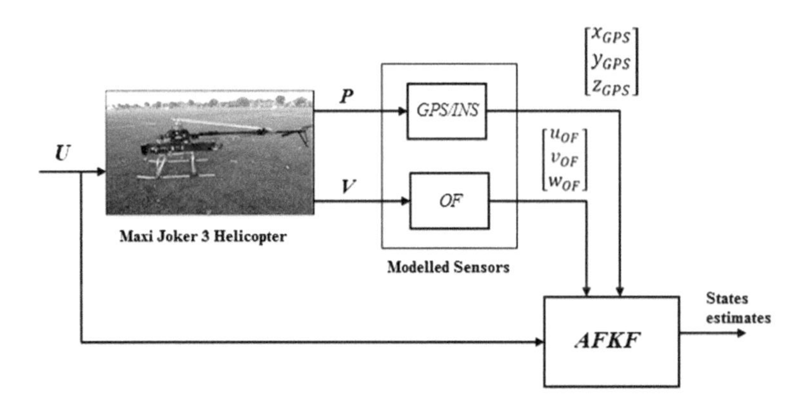

FIGURE 13.7
Sensor fusion block diagram.

estimator [3, 15]. An optimal state estimate can be achieved using its recursive state estimation algorithm. This algorithm consists of the following two sets of equations (13.9 and 13.10, and 13.11 through 13.15):

1. State and measurement prediction equations

$$\hat{x}_{k+1|k} = A_k \hat{x}_{k|k} + B_k U_k \tag{13.9}$$

$$p_{K+1|k} = A_k P_{k|k} A_K^T + Q_k \tag{13.10}$$

2. State update equations

$$v_{K+1} = z_{K+1} - H_{K+1} \hat{x}_{k+1|K} \tag{13.11}$$

$$S_{K+1} = R_{K+1} + H_{K+1} P_{K+1|k} H_{K+1}^T \tag{13.12}$$

$$W_{K+1} = P_{K+1|k} H_{K+1}^T S_{K+1}^{-1} \tag{13.13}$$

$$\hat{x}_{k+1|K+1} = \hat{x}_{k+1|K} + W_{K+1} v_{K+1} \tag{13.14}$$

$$P_{K+1|K+1} = P_{K+1|k} - W_{K+1} S_{K+1} W_{K+1}^T \tag{13.15}$$

The current state and the error covariance estimates are updated at each cycle by the innovation sequence in Equation 13.11 and its theoretical value in Equation 13.12. Equations 13.11 through 13.13 integrate the measurement updates with the a priori estimate to enhance the a posteriori estimates and covariance.

The intelligent estimation algorithm requires a linearization process for the Maxi Joker 3 state space. The linearization process is performed at near-hover. A reduced-order state vector is used as shown in Equations 13.16 and 13.17, which represent the input vector to the helicopter dynamic model.

$$X_{\text{reduced}} = \begin{bmatrix} x \\ y \\ z \\ u \\ v \\ w \end{bmatrix} \tag{13.16}$$

$$U = \begin{bmatrix} u_{lon} \\ u_{lat} \\ u_{col} \\ u_{ped} \end{bmatrix} \tag{13.17}$$

where:

u_{lon} is the longitudinal stick input
u_{lat} is the lateral stick input
u_{col} is the collective lever input
u_{ped} is the rudder pedal input

The reduced-order estimator has the following system matrix (A_{red}):

$$A_{red} = \begin{bmatrix} 0 & 0 & 0 & 1.0000 & -0.0008 & -0.0016 \\ 0 & 0 & 0 & 0.0007 & 0.9948 & -0.1018 \\ 0 & 0 & 0 & 0.0017 & 0.1018 & 0.9948 \\ 0 & 0 & 0 & -0.0190 & 0.1250 & 0.0699 \\ 0 & 0 & 0 & -0.1256 & -0.0268 & 0.5550 \\ 0 & 0 & 0 & -0.0687 & -0.5378 & -3.2826 \end{bmatrix} \tag{13.18}$$

The input matrix B_{red} is similarly calculated, as shown in Equation 13.19. The C and D matrices are diagonal matrices, and their dimensions are 6×6 and 6×4, respectively.

$$B_{red} = \begin{bmatrix} 0 & 0 & 0 & 0 \\ 0 & 0 & 0 & 0 \\ 0 & 0 & 0 & 0 \\ 0 & 0 & 0.0374 & 0 \\ 0 & 0 & 1.5680 & -17.3314 \\ 0 & 0 & -294.6074 & 0 \end{bmatrix} \tag{13.19}$$

The optimal variances of the diagonal measurement and process noise covariance matrices, R_k and Q_k, are meticulously picked out to avoid divergence of the filter. The variances of the position states are greater than the variances of the velocity states. This is due to the fact that at altitudes close to the ground, the GPS/INS position measurements become less precise than the PX4FLOW velocity measurements.

The adaptive sensor fusion technique is based on updating R_k and/or Q_k at each estimation cycle. Presuming perfect knowledge of the process noise matrix Q_k, the measurement noise matrix R_k is adapted using the innovation adaptive estimation (IAE) approach. This approach measures the matching degree (MD) between the actual covariance of the residual with its theoretical value [18]. MD can be computed as follows:

$$MD = S_{K+1} - \hat{C}_{k+1} \quad (13.20)$$

where $\hat{C}_{k+1}$ represents the actual covariance of the innovation sequence and can be evaluated as follows:

$$\hat{C}_{k+1} = (1 \setminus N) \sum_{i=i_o}^{N} v_i v_i^T \quad (13.21)$$

where:

N presents the estimation window
i_o denotes the first sample inside the estimation window

As can be noticed from Equations 13.13 and 13.14, S_{k+1} can be adjusted by varying the variances of R_k. Therefore, the discrepancy between S_{k+1} and $\hat{C}_{k+1}$ can be minimized by changing the value of S_{k+1} by adjusting the variances value of R_{k+1}. A single input–single output (SISO) fuzzy inference system (FIS) is used to adjust the variances of the measurement noise matrix R_k at each Kalman estimation cycle, as shown in Equation 13.27:

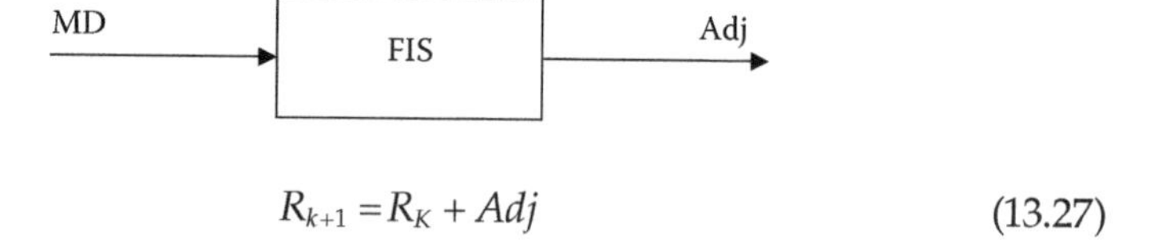

$$R_{k+1} = R_K + Adj \quad (13.27)$$

The adjustment value (Adj) is modeled using the Mamdani min–max fuzzy inference engine, with centroid defuzzification to collect the final adjustment value from the fuzzy output set. The MD (the inference engine input variable) universe of discourse is partitioned into three different fuzzy terms: negative (N), zero (Z), and positive (P). The inference engine output (the adjustment variable) universe of discourse is partitioned into three different fuzzy terms: decrease (*D*), maintain (*M*), and increase (*I*). The Gaussian bell-shaped membership functions are used to mathematically describe the fuzzy inference engine input/output fuzzy terms continuously. This nonlinear matrix adjustment is controlled by the following group of fuzzy linguistic description rules [19]:

- If *MD* is negative, then *Adj* is increased.
- If *MD* is equal to zero, then *Adj* is maintained.
- If *MD* is positive, then *Adj* is decreased.

13.5 Simulation Results

This section compares the simulation results of the intelligent AFKF's performance with the performance of our previously published classical KF in terms of the quality of the position and velocity estimations during Maxi Joker 3 terminal-phase landing [4]. The performance of the proposed AFKF estimator was tested against the classical KF using the simulation environment. The GPS/INS and OF sensors were modeled on the MIDG and PX4FLOW sensors, respectively. The measurements of these sensors were fused by the AFKF filter to obtain position and velocity estimations of the Maxi Joker 3 helicopter. After that, a simulated landing test was conducted to demonstrate the accurate performance of the proposed intelligent AFKF estimation method.

13.5.1 Position Estimation

As a precision landing operation is being performed, a highly accurate estimation of the position is essential. Figures 13.8 through 13.11 display the estimation performances of the position states. As shown in Figures 13.8 through 13.9, the helicopter was commanded to start descending from 3 m in the *Z* body frame until it reached the ground with a slope of 0.3 m/s. The AFKF exhibited an obvious superiority over the classical KF in terms of following the measurement updates. Furthermore, it can be seen from Figure 13.8 that the AFKF has a faster response than the classical KF during the transient response of the vehicle. The AFKF has estimated the altitude with a root mean square error of estimation (RMSEE) of 0.143 m, while the KF has a greater RMSEE of 0.177 m.

13.5.2 Velocity Estimation

The AFKF approach uses precise PX4FLOW measurements in m/s as the velocity measurements of the helicopter during terminal-phase landing.

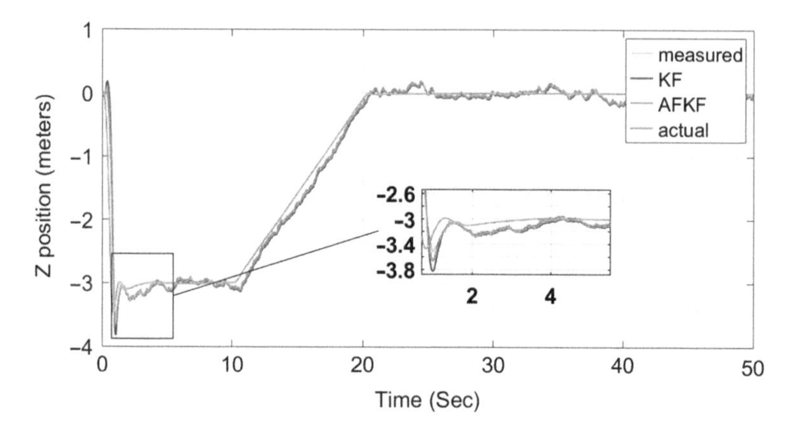

FIGURE 13.8
Altitude estimation

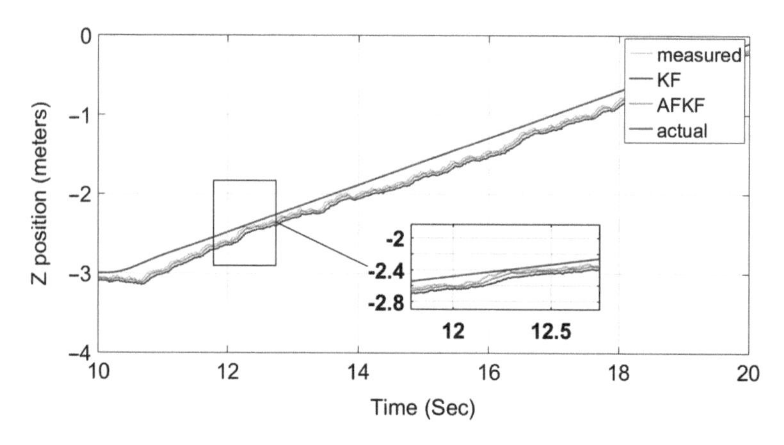

FIGURE 13.9
Zoomed altitude estimation.

Therefore, a precise velocity estimation is expected. Figures 13.12 through 13.14 compare the performance of the AFKF with the performance of the classical KF in estimating the velocity states of the helicopter. Figure 13.12 is zoomed to demonstrate the velocity estimation while the helicopter is approaching the ground. The helicopter begins to descend with 0.3 m/s Z velocity in the body frame until it reaches the ground. As can be seen, the AFKF has succeeded in rejecting the measurement noise and reducing the estimation error. The AFKF has estimated the Z velocity with an RMSEE value of 0.0249 m/s, while the KF has an RMSEE value of 0.0302 m/s. Figures 13.13 through 13.14 show that the OF sensor has captured the helicopter's translational velocities in X and Y while landing. Similarly, the AFKF is more successful than the KF in estimating the X and Y velocities. The AFKF has estimated the Z velocity with an RMSEE value of 0.0249 m/s, while the KF

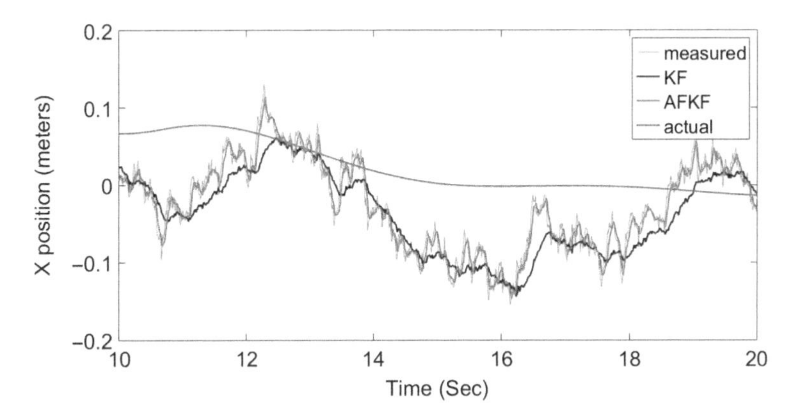

FIGURE 13.10
Estimation of X position.

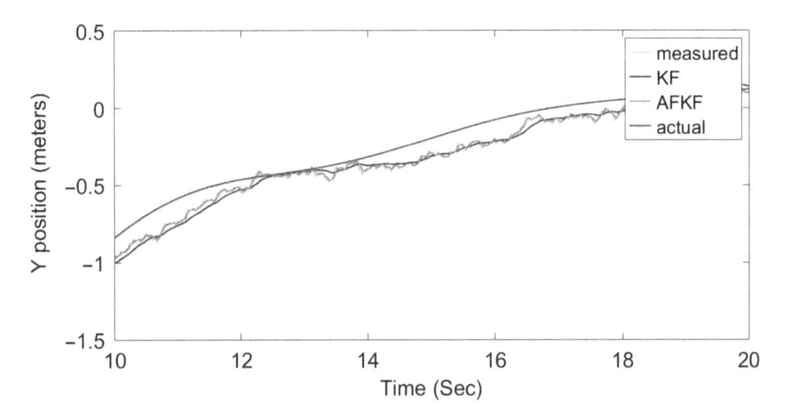

FIGURE 13.11
Estimation of Y position.

has an RMSEE value of 0.0302 m/s. As shown, the AFKF estimation algorithm achieves smaller estimation error values than the KF algorithm.

13.6 Conclusion

This chapter has studied the problem of OF sensing and its applications in unmanned small-scale vertical landings. A survey of the existing OF approaches and their uses in the field of UAV estimation and control has also been presented in this chapter. Moreover, an intelligent sensor fusion technique between the experimental model of the OF sensor and the GPS/INS solution has been proposed for precise velocity and position state estimations

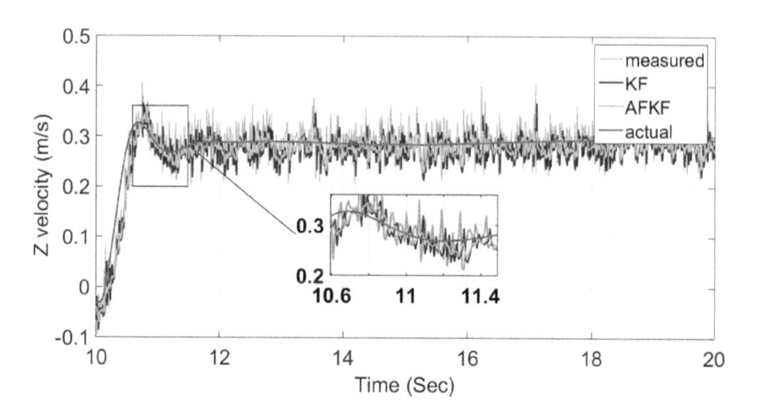

FIGURE 13.12
Estimation of Z velocity

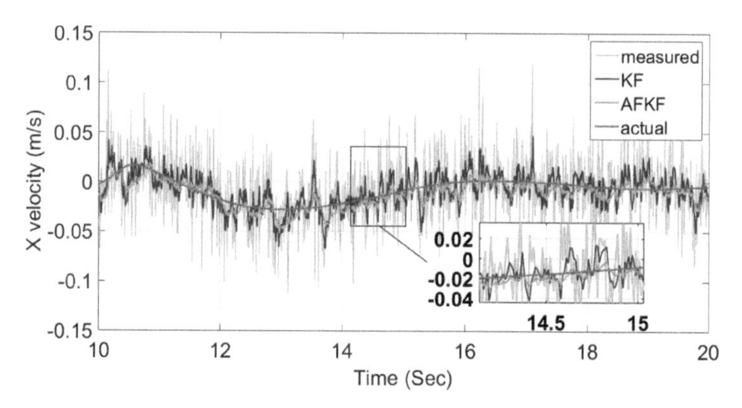

FIGURE 13.13
Estimation of X velocity

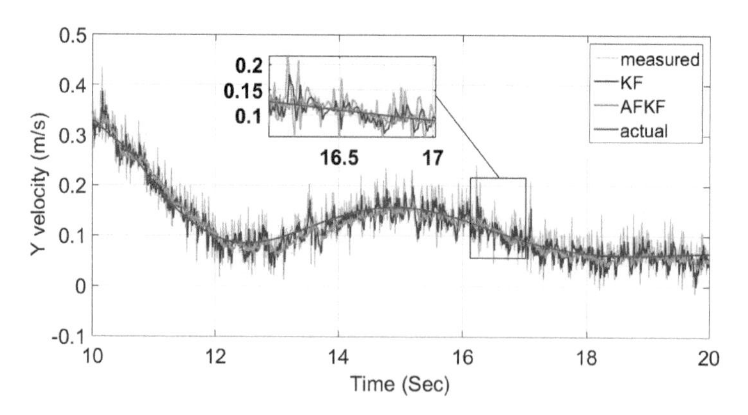

FIGURE 13.14
Estimation of Y velocity.

during the terminal phase of the helicopter. This technique is based on adapting the measurement noise covariance matrix of the KF through a SISO FIS. The simulation results of the intelligent estimation approach have been compared with the classical estimation approach. The proposed intelligent estimation technique has proved its superiority over the classical algorithm. The main goal of future research could be to incorporate intelligent optimization techniques in the fuzzy membership functions to obtain more accurate state estimates during the terminal-phase landing stage.

References

1. P. Marantos, C. P. Bechlioulis, and K. J. Kyriakopoulos, "Robust Trajectory Tracking Control for Small-Scale Unmanned Helicopters With Model Uncertainties," *IEEE Transactions on Control Systems Technology*, no. 99, pp. 1–12, 2017.
2. H. Lee, S. Jung, and D. H. Shim, "Vision-Based UAV Landing on the Moving Vehicle," *International Conference on Unmanned Aircraft Systems (ICUAS)*, pp. 1–7 2016.
3. M. Al-Sharman, M. Abdel-Hafez, and M. Al-Omari, "Attitude and Flapping Angles Estimation for a Small-Scale Flybarless Helicopter Using a Kalman Filter," *IEEE Sensors Journal*, vol. 15, no. 4, pp. 2114–2122, 2015.
4. M. Al-Sharman, "Auto Takeoff and Precision Landing Using Integrated GPS/INS/Optical Flow Solution," master's thesis, Sharjah, UAE: American University of Sharjah, 2015.
5. S. Saripalli, J. Montgomery and G. Sukhatme, "Visually-Guided Landing of an Unmanned Aerial Vehicle," *IEEE Transactions on Robotics and Automation*, vol. 19, no. 3, pp. 371–381, 2003.
6. T. Merz, S. Duranti, and G. Conte, "Autonomous Landing of an Unmanned Helicopter Based on Vision and Inertial Sensing," *Proceedings of the 9th International Symposium on Experimental Robotics (ISER 2004)*, Singapore, pp. 343–352, 2006.
7. A. Cesetti, E. Frontoni, and A. Mancini, "A Vision-Based Guidance System for UAV Navigation and Safe Landing Using Natural Landmarks," *Journal of Intelligent & Robotic Systems*, vol. 57, no. 1–4, pp. 233–257, 2010.
8. M. Meingast, C. Geyer, and S. Sastry, "Vision Based Terrain Recovery for Landing Unmanned Aerial Vehicles," *43rd IEEE Conference on Decision and Control (CDC)*, Nassau, Bahamas. pp. 1670–1675, 2004.
9. C. Sharp, O. Shakernia, and S. S. Sastry, "A Vision System for Landing an Unmanned Aerial Vehicle," *IEEE International Conference on Robotics and Automation*, Seoul, South Korea, pp. 1720–1728, 2001.
10. N. Gageik, M. Strohmeier, and S. Montenegro, "An Autonomous UAV with an Optical Flow Sensor for Positioning and Navigation," *International Journal of Advanced Robotic Systems*, vol. 10, pp. 1–9, 2013.
11. K. Sungbok and S. Lee, "Optical Mouse Array Position Calibration for Mobile Robot Velocity Estimation," *IEEE Conference on Robotics, Automation and Mechatronics*, Chengdu, China, pp. 1167–1172, 2008.

12. S. Janson and R. Welle, The NASA optical communication and sensor demonstration program. Year: 2013, *AIAA Small Satellite Conference*, Karlsruhe, Germany, 213.
13. S. Lange, N. Sünderhauf, and P. Protzel, "Incremental Smoothing vs. Filtering for Sensor Fusion on an indoor UAV," *IEEE International Conference on Robotics and Automation (ICRA)*, Karlsruhe, Germany, 2013.
14. D. Honegger, L. Meier, P. Tanskanen, and M. Pollefeys, "An Open Source and Open Hardware Embedded Metric Optical Flow CMOS Camera for Indoor and Outdoor Applications," *2013 IEEE International Conference on Robotics and Automation (ICRA)*, pp. 1736–1741, 2013.
15. M. Al-Sharman, "Attitude Estimation for a Small-Scale Flybarless Helicopter," in *Multisensor Attitude Estimation: Fundamental Concepts and Applications*, Boca Raton, FL: CRC Press, 2016.
16. D. A. Mercado, G. Flores, P. Castillo, J. Escareno, and R. Lozano, "GPS/INS/Optic Flow Data Fusion for Position and Velocity Estimation," *International Conference on Unmanned Aircraft Sysems (ICUAS)*, Atlanta, GA, pp. 486–491, 2013.
17. M. Al-Sharman, M. Abdel-Hafezand, and M. Al-Omari, "State Estimation for a Small Scale Flybar-less Helicopter," *2nd International Conference on System-Integrated Intelligence: Challenges for Product*, Bremen, Germany, pp. 258–267, 2014.
18. A. H. Mohamed and K. P. Shwarz, "Adaptive Kalman Filtering for INS/GPS," *Journal of Geodesy*, vol. 73, no. 4, pp. 193–203, 1999.
19. P. J. Escamilla-Ambrosio and N. Mort, "Development of a Fuzzy Logic-Based Adaptive Kalman Filter," in *Proceedings of the 2001 European Control Conference*, Porto, Portugal, 1768–1773, 2001.
20. A. Makni, H. Fourati, and A. Kibangou, "Energy-Aware Adaptive Attitude Estimation under External Acceleration for Pedestrian Navigation," *IEEE/ASME Transactions on Mechatronics*, vol. 21, no. 3, pp. 1366–1375, 2015.
21. D.-J. Jwo and T.-P. Weng, "An Adaptive Sensor Fusion Method with Applications in Integrated Navigation," *The Journal of Navigation*, vol. 61, no. 4, pp. 705–721, 2008.
22. H. Bian, J. Zhihua, and T. Weifeng, "IAE-Adaptive Kalman Filter for INS/GPS Integrated Navigation System," *Journal of Systems Engineering and Electronics*, vol. 17, no. 3, pp. 502–508, 2006.
23. W. Abdel-Hamid,. A. Noureldin, and. N. El-Sheimy, "Adaptive Fuzzy Prediction of Low-Cost Inertial-Based Positioning Errors," *IEEE Transactions on Fuzzy Systems*, vol. 15, no. 3, pp. 519–529, 2007.
24. C.-L. Lin, Y.-M. Chang, C.-C. Hung, C.-D. Tu, and C.-Y. Chuang, "Position Estimation and Smooth Tracking with a Fuzzy-Logic-Based Adaptive Strong Tracking Kalman Filter for Capacitive Touch Panels," *IEEE Transactions on Industrial Electronics*, vol. 62, no. 8, pp. 5097–5108, 2015.
25. M. V. Srinivasan, "An Image Interpolation Technique for the Computation of Optical Flow and Egomotion," *Biological Cybernetics*, vol. 71, p. 401–415, 1994.
26. D. Chen, H. Sheng, and Y. Chen, "Fractional Variational Optical Flow Model for Motion Estimation," in *Proceedings of the 4th IFAC Workshop on Fractional Differentiation and Its Applications (FDA '10)*, Badajoz, Spain, pp. 18–20, 2010.
27. J. Yuan, C. Schorr, and G. Steidl, "Simultaneous Higher-Order Optical Flow Estimation and Decomposition," *SIAM Journal of Scientific Computing*, vol. 29, pp. 2283–2304, 2007.

28. K. Ryan and C. J. Taylor, "Optical Flow with Geometric Occlusion Estimation and Fusion of Multiple Frames," *International Workshop on Energy Minimization Methods in Computer Vision and Pattern Recognition*, Hong Kong, China, pp. 364–377, 2015.
29. M. Kalksma and D. B. Jasper, "Choosing between Optical Flow Algorithms for UAV Position Change Measurement," *SC@RUG*, 2015.
30. S. Baker, D. Scharstein, J. Lewis, S. Roth, and M. Black, "A Database and Evaluation Methodology for Optical Flow," *International Journal of Computer Vision*, vol. 92, no. 1, pp. 1–31, 2011.
31. J. L. Barron, D. J. Fleet, and S. S. Bea, "Performance of Optical Flow Techniques," *International Journal of Computer Vision*, vol. 12(1), pp. 43–77, 1994.
32. F. Denis, P. Bouthemy, and C. Kervrann, "Optical Flow Modeling and Computation: A Survey," *Computer Vision and Image Understanding*, vol. 134, pp. 1–21, 2015.
33. M. Brendan,. B. Galvin, and K. Novins, "On the Evaluation of Optical Flow Algorithms," *Fifth International Conference on Control, Automation, Robotics and Vision*, Singapore, pp. 1563–1567, 1998.
34. H. Liu, T. Hong,. M. Herman, T. Camus, and. R. Chellappa, "Accuracy vs Efficiency Trade-Offs in Optical Flow Algorithms," *Computer Vision and Image Understanding*, vol. 72, no. 3, 1998.
35. S. Nusrat and R. Brad, "Optimal Filter Estimation for Lucas–Kanade Optical Flow," *Sensors*, vol. 12, no. 9, 2012.
36. H. Chao, Y. Gu, and M. Napolitano, "A Survey of Optical Flow Techniques for UAV Navigation Applications," *2013 International Conference on Unmanned Aircraft Systems (ICUAS)*, Atlanta, GA, pp. 710–716, 2013.
37. W. Ding, J. Wang, S. Han, A. Almagbile, M. A. Garratt,. A. Lambert, and J. J. Wang, "Adding Optical Flow into the GPS/INS Integration for UAV Navigation," *Proceedings of the International Global Navigation Satellite Systems Society IGNSS Symposium*, Queensland, Australia, pp. 1–13, 2009.
38. H. Romero, S. Salazar, and R. Lozano, "Real-Time Stabilization of an Eight-Rotor UAV Using Optical Flow," *IEEE Transactions on Robotics*, vol. 25, no. 4, p. 809–817, 2009.
39. J. Stowers, A. Bainbridge-Smith, M. Hayes, and S. Mills, "Optical Flow for Heading Estimation of a Quadrotor Helicopter," *International Journal of Micro Air*, vol. 1(4), pp. 229–239, 2009.
40. J. C. Zufferey and D. Floreano, "Toward 30-Gram Autonomous Indoor Aircraft: Vision-Based Obstacle Avoidance and Altitude Control," *Proceedings of the IEEE International Conference on Robotics and Automation*, Barcelona, Spain, pp. 2594–2599, 2005.
41. S. Griffiths, J. Saunders, A. Curtis, B. Barber, T. McLain, and R. Beard, "Maximizing Miniature Aerial Vehicles: Obstacle and Terrain Avoidance for Mavs," *IEEE Robotics & Automation Magazine*, vol. 13, no. 3, pp. 34–43, 2006.
42. B. Herissé, T. Hamel, R. Mahony, and F. X. Russotto, "Landing a VTOL Unmanned Aerial Vehicle on a Moving Platform Using Optical Flow," *IEEE Transactions on Robotics*, vol. 28, no. 1, pp. 77–89, 2012.
43. H. Chao, Y. Gu, J. Gross, G. Guo, M. L. Fravolini, and M. R. Napolitano, "A Comparative Study of Optical Flow and Traditional Sensors in UAV Navigation," *Proceedings of the American Control Conference*, Washington, DC, pp. 3858–3863, 2013.

44. Agilent Technologies, "Agilent ADNS-3080 High-Performance Optical Mouse Sensor Data Sheet." Online: http://www.dexsilicium.com/Agilent_188515_ADNB-3081.pdf (accessed 10.03.2017).
45. Centeye TinyTam. Online: http://www.centeye.com/technology/67-2 (accessed 10.03.2017).
46. D. Watman and H. Murayama, "Design of a Miniature, Multidirectional Optical Flow Sensor for Micro Aerial Vehicles," *Proceedings of the IEEE International Conference on Robotics and Automation*, Shanghai, China, pp. 2986–2991, 2011.
47. OmniVision, "OV7670/OV7171 CMOS VGA (640x480) Camera Implementation Guide." Online: http://www.arducam.com/downloads/modules/OV7670/OV7670_CMOS_Camera_Module_REVC_DS.pdf (accessed 10.03.2017).
48. CMUcam, "CMUcam: Open Source Programmable Embedded Color Vision Sensors." Online: http://www.cmucam.org/projects/cmucam5 (accessed 10.03.2017).
49. H. Chao, Y. Gu, J. Gross, G. Guo, and M. L. Fravolini, "A Comparative Study of Optical Flow and Traditional Sensors in UAV Navigation," *Proceedings of the American Control*, Washington, DC, pp. 3858–3863, 2013.
50. J. Fang and X. Gong, "Predictive Iterated Kalman Filter for INS/GPS Integration and Its Application to SAR Motion Compensation," *IEEE Transactions on Instrumentation and Measurement*, vol. 59, no. 4, pp. 909–915, 2010.

Index

A

Acoustic optic effect, 28
Active concentration assays, 198
Active pixel, 131
Affinity analysis, 198
Aldehyde coupling, 207
Amine coupling, 206
APDs, *see* avalanche photodiodes
ASIC readout electronics, 117–124
overview, 117
noise-recurring engineering (NRE), 117
photon counting, 118–120
charge-sensitive amplifier (CSA), 119
ChromAIX, 121–122
Medipix3, 120
Timepix, 120
sensor attachment, 117–118
industry-standard bump-bonding, 118
spectroscopic ASICs, 122–124
HEXITEC, 124
IDeF-X, 122–123
VAS UM/TAT4, 123–124
Avalanche photodiodes (APDs), 103–104
avalanche effect, 103
breakdown voltage, 103
AW, *see* acoustic wave

B

Baseline buffers, 204
BAW, *see* bulk acoustic wave
BDSe, *see* Brillouin distributed sensors
BFS, *see* Brillouin frequency shift
Binding stoichiometry, 198
Bit energy, 273
BOCDA, *see* Brillouin optical correlation-domain analysis
BOFDA, *see* Brillouin optical frequency-domain analysis
Bolometric THz detection, 2–3
CNT film, 2
THz-induced electrical-gate effect, 3
THz-thermoelectric effect, 3
BOTDA, *see* Brillouin optical time-domain analysis
Bragg diffraction, 28
Bragg reflector, 74
Bridgman technique, 112
Brillouin threshold, 36
Buffer preparation, 204–205
baseline buffers, 204
regeneration buffers, 204–205
running buffers, 205

C

Cadmium telluride sensors, 111–112
Cadmium zinc telluride sensors, 112–114
Bridgman technique, 112
Candida antarctica B lipase, 194
Capacitive coupling, 12
Capacitive tactile sensors, 153–155
microcellular polyurethane, 154
polydimethylsiloxane (PDMS), 155
shear forces, 153
Chemical modifications, 192–193
gold-sulfur interaction, 193
self-assembled monolayers (SAMs), 192
silanization, 193
Chrom AIX, 121–122
CMOS, *see* mixed-signal complementary metal oxide semiconductor technology
CMOs, *see* post-complementary metal oxide semiconductor
CNT-based THz detector, 2–9
Coaxial cable, 15

Covalent immobilization, 206
Cross-talk phenomena, 131
Crystal structures, 188–193
chemical modifications, 192–193
Langmuir-Blodgett (LB), 191–192
polymer coating, 189–191
CSA, *see* charge-sensitive amplifier
CW, *see* continuous effect
Cyclotron resonance, 5
Czochralski process, 110

D

DDA, *see* discrete dipole approximation
Dichroic Lycurgus Cup, 45
Dirac fermions, 11
Direct conversion detectors, 105–109
CZT direct detection in baggage scanning, 107–108
CZT direct detection in computed tomography, 108–109
gadolinium oxysulfide (GOS), 108
Direct X-ray conversion
ASIC readout electronics, 117–124
scintillators *vs.* direct-conversion sensors, 98–109
direct-conversion detectors, 105–109
overview, 98–100
photodiode technology, 102–104
technology, 100–101
sensor selection, 109–117
cadmium telluride sensors, 111–112
cadmium zinc telluride sensors, 112–114
gallium arsenide sensors, 111
germanium sensors, 110–111
high-z materials, 109–110
other semiconductor material, 114
photon energy range, 115–117
DNA biosensors, 195–196
DUV, *see* deep ultraviolet irradiation
Dynamic measurements, 31–35
multiple frequency components, 35
phase-modulated (PM), 33
self-heterodyne detection, 33
signal-to-noise ratio (SNR), 35
slope-assisted method, 32
Dynodes, 102

E

EBL, *see* electron beam lithography
Electroless approaches, 56–57
Electronic skin, 151–153
epidermal electronic system, 152
Energy binning, 100
Energy harvesting
application, 229–239
bridge monitoring to medical implants, 229–236
human wearable and implantable devices, 236–239
types and power ranges, 218–223
MPPT algorithms, 223–229
pervasive monitoring, 221
E-nose detectors, 292–293
Enzyme biosensors, 194
Candida antarctica B lipase, 194
poly(lactic-co-glycolic acid) (PLGA), 194
Epidermal electronic system, 152
EPSRC, *see* Engineering and Physical Science Research council
EQE, *see* external quantum efficiency

F

FDM, *see* finite difference method
FDTD, *see* finite-difference time domain
FEM, *see* finite element method
FET, *see* field-effective transistors
FIA, *see* flow injection analysis
FIB, *see* focused ion beam
Fibre-optic Brillouin distributed sensors
BOTDA sensors, 31–39
dynamic measurements, 31–35
long-range measurements, 35–39
fundamentals, 28–30
acousto-optic effect, 28
Bragg differaction, 28
Brillouin frequency shift (BFS), 28
Brillouin gain spectrum (BGS), 29
Brillouin optical correlation-domain analysis (BOCDA), 30
Brillouin optical frequency-domain analysis (BOFDA), 30
continuous wave (CW), 30

nonlinear Brillouin scattering effect, 28
stimulated Brillouin scattering (SBS), 28
overview, 27–28
Brillouin distributed sensors (BDSe), 27
Brillouin optical time-domain analysis (BOTDA), 28
Field-effective transistors (FET), 165–172
pressure modulated organic field-effect transistor (PMOFET), 171
FPW, *see* flexural plate wave
FTS, *see* Fourier transform spectroscopy

G

GaAs-based UTC-PDs and LPCs, 73–79
Gallium arsenide sensors, 111
GBL, *see* graded bandgap layer
Germanium sensors, 110–111
Czochralski process, 110
Gold-sulfur interaction, 193
Graphene-based THz detector, 10–13
Fourier transform spectroscopy (FTS), 10
plasmonic detection, 10
silicon-based bolometers, 10
time domain spectroscopy (TDS), 10

H

HEXITEC, 124
Engineering and Physical Science Research council (EPSRC), 124
Research Council UK (RCUK), 124
High-speed Ge-on-Si-based PDs, 87–90
High-z materials, 109–110
Hot electron-based plasmonic sensors, 52–54
Hybrid imagers, 137

I

IDeF-X, 122–123
baseline holder (BLH), 123
Identify medications, 255
Identify patients and prevent adverse effects, 255
Immunosensors, 194–195
Industry-standard bump-bonding, 118
Information rate of biosignal amplifier, 278–280
Information rate using first-order model, 269–273
InP-based uni-traveling carrier, 68–73
Integrated near-field THz nano-imager, 16–20
Interaction mechanism, 198
Inverse piezoelectric effect, 183
Ion mobility spectrometry, 288–290
IoT, *see* Internet of Things
IR spectrometer, 11–13
capacitive coupling, 12
Dirac fermions, 11

K

Kinetic analysis, 198

L

Label-free detection of biomolecular interactions
overview, 181–183
acoustic wave (AW), 181
surface plasmon resonance (SPR), 181
Quartz crystal microbalance (QCM), 183–198
biosensor application, 193–198
bulk acoustic wave (BAW), 183
crystal structures, 188–193
electrochemical QCM, 186–188
flexural plate wave (FPW), 183
inverse piezoelectric effect, 183
measuring principle, 185–186
QCM-FIA, 188
shear horizontal surface acoustic wave (SH-SAW), 183
surface acoustic wave (BAW), 183
surface plasmonics resonance, 198–207
active concentration assays, 198
affinity analysis, 198
binding stoichiometry, 198
interaction mechanism, 198

kinetic analysis, 198
optical biosensors, 198
p-polarization, 199
thermodynamic analysis, 198
ultrasensitive photodiode, 199
working principle, 200–201
Long-range measurements, 35–39
Brillouin threshold, 36
dithering, 37
nonlocal effects, 36
Raman scattering, 35
time-multiplexing, 36
Low-power biosensor design technique
bit energy, 273
information rate of biosignal amplifier, 278–280
information rate using first-order model, 269–273
noise and information rates in amplifiers, 268–269
noise efficiency factor, 273
overview, 267
Mixed-signal complementary metal oxide semiconductor (CMOS) technology, 267
trade-offs in amplifier design, 274–278
Low-powered plasmonic sensors
biomedical technology, 49–50
communication industries, 50–51
localized surface plasmons resonant (LSPR), 50
post-complementary metal oxide semiconductor (CMOs), 50
surface plasmons polaritons (SPPs), 50
surface plasmons (SPs), 50
very-large-scale integration (VLSI), 50
electroless approaches, 56–57
ultra-smooth plasmonic metal surfaces, 57
hot electron-based plasmonic sensors, 52–54
locally resonant surface plasmons (LRSPs), 53
planar Schottky, 53
plasmonic resonant energy transfer (PRET), 53
polycrystalline nanostructures, 54
prism-coupling method, 53
nanostructure and local field enhancements, 54–56
Mie scattering theory, 54
overview, 43–48
anodized alumina, 48
dichroic Lycurgus Cup, 45
diffraction limit of light, 44
discrete dipole approximation (DDA), 48
electron beam lithography (EBL), 47
fabrication, 46–48
finite difference method (FDM), 48
finite-difference time domain (FDTD), 48
finite element method (FEM), 48
focused ion beam (FIB), 47
history and development, 44–46
mathematical simulation, 48
metamaterials, 44
physical vapor deposition (PVD), 48
structured conductor-dielectric interfaces, 44
surface-enhanced Raman scattering (SERS), 45
research and development, 51–52
silicon-based micro-devices, 49
wavelength selectivity with nanostructures, 55–56
opto-plasmic sensors, 56
Low-power energy harvesting solutions
energy harvesting, 218–239
application, 229–239
MPPT algorithms, 223–229
types and power ranges, 218–223
multi-source self-powered device conception, 239–242
multi-harvesting power chip (MHPC), 241
overview, 217–218
LRSPs, *see* locally resonant surface plasmons
LSPR, *see* localized surface plasmons resonant

M

Maleimide coupling, 207
Maximum power point tracking (MPPT) algorithms, 218, 223–229
Medipix3, 120
Metal-coated optical fibre, 15
Metal oxide sensors, 290–291
Metamaterials, 44
MHPC, *see* multi-harvesting power chip
Microcellular polyurethane, 154
Micro-GC detectors, 291–292
Mie scattering theory, 54
Modern application potential of miniature chemical sensors
  applications, 285–288
    agriculture, 288
    medical, 286–287
    military and security, 287–288
  examples, 288–293
    E-nose detectors, 292–293
    ion mobility spectrometry, 288–290
    metal oxide sensors, 290–291
    micro-GC detectors, 291–292
  overview, 283–285
MPPT, *see* maximum power point tracking (MPPT) algorithms
MPPT algorithms, 223–229
  Internet of Things (IoT), 225
  solar tracking, 224
  Trillion sensor (TSensor), 225
  wireless sensor networks (WSNs), 225
Multi-channel digital acquisition system, 102

N

Nano-carbon terahertz sensors and images
  CNT-based THz detector, 2–9
    bolometric THz detection, 2–3
    THz-induced electrical-gate effect, 5–7
    THz photon-assisted tunneling (PAT), 3–5
    THz-thermoelectric effect, 7–9
  graphene-based THz detector, 10–13
    IR spectrometer, 11–13
  near-field THz imager, 13–20
    imaging, 14–16
    integrated near-field THz nano-imager, 16–20
  overview, 1–2
Nanostructure and local field enhancements, 54–56
Near-field THz imager, 13–20
  coaxial cable, 15
  metal-coated optical fibre, 15
NEP, *see* noise-equivalent power
Noise and information rates in amplifiers, 268–269
Noise efficiency factor, 273
Non-covalent capture, 207
Nonlinear Brillouin scattering effect, 28
NRE, *see* noise-recurring engineering

O

OF algorithms, 310–312
  block-matching methods, 310
  differential method, 311
  energy-based methods, 311–312
  feature-based methods, 311
  fractional-order operator-based method, 312
  fusion-based method, 312
  image interpolation method, 311
  phase-based methods, 312
OF computation, 308–310
OLEDs, *see* organic light-emitting diodes
Optical biosensors, 198
Optical flow sensing and applications
  evaluation methodology, 312–317
  overview, 304–306
  sensing techniques, 307–312
    OF algorithms, 310–312
    OF computation, 308–310
  OF sensor model, 317–323
  simulation results, 323–325
  small-scale flybarless helicopter modeling, 306–307
Opto-plasmic sensors, 56
Organic imagers
  active, 133–139

hybrid imagers, 137
single all-organic pixels, 133
overview, 129–130
organic light-emitting diodes (OLEDs), 130
organic semiconductors (OSCs), 129
thin-film transistors, 130
passive, 131–133
external quantum efficiency (EQE), 133
structure, 130–131
active pixel, 131
cross-talk phenomena, 131
passive/active matrices, 130
passive pixel, 131
sneak path problem, 130
unconventional architectures, 140–144
deep ultraviolet irradiation (DUV), 143
organic phototransistors, 140
photosensitive voltage divider, 142
OSCs, *see* organic semiconductors

P

Passive/active matrices, 130
Passive pixel, 131
PAT, *see* THz photon-assisted tunneling
PDMS, *see* polydimethylsiloxane
Pervasive monitoring, 221
Photo-absorption, 68
Photodiode technology
avalanche photodiodes (APDs), 103–104
avalanche effect, 103
breakdown voltage, 103
photomultiplier tubes (PMT), 102–103
dynodes, 102
solid-state photomultipliers, 104
Geiger mode, 104
Photon counting, 118–120
Photon energy range, 115–117
Photo-receiver array, 87
PICs, See photonic integrated circuits
Piezoelectric and pyroelectric tactile sensors, 162–165
paraelectric state, 163
polyvinylidene fluoride-trifluroethylene (PVDF-TrFE), 164
pyroelectric nanogenerator (PNG), 164
Piezoresistive sensors, 158–160
Planar Schottky, 53
Plasmonic detection, 10
PM, *see* phase-odulation
PMT, *see* photomultiplier tubes
PNG, *see* pyroelectric nanogenerator
Polycrystalline nanostructures, 54
Polymer coating, 189–191
drop coating, 189
electropolymerized films, 191
electrospinning, 190
spin coating, 190
spray coating, 190
P-polarization, 199
PRET, *see* plasmonic resonant energy transfer
Prism-coupling method, 53
p-type GaAs-based photo-absorption layer, 74
PVD, *see* physical vapor deposition
PVDF-TrFE, *see* polyvinylidene fluoride-trifluroethylene
PVT, *see* polyvinyltoluene

Q

QCM-FIA
flow injection analysis (FIA), 188
Quantum detection, 3
Quartz crystal microbalance (QCM)
biosensor application, 193–198
cell, 196–198
DNA, 195–196
enzyme, 194
immunosensors, 194–195
crystal structures, 188–193
chemical modifications, 192–193
Langmuir-Blodgett (LB), 191–192
polymer coating, 189–191
electrochemical QCM, 186–188
measuring principle, 185–186
active oscillator mode, 185

passive mode, 185–186
QCM-DTM, 186
QCM-FIA, 188
flow injection analysis (FIA), 188

R

RA, *see* rapid adaptation
Raman scattering, 35
RCUK, *see* Research Council UK
Regeneration buffers, 204–205
Resistive tactile sensors, 155–162
piezoresistive sensors, 158–160
pyroresistive tactile sensors, 160–162
resistance temperature detectors (RTDs), 157
RFID supporting IoT in health
access control, 254–255
big data handling, 258–260
case study, 260–261
identify medications, 255
identify patients and prevent adverse effects, 255
overview, 253–254
radio-frequency identification (RFID), 253
sensing and monitoring, 255
smart home appliances, 256
track objects and people, 255
wearable passive RFID tags, 256–258
RTDs, *see* resistance temperature detectors

S

SA, *see* slow adaptation
SAMs, *see* self-assembled monolayers
Scintillator crystal, 102
Scintillators *vs.* direct-conversion sensors, 98–109
direct conversion detectors, 105–109
CZT direct detection in baggage scanning, 107–108
CZT direct detection in computed tomography, 108–109
energy binning, 100
inorganic scintillators, 101
limitations, 101
organic scintillators, 101
polyvinyltoluene (PVT), 101
photodiode technology, 102–104
avalanche photodiodes (APDs), 103–104
multi-channel digital acquisition system, 102
photodetectors, 102
photomultiplier tubes (PMT), 102–103
scintillator crystal, 102
solid-state photomultipliers, 104
technology, 100–101
full-energy interaction, 100
SCS, *see* space-charge screening
Self-heterodyne detection, 33
Semiconductor-based photo-absorption layer, 88
Sensing and monitoring, 255
Sensor chip immobilization, 206–207
aldehyde coupling, 207
amine coupling, 206
covalent, 206
maleimide coupling, 207
non-covalent capture, 207
thiol coupling, 206
Sensor selection, 109–117
cadmium telluride sensors, 111–112
travelling heater method (THM), 111
cadmium zinc telluride sensors, 112–114
Bridgman technique, 112
gallium arsenide sensors, 111
germanium sensors, 110–111
Czochralski process, 110
high-z materials, 109–110
other semiconductor material, 114
photon energy range, 115–117
extension to higher energies, 116–117
maximum, 115–116
minimum, 115
SERS, *see* surface-enhanced Raman scattering
SH-SAW, *see* shear horizontal surface acoustic wave
Si-based CMOS ICs, 88
Silanization, 193
Silicon-based bolometers, 10

Single all-organic pixels, 133
Slope-assisted method, 32
Sneak path problem, 130
SNR, *see* signal-to-noise ratio
SOI, *see* Silicon-on-insulator substrate
Solar tracking, 224
Somatosensory system, 149–151
  mechanoreceptors, 150
  nociceptors, 150
  rapid adaptation (RA), 150
  slow adaptation (SA), 150
  thermoreceptors, 150
Spectroscopic ASICs, 122–124
  HEXITEC, 124
  IDeF-X, 122–123
  VAS UM/TAT4, 123–124
SPPs, *see* surface plasmons polaritons
SPRi, *see* surface plasmon resonance imaging
SPR-MS, *see* surface plasmon resonance-mass spectrometry
SPR-QCM, *see* Surface plasmon resonance-Quartz crystal microbalance
SPs, *see* surface plasmons
Structural health monitoring (SHM), 218
Surface plasmonics resonance, 198–207
  protocols, 203–207
    buffer preparation, 204–205
    data analysis, 207
    sample preparation, 204
    sensor chip immobilization, 206–207
    sensor chip priming, 203–204
    suitable chip selection, 205–206
  surface-enhanced Raman scattering (SERS), 202
  surface plasmon resonance imaging (SPRi), 201–202
  surface plasmon resonance–mass spectrometry (SPR-MS), 201
  Surface plasmon resonance–Quartz crystal microbalance (SPR-QCM), 202–203

T

Tactile sensors for electronic skin
  capacitive, 153–155
    microcellular polyurethane, 154
    polydimethylsiloxane (PDMS), 155
    shear forces, 153
  electronic skin, 151–153
    epidermal electronic system, 152
  field-effective transistors (FET), 165–172
    pressure modulated organic field-effect transistor (PMOFET), 171
  piezoelectric and pyroelectric tactile sensors, 162–165
    paraelectric state, 163
    polyvinylidene fluoride-trifluroethylene (PVDF-TrFE), 164
    pyroelectric nanogenerator (PNG), 164
  resistive, 155–162
    piezoresistive sensors, 158–160
    pyroresistive tactile sensors, 160–162
    resistance temperature detectors (RTDs), 157
  somatosensory system, 149–151
    mechanoreceptors, 150
    nociceptors, 150
    rapid adaptation (RA), 150
    slow adaptation (SA), 150
    thermoreceptors, 150
TDS, *see* time domain spectroscopy
Thermodynamic analysis, 198
Thin-film transistors, 130
Thiol coupling, 206
THM, *see* travelling heater method
THz-induced electrical-gate effect, 5–7
  cyclotron resonance, 5
  ultra-sensitive readout, 7
THz photon-assisted tunneling (PAT), 3–5
  Mylar sheet, 3
  noise-equivalent power (NEP), 5
  photon sidebands, 3
  quantum detection, 3
THz-thermoelectric effect, 7–9
  Golay cell detectors, 7
  polarization nisotrophy, 8
  pyroelectric detectors, 7
  Schottky barrier diodes, 7
  THz tomography, 9

ultra-broad band photodetectors, 8
THz tomography, 9
Timepix, 120
Track objects and people, 255
Trade-offs in amplifier design, 274–278
TSensor, *see* Trillion sensor

U

Ultra-fast photodiodes
GaAs-based UTC-PDs and LPCs, 73–79
Bragg reflector, 74
p-type GaAs-based photo-absorption layer, 74
high-speed Ge-on-Si-based PDs, 87–90
photonic integrated circuits (PICs), 89
semiconductor-based photo-absorption layer, 88
Si-based CMOS ICs, 88
silicon-on-insulator (SOI) substrate, 89
InP-based uni-traveling carrier, 68–73
photo-absorption, 68
ultra-low background doping, 69
limitation in speed, 67–68
large junction capacitance (CJ), 68
space-charge screening (SCS), 67
overview, 65–67
type II InP-based UTC-PDs, 79–87
absorption-collection (A-C) junction, 80
graded bandgap layer (GBL), 81
photo-receiver array, 87
Ultra-low background doping, 69
Ultrasensitive photodiode, 199
Ultra-sensitive readout, 7
Ultra-smooth plasmonic metal surfaces, 57
Unconventional architectures, 140–144

V

VAS UM/TAT4, 123–124

W

Wavelength selectivity with nanostructures, 55–56
Wearable passive RFID tags, 256–258